LONG WAVES IN THE WORLD ECONOMY

LONG WAVES IN THE WORLD ECONOMY

edited by

Christopher Freeman
Deputy Director of Science Policy Research Unit, University of Sussex

Butterworths
London Boston Durban Singapore Sydney Toronto Wellington

First published 1983

British Library Cataloguing in Publication Data

Long waves in the world economy
1. Economic history – 1945-
2. Economics – History – 20th century
I. Freeman, Christopher
330.9181'2 HC59

ISBN 0-408-22156-9

Photoset by Butterworths Litho Preparation Department
Printed and Bound in Great Britain by Page Bros Ltd, Norwich, Norfolk

PREFACE

C. Freeman

Most of this collection of papers was first published in a special issue of *Futures* in 1981, but five new papers have been added that greatly enlarge the scope of this book. The original publication did not really cover the empirical evidence relating to the existence and characteristics of the "long waves" themselves. In this book there are two major papers by a group of Kiel economists (Glismann, Rodemer and Wolter) and by Cleary and Hobbs, which do much to remedy this deficiency. Two important contributions by Italian economists, Salvati and Dosi, greatly widen the interest of the book by their original insights. Michele Salvati takes up the fundamental issue of the relationship between political and economic factors in long (and short) business cycles. Giovanni Dosi develops the notion of "technological paradigms" analogous to the scientific paradigms of Thomas Kuhn. Finally, a paper by George Ray takes up the issue of the national origins of major innovations and the possible characteristics of a new economic upswing. There are still of course important gaps in the coverage of this volume—in particular the role of the Third World is not discussed at all. Nevertheless, it is to be hoped that with the new papers it will form an important source of reference for future debates and further research in the 1980s.

The book reflects the rapidly growing interest in the topic of "long waves" in the development of the economy or, as they are often called, "Kondratiev" cycles. As Tinbergen's paper (chapter 2) points out, Kondratiev was not the originator of the idea, which was already fully formulated by van Gelderen in 1913. It was van Gelderen who coined the metaphor that is still perhaps the most apt description of the phenomenon—"spring tides" of rapid expansion, followed by "ebb tides" of relative stagnation and depression, with each complete cycle lasting approximately half a century. As the contrast between the 1970s (and early 1980s) and the "spring tide" of the 1950s and 1960s became more and more apparent, it was obvious that there would be a revival of long-wave theories, and Mandel is entitled to observe that, whereas in the mid-1960s he stood on somewhat lonely ground in predicting the end of the long post-war boom, there are now many who are thinking in terms of a rather prolonged period of slow growth or more serious depression.

As he points out, one of the eminent economists who has recently spoken of the last quarter of the 20th century in terms of this contrast is Paul Samuelson:[1]

> No one can predict the future with confidence. Still, it is my considered guess that the final quarter of the twentieth century will fall far short of the third quarter in its achieved rate of economic progress.

> The dark horoscope of my old teacher Joseph Schumpeter may have particular relevance here. When I was a precocious student, I didn't think much of Joseph Schumpeter's futurology. But, like Mark Twain, who said that when he was fourteen he thought his father was awful dumb, but by the time of reaching twenty-one he was surprised at how much the old boy had smartened up, when I reread Schumpeter's 1942 *Capitalism, Socialism and Democracy* I find new meanings in it.
>
> To be sure, some of those meanings were not put there by Joseph Schumpeter. But what was Harvard paying him for if not to stimulate students like me to new theories?

Samuelson's reference to Schumpeter is of particular interest, as it was Schumpeter who gave a new twist to the whole long-wave debate in 1939, when he explained Kondratiev cycles in terms of major innovations providing at intervals the basis for the spring tides of expansion. This connection, or possible connection, between technical innovation and long waves of economic development is one of the main issues discussed in this book. This does not mean that any of the papers are advocating a simple-minded technological determinism. Indeed, as we shall see, several of them explicitly repudiate the view that innovation is the major factor in any satisfactory theoretical explanation. But all of them recognise that innovation is an important influence that certainly cannot be ignored.

The first two papers introduce the whole set, although in very different ways. Delbeke seeks to classify some of the recent contributions to the long-wave debate in terms of their relative emphasis on the various factors involved—entrepreneurship and innovation (Schumpeter, Mensch, van Duijn), capital accumulation, lags, shortages and rate of profit (Forrester and Mandel), employment (Freeman), and raw materials and foodstuffs (Rostow). With the exception of Rostow, the authors discussed are represented in this book either directly or through others with similar ideas, so that the reader will have the opportunity both to find a convenient summary of their earlier work and some useful critical comment by Delbeke, and to become acquainted with the most recent developments of their ideas. It will, of course, be for readers to decide how far they agree with Delbeke's criticism (or indeed his classification). But in any case I believe they will find his paper a useful clarification of a complex field, and a valuable and stimulating introduction to the book.

The second introductory paper, by Tinbergen, is very different. It serves to remind us, through a brief historical retrospect, of the outstanding contribution made by Dutch economists to the development of long-wave theories. Many of the ideas that we are now debating were anticipated in this earlier discussion, and it is a source of particular pleasure to me that Tinbergen himself was able to contribute this short paper, providing a direct personal link with an earlier "long-wave" debate and a set of ideas to which he himself made such an outstanding contribution.

Reference

1. P. A. Samuelson, "The world's economy at century's end", *Japan Economic Journal*, 10 March 1981, page 20.

CONTENTS

CHAPTER 1

RECENT LONG-WAVE THEORIES

A critical survey

Jos Delbeke

The author gives a critical overview of some recent long-wave theories, which, although not complete, does seek to diminish the present confusion in the literature and to emphasise the complementary character of the different approaches. A classification scheme is developed, based on the alternating scarcity and abundance of production factors in the upswing and downswing phases of the long wave. In this way theories have been classified as: those which consider the role of innovation (ie entrepreneurship) as crucial (Schumpeter and Mensch); the capital theories (Mandel and Forrester); the labour theory (Freeman); and the raw materials and foodstuffs theory (Rostow). The importance of institutional and social changes is emphasised.

A REMARKABLE revival of interest in economic fluctuations and especially in long waves has occurred since the second half of the 1970s. This renewed interest may be attributed to the present world economic crisis and the failure of short- and medium-term analysis to account for it.

Although there is widespread scepticism about any long-wave 'law', there are authors who forecast the present crisis on a theoretical basis. In 1944, Colin Clark[1] wrote in *The Economics of 1960:* "at some date in the 1960s the world will probably again reach an abrupt termination of this capital-hungry phase and another capital-sated phase will follow". In 1964, E. Mandel[2] was expecting a slowdown of the post-war boom to occur in the following few years. It is also misleading to presume that the present economic experiences are unique in history. The first sentences van Gelderen[3] wrote in 1913 in his study *Springvloed* can be read in any newspaper today: "It is almost a commonplace to say that we live in an expensive time. No economic phenomenon of an international nature is so much the subject of public discussion as the continual rise of prices".

However, most economists have not been flexible enough to change their analytical framework. Too quickly, many seemed to forget the difficult events of

J. Delbeke is a research assistant at the Centrum voor Economische Studien, Katholieke Universiteit Leuven, Van Evenstraat 2 b, B-3000 Leuven, Belgium.

the 1930s, and from the 1950s the theory of economic fluctuations was gradually replaced by growth theory. The neoclassical model, stressing the equilibrating tendencies of the system, became the cornerstone of economic thinking and analysis. The conditions for convergence to steady-state growth have been investigated in detail: identical growth rates of total output and capital stock, a constant profit rate, identical growth rates of real wages and labour productivity, and constant income distribution between labour and capital.[4] There was a tendency to assume in the 1960s that real world growth approximated this ideal model.

Against this, van Duijn argued that rational economic behaviour *ex ante* appears to generate uneven economic development *ex post*, and that therefore equilibrium growth is a mental construct, an ideal image that is never achieved because we do not act that way.[5]

Widespread belief in neoclassical economics, however, is not the only cause of the recent neglect of economic fluctuations. Other reasons are that the long-wave theories are dispersed over a vast amount of literature, and that they show tendencies toward monocausality, and to confusion between cause and effect in explaining the phenomena. Many authors construct theories without empirical testing or simply present time series of fluctuations without a theory. One reason for this is that empirical testing and verification are inherently difficult because the sample contains only 3·5 long waves. Therefore, a more intensive application of mathematical and statistical methods may not eliminate the existing confusion between the various theories.[6]

The long-wave pioneers—Parvus, van Gelderen, de Wolff and Kondratiev—observed the long wave by means of price data, and considered 1815, 1849, 1873 and 1896 as years of crucial turning points. They held that a relationship existed between price fluctuations and economic development in a broad sense. They also considered the long wave as a phenomenon inherent in the capitalist way of production, and they sought to explain it as a cyclical and endogenous process.[7]

Kondratiev, the best known long-wave pioneer, emphasised the discontinuity in the production and duration of long-term fixed capital as the most powerful force behind the long wave.[8] Karl Marx's conjunctural theory was clearly the starting point for his explanation. But he overestimated the role of price fluctuations and he was unable to take full account of fluctuations in industrial production and changes in productivity. His theory has also been discredited because of inability to incorporate consistently endogenous variables in the explanation of turning points. Although he considered technological development as an important factor, he underestimated its powerful influence.

Building further on the work of Kondratiev, several authors refined his analysis, and dated the long waves in slightly different ways using the four phases—prosperity, recession, depression, recovery—which Schumpeter[9] developed (see chapter 3 by van Duijn in this book, Table 3.2).

Since the work of Schumpeter, there have been several theories which have taken into account his increased emphasis on the role of innovations in long waves. These theories are the subject of this survey, as of this book as a whole. This survey compares and criticises the theories, classifying them according to the relative scarcity or abundance of the various production factors—entrepreneurship, capital, labour and materials. The analysis concentrates on the first three of these as they are the ones represented in this collection of papers. Moreover, we do not intend to discuss theories that consider the outbreak of wars and monetary factors as important variables. We recognise, however, that the money and credit supply, even if it does not influence the very long-term development of the economy, seems to be important for the determination of the turning points.[10]

The scarcity and abundance of entrepreneurship

In this section, we analyse the relationship between the long-wave movement and the frequency and intensity of innovational activity. Some authors, such as Schumpeter and more recently Mensch and Kleinknecht, assign technological innovation an almost exclusive role as the engine of economic development.[11] Theories that do not consider technological change to be of such primary importance are discussed later.

Schumpeter considered the irregular clusters of innovation to be crucial for development. His theory can be summarised as follows: economic development appears in the form of innovations; it occurs in cycles; and entrepreneurship is an innovative action while management is not.[12] The upper turning point of the cycle is caused by the disappearance of profit (the shift from monopolistic to pure competition), the lower occurs after the creative destruction of a depression. The fundamental solution arises when new consumer goods and new methods of production and transport appear, when new markets are created, and when there are breakthroughs in the forms of industrial organisation. However, Schumpeter is unclear about why clusters of innovation take place every 45 to 60 years.

The theory of Mensch

Mensch updated Schumpeter's theory and gave it an empirical base. Around the years 1825, 1886, and 1935, new clusters of basic innovations took place and generated completely new sectors. Massive new markets allowed these sectors to grow rapidly and to improve the products and the production processes. Competition, rationalisation, and concentration raise the capacity of the new industrial sectors. However, at a certain point, the capacity becomes too great for the domestic market so the export market becomes vital. Since foreign industrial countries have also gone through this same process, competition grows among a diminishing number of large concerns on the world market. Aware of the growing saturation, these concerns invest less money in the industrial sectors and more on the international money and capital markets. In the next stage, the monopolies try to sell their surplus in the LDCs with the help of massive credits. Ultimately, these countries reach their credit limits, and the saturation of demand is complete.[13]

During this critical period, which begins with recession and lasts until the end of the depression, the economy has no fundamental solution other than the creation of basic innovations and new sectors. Only "innovations will overcome depression" ("Innovation überwinden die Depression").[14] Mensch calls the situation in which society is forced to innovate the 'technological stalemate' (*das technologische Patt*). "Der menschliche Drang zur Neuerung wird im technologischen Patt zur wirtschaftlichen Notwendigkeit, zur Sachzwang".[15] The technological stalemate is a period of reorientation, of crisis in the sense of the Greek κρίσις, which means turn, change. We can observe such periods in 1814–27, 1870–85, 1925–39, 1975–?.

In the technological stalemate, economic policy cannot offer fundamental solutions to the crisis, at least in the short run. The government can only *finessieren*, ie play for time and avoid a collapse of domestic and foreign demand. In the long term, it has to stimulate the search for new and basic innovations by supporting innovative projects with massive amounts of money (which can be found on the international markets).[16] Since fundamental changes occur only in the micro-economic cell, government cannot control the outcome and nature of the basic innovations, it can only create a favourable climate. In any event, Mensch argues that the economic policy of deficit spending is not the most satisfactory. *Schubladenprojecte* ('drawer projects') do not have a positive multiplier effect, because they do not affect the essential problems of stagnation, which are the lack of basic innovations and the saturation of demand in the leading sectors.[17] In short, the only fundamental solution to demand saturation is an aggressive innovation policy.

Mensch makes interesting distinctions between different kinds of innovations. The basic innovations change essentially the direction of economic development and create new leading sectors, which will form the basis of the prosperity of the next upswing. They only occur under the pressure of demand saturation of the existing leading sectors when entrepreneurs are looking for new and profitable investment opportunities. Once entrepreneurs invest in these new and risky sectors, turbulent technological development begins in the same direction: these are the improvement innovations of products and production processes. The new products start a development with an S-shaped life cycle. Along with the maturing of the new industry and the increasing demand saturation, a concentration movement develops, and the monopoly firms try to protect their market share by means of *Schein-Innovationen* (pseudo-innovations), which also cause price rises. An important characteristic of the technological stalemate is that pseudo-innovations become more important than improvement innovations.[18]

Comment

Mensch's analysis is interesting because of its updating and statistical testing of Schumpeter's theory and the emphasis it places on demand and innovation as crucial factors for economic development. But Mensch skirts dangerously close to monocausality and neglects other crucial factors, which we consider in the following sections.

Mensch does remedy a major lacuna of most theories, namely the neglect of demand as the engine for long-wave movement. However, we do not agree with the position that expansionary innovations are always more demand-inducing than rationalising innovations because of the different income effects they generate.[19] Rationalising innovations surely can induce a larger demand if the more efficient use of production factors is translated into increased purchasing power and unemployment is neutralised by job creation in other sectors, eg the tertiary sector. However, when there is a technological stalemate, we do agree that rationalising innovations will be less demand-inducing than expansionary ones because of the difficulties in creating alternative employment because of the slowdown of economic growth. In short, Mensch's comparative statics do not always give the same dynamic results.

A major weakness in Mensch's empirical research is the lack of a good criterion for identifying the basic and improvement innovations. Although the difference is crucial, the interpretation of a concrete innovation is difficult and risky. For this reason, it is difficult to integrate into Mensch's theory those innovations that needed two long waves to exert their full effect on economic growth, eg automobile construction. Because these sectors have undergone intensive changes, it is uncertain whether they should be classified in terms of basic or of improvement innovations.

An important feature of innovation theory is the emphasis on the discontinuity, the clustering, of basic innovative activity. Periods of scarcity and abundance of entrepreneurial activity seem to alternate in economic development. But there is a sharp contrast between the Schumpeterian way of treating innovations, and the way macro-economic and management literature deal with them. In macro-economics, technological innovation is considered as a continual and even process, so only improvement innovations are dealt with. Such an approach fails to take account of the revolutionary character of basic innovations. The same idea of continuity is paramount in management literature, but with a more normative character: "Innovations do not happen, they are made to happen".[20]

In conclusion, Mensch provides a fruitful analysis of the current situation and implications for the study of the long wave, and the theory's importance is increasing as part of a multicausal explanation.

Scarcity and abundance of capital

Many authors who have dealt with the long wave, and particularly the pioneers because of their Marxist orientation, have considered varying intensity in capital goods production as the most important characteristic of the long wave. We discuss here the approaches used by Mandel and Forrester.

The theory of Mandel

Mandel defines the industrial cycle as the successive acceleration and deceleration of capital accumulation. In the upswing period, the mass and rate of profit, and the volume and rhythm of capital accumulation all increase. Conversely, during a crisis and the subsequent period of depression, these factors all decrease.[21]

When the acceleration of capital accumulation reaches a certain point during the upswing, it becomes impossible to invest the total mass of accumulated capital at an adequate rate of profit. This is the situation of overaccumulation, and it must be understood as a relative phenomenon: there is simply too much capital available to attain the expected social average rate of profit. Consequently, the average rate of profit of the total accumulated mass of capital declines. Conversely, in the crisis and depression phase, capital is devalorised. Underinvestment now occurs: less capital is invested than would be expected at a given level of surplus-value production and at the given (rising) average rate of profit. This periodical devalorisation and underinvestment has the function of again raising the average rate of profit of the entire mass of accumulated capital, which in turn allows the intensification of production and capital accumulation. The capitalist industrial cycle thus appears to be the consequence of accelerated capital accumulation, overaccumulation, decelerated capital accumulation, and underinvestment. The rise, fall, and revitalisation of the rate of profit correspond to and determine the successive movements of capital accumulation.[22]

The periodical underinvestment of capital also makes the introduction of fundamental innovations possible: "It [the periodical underinvestment] ... creates a historical reserve fund of capital, from which can be drawn the means for additional accumulation needed over and above 'normal' extended reproduction to allow a fundamental renewal of productive technology ...".[23] The fall in the rate of profit indicates the watershed from which idle capital becomes available, first relatively and then absolutely. At a particular point, however, this additional capital is expended on a massive scale because there is a sudden increase in the rate of profit, because of sociopolitical changes and the breakthrough of technological and institutional innovations.

Comment

Mandel's theory is essentially supply-oriented and tends to neglect the role of final demand. Actually, one reason for the fall in the rate of profit is that the final demand is saturated, and this situation can only be overcome by technological and/or institutional innovations.

Furthermore, it is not clear why the average rate of profit has to increase suddenly, which Mandel repeatedly emphasises. It would be reasonable to assume that the rate of profit at first becomes high in only a few small sectors, and that it takes years before massive amounts of capital are invested in new fields because of their risky nature. In contrast, other authors, such as Forrester, consider time lags involved in investment as the essential engine of the dynamics of the long wave.

Along with the supply side and the investment goods sector, Mandel emphasises fundamental revolutions in technology—especially the machine production of steam-driven motors since 1848, of electric and combustion motors since the 1890s, and of electronic and nuclear-powered apparatus since the 1950s. Technological innovation spreads through the economy in two distinct phases. During the upswing the technology itself undergoes a revolution (eg the creation of production sites for the new means of production), while in the downswing the new production means become adopted generally in all branches of industry and the economy. Although Mandel does not investigate thoroughly the pressures that generate innovation, the distinction between the effects of technological innovation in the capital goods sector and the spin-off in the other sectors is extremely important.

The theory of Forrester

Forrester has developed a completely new method of analysing growth dynamics in which the capital goods sector plays the crucial role.[25] Using system dynamics computer simulation, he is able to combine masses of descriptive information with the inherent complexity of economic dynamics. Forrester's theory is based on two important pillars that create endogenous fluctuations.[26] The first is a multiplier accelerator mechanism that transmits the fluctuations in investment to the rest of the economy and works on the long term. The second is rational behaviour along a capital–stock adjustment model, which causes alternating phases of capital hunger and capital saturation. The economic agents continually exaggerate: they expand the capacity of the capital sector too much in the upswing to catch up with demand, and they let the capital stock collapse to a level below the long-term average in the downswing. The entrepreneur is rational *ex ante* but not *ex post* because of perception delays.

'Bootstraps' and lags in the construction of industrial infrastructure are the cornerstone of Forrester's theory. Bootstrap structures occur when a sector uses a part of its own output as a production factor. If consumer demand is increasing fast enough, the capital sector has to expand its own production capacity, but the only way to acquire capital equipment is to divert it first from the consumer goods sector that is generating the higher demand. Therefore, there is a tendency to supply less capital equipment in response to the demand for more. "The bootstrap structures tend to be highly destabilizing and to lengthen substantially the periods of fluctuation that might otherwise prevail ... In the model, the bootstrap linkage around capital creates the 50-year cycle out of what would otherwise be a 20-year medium cycle in capital acquisition".[27]

Forrester claims that the central generating structure of the cycle has changed very little over two centuries because the "fundamental causes revolve around the life of capital plant, the bootstrap structure around the capital sector, and the perception delays in human decision making".[28]

Comment

Forrester's analysis has some interesting features, not the least of which is that it is a quasi-complete endogenous model of growth dynamics. He emphasizes the importance of destabilising constraints of human actions such as perception delays and bootstrap structures. The role he assigns to final demand, however, is too passive, certainly for the interpretation of the present crisis. Moreover, although the results are spectacular, the method is very mechanistic and is based heavily on capital formation data, which are difficult to assemble accurately.

The most surprising feature of his analysis, however, is its neglect of the special dynamics that technological innovation can generate. He simply consolidates the new technology into the successive upswings, each with its compatible infrastructure. After a major depression, during which the new plant construction rate has been at a minimum for several years, the country has to replace its aging plant, and it is unlikely that the technology of 20 years before will be revived because two or three decades of unused inventions and innovations are available. Industry begins to develop a whole new kind of infrastructure. When the upswing has progressed half way, the style and pattern rigidify, and only minor improvements take place. Here Forrester's theory resembles that of Mensch. Radically different ideas and inventions must wait until the next upswing. Still, Forrester neglects the non-continuous nature of the birth of technological innovation, despite continuing R and D activities. It is the result of pressure. The system dynamics model clearly needs a more dynamic demand function because, with the assumption of passive consumer demand, one cannot expect saturation pressures with which growth sectors could be confronted. Moreover, recent experience (eg electronics) shows that technological innovation itself has a more active influence on investment activity by rendering the production infrastructure obsolete more rapidly. Therefore, we are inclined to assign a greater role to technological innovation in the behaviour of the long wave than does Forrester.

Scarcity and abundance of labour

It is surprising that the long wave has only recently been studied by means of an explicit employment approach. The Keynesian interest in full employment and the recent failure of this policy have stimulated research in this direction: "Neither Kondratiev nor Schumpeter explicitly discuss the possibility that a particular wave of innovation might first of all have employment generating effects but at a later stage employment displacing effects. They were apparently thinking mainly in terms of the general 'climate of investment' and accumulation and fluctuations in the overall level of investment".[29]

The theory of Freeman

During the upswing, after decades of preliminary scientific and technical work, major new technologies generate both new investment and new employment on a large scale, thus becoming substantial new branches of activity. As a result of rapid technical change and increasing competitive pressures, relative prices of new goods fall and their cost advantage increases. The main employment effect during the upswing is a steep increase in employment. There will probably be shortages of specific types of labour which create an upward pressure on wages and induce ultimately labour-saving innovations.

After about a quarter of a century, the new branches of industry are firmly established, and their role as generators of additional new employment diminishes and eventually disappears. During the downswing, competitive pressures within industry become stronger, capital intensity grows, and investment continues, but labour-saving and material-saving technical changes become increasingly important. For example, employment in the US automobile industry reached a peak in the 1920s, and although output expanded, the labour force did not.

The loss of impetus for the growth of employment in the new industries is reinforced by the acceleration of labour-displacing technical changes, and economies of scale which arise from the success of the new technology itself. This process is the most intense during the Kondratiev downswing.[30]

Freeman applied his theory to the present observable evolution in the electronics industry. In the 1960s, the importance of the sector shifted from data processing towards control, monitoring, regulation and automation. After the product innovations of the upswing, process innovations prevailed in the downswing. The applications of the electronics industry spread out increasingly through the rest of the economy, and affected production methods in a revolutionary way. This process was stimulated intensively by astonishing improvements in quality, accompanied by drastic reductions in the production costs of electronic components. Therefore, "the competitive substitution pressure in a wide range of industries and services will become intense in the last quarter of this century".[31]

Comment

Freeman's theory is of importance for the present because of its ability to deal with the intensive job destruction effects of the electronics sector. However, it may be queried whether the electronics sector is having a more devastating effect on jobs than did the earlier leading sectors. An empirical test would be complicated, because one would have to know how much unemployment is caused by the process innovation, and how much by the saturation of consumption demand. Still, research in this field could be fruitful.

The distinction between process and product innovations is useful and more adequate than the distinctions between basic and improvement innovations developed by Mensch. The most interesting features of Freeman's theory, however, are those that accord with the capital-oriented long-wave theories. The key aspect of the long wave lies in investment behaviour, for example, in the production of electronic capital goods and its influence on other sectors via automation. A major difference is that capital intensity grows in the downswing in the labour approach, but declines in Forrester's capital approach. Furthermore, both long-wave theories emphasise the growing unemployment of production factors in industry during the downswing: labour for Freeman, and capital for Mandel and Forrester.

Scarcity and abundance of foodstuffs and raw materials

The theory of Rostow

More than the other authors considered here, Rostow emphasises the dynamics of growth in a world perspective.[32] He sharply distinguishes and relates three distinct phenomena: the forces set in motion by a leading sector in growth, by changes in the profitability of producing foodstuffs and raw materials, and by large waves of international or domestic migration. The second factor is particularly important because it generates a price revolution, ie a short period of a sharp rise in the relative prices of food and raw materials at the expense of the prices of industrial goods. If the producers of the former are the LDCs, and of the latter the developed countries, then a price revolution means a period of deteriorating terms of trade for the developed countries.

The price revolution reflects the disequilibrium between supply and demand of foodstuffs and raw materials at the end of the period of declining prices. Growing population and rising real income create an increasing pressure on the supply of food. In the same way, developing industrialisation creates a pressure on the available supply of raw materials. Within the existing structures, the supply of these goods becomes less flexible because of the declining marginal returns of improvement in exploitation. This disequilibrium is reflected in an enormous price rise, measured by historical standards. The opportunities for profit in these sectors change radically, so growing numbers of people and large amounts of capital are attracted to them. People search for fundamental solutions to this relative scarcity, and leading sectors, with important secondary effects, can develop. Once the scarcity of foodstuffs and raw materials is overcome, a new period of falling prices begins. During the downswing, the price decline of raw materials accelerates the diffusion of the new sectors through the economy.

Comment

Rostow has introduced some interesting elements into long-wave theory, although his analysis does not adequately integrate the post-war situation. At first sight, perversely, the 1951–72 period is considered as a downswing, with the upswing starting in 1972. However, Rostow's theory seems to be enlightening for the analysis of the present price revolution. The new shortages of primary goods and especially energy fit well into the adjustment theory and are likely to create new leading sectors on the input side of the economy: new energy sources (nuclear, solar, wind), new raw material sources (ocean reclamation), new foodstuffs industries (bio and genetic industries), environment industries, and so on.[33]

Conclusions

The theories we have dealt with emphasise different aspects of the long-wave movement, and have different starting points and assumptions. In reality, they are more complementary than we have so far suggested, multicausality being clearly a better approach to the complex phenomenon of the long wave. Along with Rostow and van Duijn, who integrated several theories, we also prefer an eclectic approach.

Van Duijn[34] constructs his synthesis by combining the innovation theory of Schumpeter and Mensch, the hypothesis about the life cycle of industry along an S-curve, and Forrester's multiplier accelerator mechanism of investment, which intensifies the growth and saturation of basic innovations. The role of time lags is considered essential for the particular behaviour and length of the long wave.

It is likely that the newest leading sectors will be reactions to the present scarcity of primary materials. This corresponds with recent research on new forms of energy and raw materials.[35] In this context, we are convinced that long-wave researchers have to investigate further the link between, on the one hand, the relative prices and the substitution of production factors, and on the other, the behaviour of industry and the economy during the long-wave movement. Rostow's theory about changing relative prices should be extended to include wages.

It is important to take into account how much the social and institutional framework has changed since the industrial revolution. It is valuable, therefore, to study the different long waves more in their historical context. Often, institutional changes took place unevenly and in step with the long wave, and technological innovations seemed to break through accompanied by them. According to Freeman, "It is important to remember that a major technology requires not one or two but a whole cluster of related inventions and institutional changes, which in combination enable it to take off ...".[36] And Mandel asserts: "...it is plain that this ascent and decline (of the rate of profit) is not determined by one single factor but must be explained by a series of social changes...".[37]

Moreover, many authors interpret the major economic crisis from the viewpoint of institutional innovations. Referring to the downswing in the last quarter of the 19th century, Chandler has stated: "In that period, the basic innovations were more in the creation of new forms of organisation and new ways of marketing. The great modern corporation ... had its beginnings in that period. Such organisations hardly existed, outside of the railroads, before the 1880s. By 1900 they had become the basic business unit in American industry".[38] Analogously, Akerman presents an institutional interpretation for the crisis of the 1930s, although he considers technological factors primary for the 1870s: "Si cette frontière structurelle [ie the 1870s] a des causes avant tout techniques, celle de 1920–29 est due essentiellement à raisons politiques: le passage du principe du profit à une structure d'économie dirigée".[39]

The question arises whether future technological innovations will be accompanied by institutional changes, and, if so, what their nature will be. While the McCracken report forecast further economic growth based on the institutional pillars of the past, the more recent INTERFUTURES report emphasises important social and institutional changes that are now breaking through.[40] These consist of changes in the ideas of the post-war generation *vis-à-vis* economic life and especially labour; the growing economic and political power of the LDCs; the increasing political pressure for better income distribution, both domestically and internationally; and the growing interdependence of the world economy.

If there is such a close link between economic and institutional change, a fundamental analysis of the long wave has to deal with it. And particularly in empirical testing, attention has to be paid to important changes in the parameters and the specification of the model. Perhaps the most fundamental problem is the question of whether our industrial society is itself following a life cycle of economic development. Should this be the case, we are the witnesses of the transition from a growth economy to an equilibrium economy.[41] Such a possible change presents a challenge to aggressive research on economic and industrial growth dynamics in the very long term.

Notes and references

1. C. Clark, *The Economics of 1960* (1944).
2. E. Mandel, "The heyday of neo-capitalism and its aftermath", *Socialist Register,* 1964.
3. J. van Gelderen and J. Fedder, "Springvloed", *De Nieuwe Tijd,* April–June 1913, *18,* 4, 5, 6.
4. J.J. van Duijn, *De Lange Golf in de Economie* (Assen, Van Gorcum, 1979), page 3. See also English edition: *The Long Wave in Economic Life* (London, Allen & Unwin, 1983).
5. Van Duijn, reference 4, page 6.
6. T.J. Broersma, *De Lange Golf in het Economisch Leven: Empirische en Theoretische Onderzoekingen* (Groningen, VRB, 1978); and T. Kuczynski, "Kondratiev cycles—appearance or reality?", *Proceedings of the 7th International Economic History Congress,* Vol 2 (Edinburgh, 1978), pages 79–86.

7. Van Duijn, reference 4, page 86; see also chapter 2 by J. Tinbergen in this book.
8. N.D. Kondratiev, "The long waves in economic life", *Review of Economic Statistics*, 1935, pages 105–115.
9. J.A. Schumpeter, *Business Cycles*, 2 vols (McGraw-Hill, 1939).
10. L.H. Dupriez, *Des Mouvements économiques généraux*, 3rd edition (Leuven, Nauwelaerts, 1966); and P. Temin, *Did Monetary Forces Cause the Great Depression* (New York, 1976).
11. G. Mensch, *Das technologische Patt* (Umschau Verlag, 1975); A. Kleinknecht, "Basisinnovationen und Wachstumsschübe: das Beispiel der westdeutschen Industrie", *Konjunkturpolitik*, 1979, 5/6, pages 320–343.
12. Van Duijn, reference 4, pages 50–51.
13. Mensch, reference 11, pages 17–27.
14. *Ibid*, subtitle.
15. *Ibid*, page 16.
16. *Ibid*, pages 29–43.
17. *Ibid*, page 108.
18. *Ibid*, pages 54–58
19. G. Mensch, *Stalemate in Technology, 1925–1935: The Interplay of Stagnation and Innovation*, Discussion paper (Berlin, International Institute of Management, 1977).
20. J.J. van Duijn, reference 4, pages 46–48.
21. E. Mandel, *Het Laatkapitalisme* (Amsterdam, Van Gennep, 1975).
22. *Ibid*, pages 85–86; see also chapter 14 by Mandel in this book.
23. *Ibid*, page 89.
24. *Ibid*, pages 92–95.
25. J.W. Forrester, "Growth cycles", *De Economist*, 1977, *4*, pages 525–543.
26. J.J. van Duijn, "The long wave in economic life", *De Economist*, 1977, *125*, 4, pages 544–575.
27. Forrester, reference 25, page 534.
28. *Ibid*, page 536.
29. C. Freeman, "The Kondratiev long waves, technical change and unemployment", in *Structural Determinants of Employment and Unemployment*, Vol II (Paris, OECD, 1977), page 183. For a more recent statement, see C. Freeman, J. Clark and L. Soete, *Unemployment and Technical Innovation* (London, Frances Pinter, 1982).
30. *Ibid*, pages 188–190.
31. *Ibid*, page 193.
32. W.W. Rostow, "Kondratiev, Schumpeter and Kuznets: trend periods revisited", *Journal of Economic History*, 1975, *4*, pages 719–753; and W.W. Rostow, *The World Economy, History and Prospect* (Macmillan, 1978).
33. OECD, *Facing the Future, Mastering the Probable, and Managing the Unpredictable*, INTERFUTURES Report (Paris, OECD, 1979).
34. Van Duijn, reference 26; J.J. van Duijn, "Actualiteit en Beleidsimplicaties van de Kondratiev Cyclus", *Economische en Statistische Berichten*, 1978, pages 76–82; and Van Duijn, reference 4.
35. OECD, reference 33.
36. Freeman, reference 29, page 188.
37. Mandel, reference 21, page 102.
38. A.D. Chandler, "The beginnings of 'Big Business' in American industry", *Business History Review*, 1959, pages 1–31.
39. J. Akerman, "L'analyse structurelle des variations économiques", *Bulletin de l'Ires*, 1948, 3/5, page 517.
40. P. McCracken *et al*, *Towards Full Employment and Price Stability* (Paris, OECD, 1977); and OECD, reference 33.
41. Forrester, reference 25; R.L. Heilbroner, *Business Civilization in Decline* (New York, Norton and Company, 1976).

CHAPTER 2

KONDRATIEV CYCLES AND SO-CALLED LONG WAVES

The early research

J. Tinbergen

This paper recalls some early work of the Dutch pioneers of long-wave research which anticipated many of the contemporary debates. Various explanations which have been advanced for the existence of long waves are reviewed, and the applicability of long-wave theories in a number of contexts is examined.

IN THE 19th and 20th centuries—ie from the beginning of industrialisation—several statistical time series have shown rising and declining movements, which could be interpreted as waves with a period of about 40 years. This applies clearly to the index numbers of wholesale prices, and also to the prices of wheat. Deviations from trends in many production series and interest rates are other examples of series showing wave-like movements. These fluctuations reached their highest points about 1870, 1920, and 1970, and their lowest points about 1850 and 1895.

Precise identification of these turning points is not possible because of disturbing influences caused by World Wars 1 and 2. The Dutch economist J. van Gelderen[1] (writing under the pen-name J. Fedder) was the first author to draw attention to these wave-like movements. In the international literature the Russian economist N.D. Kondratiev[2] is generally credited with the discovery of the long waves.[3] The register of Schumpeter's book does not even mention the names of Fedder or van Gelderen. On the other hand, Wesley C. Mitchell[4] gives full credit to van Gelderen and S. de Wolff for their pioneering research into long waves, which he explains in considerable detail.

Waves

The alternation of rising and declining movements can be regarded as waves,

Professor J. Tinbergen is Emeritus Professor of Economics at Erasmus Universiteit, Rotterdam, Postbus 1738, 3000 DR, Rotterdam, The Netherlands.

meaning that successive phases are connected by a causal mechanism or a logical interrelationship. Those who adhere to this interpretation prefer to speak of 'waves', or more precisely 'long waves', because the existence of two or three shorter cycles has been established, including the so-called business cycle, also called a 'Juglar' after the French economist who discovered it.

Those who are not yet convinced of the existence of a causal relationship between the various phases are inclined to avoid the use of the term 'waves', and prefer to speak instead of 'so-called long waves'.

Monetary theories

Various explanations of the long waves have been given. The first attempts at explanation were based on monetary theory. They emphasise the parallelism between the fluctuations in prices and in the total stock of monetary gold.[5] Movements in the gold stock have been influenced by various factors and by the discovery of new gold fields. If one agrees that these discoveries were largely a matter of chance, then the concept of a 'wave' would not play a role in this context. If, on the other hand, one takes into account that the production of gold is influenced by its price, then there is room for a systematic factor in the movements of the total stock of gold.

Whatever the case may be, the explanation is no longer of importance for practical policy, because the members of the International Monetary Fund (IMF) have agreed that in principle total money circulation should not depend solely on the gold stock (and according to others not at all). Instead it depends partly, and in future wholly, on the Special Drawing Rights (SDRs) allotted by the IMF to the various central banks. We have freed ourselves from the discoveries of gold fields and production plans (mainly of South Africa and the USSR) as determining factors, or at least we are on the road to greater freedom.

Investments for replacement

Other economists have tried to explain 'long waves' on the basis of the investment decisions of business enterprises. Two different theories can be distinguished which are important for the choice of a policy aimed at the gradual elimination of the long waves.

In the analysis of investment a distinction is made between replacement of existing capital goods, and new or autonomous investments. The first occurs when an existing capital good (building, machinery, vehicle, etc) goes out of use and must be replaced by a newly produced unit. The second type of investment (new or autonomous) occurs when someone starts a new production, ie enlarging the existing capacity for the output of known products, or when he starts producing entirely new products, such as the first vacuum cleaners, radio sets, television sets, or satellites.

Replacement investment will occur when there is no longer a profitable use for existing equipment. In an extreme technical sense this is the case, for instance, when a truck is destroyed in a road accident; in a less extreme sense when a machine is used up through wear and tear, or when a building has become dilapidated beyond repair.

In an economic sense replacement will be necessary also if the equipment has become obsolete, ie when it can no longer compete with equipment from more modern 'vintage years'. The new 'vintages' are usually cheaper in the productive process either because of the introduction of improvements, or because less labour is required. The Central Planning Bureau of the Netherlands attributes considerable importance to this type of economic obsolescence.

Reinvestments build a link with the remaining stock of capital goods from the past. Thus they may contribute to a theory of a wave. The theory which explains the long waves (and also the Juglar) with the help of investments for replacement is called the theory of the echo-principle, and has been advanced by S. de Wolff.[6] De Wolff distinguishes two types of capital goods, with an average length of life, respectively, of 40 years and about eight years. If, because of an accidental cause in a period t (of several years duration), a large number of capital goods with a life expectancy of T years is produced, then demand for replacement will be larger in the period $t + T$, and again T years later. Upper levels of reinvestment will be reached after periods of T years. Capital goods with a length of life of about 40 years include rolling stock of the railways according to figures published by the Netherlands Central Bureau of Statistics.[7]

Technical and economic length of life

The theory of the echo-principle has been discussed in several works, notably in the volume of essays published in honour of de Wolff in 1938.[8] The discussion was concerned mainly with the possible effects of the large variations in technical lifetimes according to figures assembled by H. Vos.[9]

Starting from one year t with a large production of the category of capital goods considered, and assuming replacement after a technical lifetime of T' years, the first echo shows a very flat curve only (around the year $t + T'$), and hardly a second echo. If, on the other hand, one takes into account that not the technical but the economic lifetime is decisive for replacement, it can be shown that waves with unchanged amplitudes are possible.[10]

It should be mentioned that de Wolff did not consider the initial high output in the year t as accidental, but as a result of a preceding more agrarian phase in the development of society which experienced years of large crops and of smaller crops, and consequently of low and high prices of cereals. These waves were probably influenced by the periodicity of sun spots. Although many questions about these cosmic influences on crops have remained unanswered, the elegance of the theory is not to be denied.

Life cycles or lag cycles?

Compared to competing business cycle theories, the echo principle has another advantage which applies to the Juglar with a possible extension to the long waves. The competing theories may be summarised as based on lag cycles. They have in common that a cyclical movement may develop if a time lag exists between the reaction of supply of a commodity to changes in its price. Well known examples which have been widely discussed are the markets for pigs, coffee and ships.

With respect to the market for pigs, it could be concluded, briefly, that after a period of high prices (or low prices) the supply of pigs would be high (low) about 1½–2 years later. This time lag is due to psychological and commercial reactions of the farmers, and the length of the biological process to 'produce' pigs. Thus a high price was followed after 1½–2 years by an increased supply, which could be sold only at reduced prices. Again 1½–2 years later supply was at a reduced level, and prices went up. The length of the cycle was 3–4 years, ie twice the time lag between price and supply.[11]

On the market for coffee,[12] a high price resulted after 6–7 years in an increased supply, because it encouraged increased planting of coffee trees. This development depressed coffee prices, but the trend continued because a coffee tree remains productive during a number of years, contrary to what happens in the case of pigs.

In the case of ships the situation was roughly the same. After a year of high freight rates more new ships were ordered, and after about a year these vessels were launched. Thus they tended to depress freights, and would continue to do so as long as they kept running—on an average about 17 years for a Dutch shipowner, often succeeded by low-cost operators for another 17 years. A mathematical analysis reveals that the resulting waves have a length equal to about four times the time lag involved.[13]

In more complete models of a country's economy as a whole, which include variables such as total production and total consumption, imports and exports, their prices etc, other time lags occur, for instance between income and consumption expenditure, and between profits and investment. Usually these time lags do not exceed about one year, but they can help to explain cyclical movements with a wave length of about eight years.

A disadvantage of this explanation of the ordinary business cycle is that small changes in the time lags mentioned may entail large changes in the length of the waves. However, the statistically observed waves, though far from regular, show wave lengths which are rather constant. The sensitivity of the waves determined by the length of life of capital goods is less, and this is an argument in favour of the theory based on the reinvestments.

Waves by autonomous new investments?

As was mentioned above, new investments are frequently caused by the discovery of new products and new methods of production. A number of economists have interpreted these developments as the causes of the long waves. They claim that important discoveries have occurred in certain periods, but much less so in between. As an example they mention the discovery and thereafter the application (this is what matters) of the steam engine, which around 1800 made industrialisation possible in England. Other examples are the introduction of steamships which after 1870 contributed largely to the transport of cereals across the Atlantic Ocean, of electricity, which caused steam engines to be replaced by electromotors, and a number of discoveries and innovations in chemistry, which greatly stimulated the growth of chemical industries. The discovery of the internal combustion engine—and therefore of the automobile—should also be mentioned.

The difficulty with this approach is, first, that it is not easy to define a measure of the importance of discoveries. Moreover, the question may be put whether accidental factors play a role here, and then we could not speak of true waves. With regard to the present, the economists mentioned hope for a new 'important' discovery to help us out of the deep depression which has been dragging on. Attention is focused on microprocessors: although it is not possible to make predictions, clearly they could exert an important influence on production and employment. On the other hand, there is every reason to question whether only technical innovations can bring about large-scale economic recovery. Could not social changes on a worldwide scale have a similar effect? Are we not able to control these social innovations ourselves, as in the case of the SDRs, which have replaced gold? There could be major advantages in, for instance, a much more vigorous development policy for the Third World.

Furthermore, strong arguments could be advanced in favour of accelerated programmes for scientific research. The projects of vital importance for the future of all include the following:

- How will the use of fossil fuel affect the climate?
- Which part of the total area of the globe should remain free and unobstructed nature?
- What risks are involved in the production of various types of energy?
- How could labour satisfaction be improved by alternative types of work?

At a large number of institutes research is underway on these and many other projects. How could this research be accelerated, coordinated, and financed? In doing this, how could we avoid more inflation?

Is there a causal link between new investments and preceding phases?

Returning once again to the new investments, should one consider them after all as being caused by the preceding long depression, instead of being autonomous and accidental? The longer the depression lasts, the more certain it is that enterprises and technicians will study new possibilities in an effort to set the industries on their feet again. If indeed this should be the case, then this explanation of rising and declining long-term movements would result in a wave, and therefore necessarily in a return to prosperity. Perhaps these aspects of the problem deserve more attention and can be added to our list of vital projects.

The most modern methods of research into the long-term movements include the work of J. Forrester.[14] In his case (and that of many other modern quantitative research workers) rather complicated, interdisciplinary, and mathematically formulated models are used, which cannot be considered fully within the context of a short article. Those authors should be considered as pioneers in a new and largely unknown domain. Interested readers are referred, for instance, to the work by P. Nijkamp *et al.*[15]

Notes and references

1. J. van Gelderen, "Springvloed: beschouwingen over industrieële ontwikkeling en prijsbeweging", in *De Nieuwe Tijd*, 1913, *18*.
2. N.D. Kondratiev, "Die langen Wellen der Konjunktur", *Archiv für Sozialwissenschaft*, December 1926, *LVI*, pages 573–609.
3. J.A. Schumpeter, *Business Cycles* (McGraw-Hill, 1939), page 164.
4. Wesley C. Mitchell, in *Business Cycles*, Volume I (National Bureau of Economic Research, 12th imprint, 1936), pages 227–229.
5. S. de Wolff, *Het economisch getij* (Amsterdam, J. Emmering, 1929, page 431.
6. *Ibid.*
7. Netherlands, Central Bureau of Statistics, *De Nederlandsche Conjunctuur* (1934), pages 22 and 37.
8. J. van der Wijk *et al*, eds, *Strijdenskracht door Wetensmacht* (Amsterdam, 1938).
9. H. Vos, "Over de levensduur van machines en gebruiksgoederen", in *ibid*, pages 117–132.
10. J. Tinbergen, *Selected Papers* (Amsterdam, North Holland, 1959), pages 85–92; see also J. Tinbergen and J.J. Polak, *The Dynamics of Business Cycles* (Chicago, 1950), pages 155–158, 178–180, 202, and 230.
11. H. Hanau, "Die Prognose der Schweinepreise", Sonderheft 2, zu den *Vierteljahresheften zur Konjunkturforschung* (Berlin, 1928).
12. H. Roth, *Die Übererzeugung in der Welthandelsware Kaffee im Zeitraum von 1790–1929* (Jena, Gustav Fischer, 1929).
13. J. Tinbergen, "Tonnage and freight", in Tinbergen, reference 10, pages 93–111.
14. J.W. Forrester, *World Dynamics* (Cambridge, MA, Wright-Allen, 1971).
15. P. Nijkamp *et al*, *Multidimensional Spatial Data and Decision Analysis* (New York, Wiley, 1979).

CHAPTER 3

FLUCTUATIONS IN INNOVATIONS OVER TIME

Jacob J. van Duijn

This paper first explains the concept of innovation life cycles and their relationship to leading sectors of the economy. It then briefly describes the relationship between innovation life cycles and long waves in macro-economic development, and presents a chronology and characterisation of the four phases of the long wave derived from Schumpeter. Various phases of the long wave could be expected to have different effects on the propensity to innovate, and this propensity probably varies also between new industries and existing established industries, as well as between product and process innovations. A list of 80 major innovations is classified by sector and date, to provide a preliminary empirical validation of the hypotheses advanced on the varying influence of the successive phases of the long wave on different types of innovation. Finally, the author's findings are compared with those of Mensch and Schmookler.

IN AN EARLIER study,[1] I have developed a long-wave theory in which the clustered occurrence of major or basic innovations is viewed as the major force which triggers economic fluctuations of the long-wave type. Some of these basic innovations will give rise to new industrial sectors. These sectors develop according to the familiar S-shaped life cycle pattern. New sectors require their own infrastructure. This demand for capital equipment follows output demand in the innovation-incorporating sectors. Through the operation of a multiplier-accelerator mechanism, excess accumulation of physical stock will eventually occur. The levelling off of demand in the innovation-incorporating sectors will accentuate the overexpansion of the capital sector. The combined effect of those two forces is a long-wave downturn, such as occurred in the early 1970s.

A long-wave decline means a period of depressed investment. A recovery sets in when replacement investment picks up and the multiplier-accelerator mechanism starts operating in an upward direction again. The stimulus of invest-

Jaap van Duijn is Professor of Economics at the Graduate School of Management, Poortweg 6, 2612 PA Delft, The Netherlands. This article is based on a paper written for the 1979 meeting of the Dutch Economic Association, entitled "De fluctuaties van innovaties in de tijd", *Preadviezen van de Vereniging voor de Staathuishoudkunde* (Leiden, Stenfert Kroese, 1979), pages 25–59. The author would like to thank Chris Freeman, Alfred Kleinknecht, and Gerhard Mensch for their comments on an earlier version of this article.

ment demand will change the overall economic outlook, setting the scene for a new burst of innovations.

In this paper, the issue of the timing of major innovations is further explored. I will do this at the meso-level, using the concept of the innovation life cycle; and at the macro-level, by dividing up the long wave into four phases: prosperity, recession, depression, and recovery.

Innovation life cycles

The perception of the economic life of a basic or major innovation as an innovation life cycle is not a novel one. I used it in earlier publications,[2] but as it turns out, it was developed earlier by Kuznets, and even, in a rudimentary form, by Schmookler.[3] The life cycle of a major innovation indicates how an innovation develops over time, measured as the output accounted for by that innovation.

Life cycles are usually assumed to be S-shaped, up to their decline phase, with gradually decreasing growth rates of output. While various interpretations and applications of the S-shaped growth curve exist, they can be reduced to two main types: the limited possibilities for further technical improvements given a certain state of technology, versus the limited possibilities for further market penetration given a certain rate of penetration. Both interpretations can be applicable to an innovation life cycle. Usually the two aspects will be interrelated: cost-reducing improvement innovations can increase the rate of penetration of a product; market saturation will be a strong incentive for product improvements, such that a decline in sales can be prevented or postponed. The strongest force, however, would seem to be the market.

A simple four-phase model can be used to illustrate the course of an innovation life cycle. The different phases are characterised by demand structure and type of innovative activity:

Introduction: there is a large number of product innovations, as different technological options exist, and little is known about the nature of demand.

Growth: there is increasing acceptance by customers, with a decreasing number of product innovations. Sales growth leads to standardisation of technology, and there are cost-reducing process innovations.

Maturity: the output rate slows down, and competition through product differentiation increases; innovations concern improvements. Process innovations are labour saving.

Decline: declining sales. Attempts are made to escape saturation through changes in technology, and the use of labour-saving process innovations continues.

Actually, the decline phase should be considered an open-ended phase. Absolute decline, resulting in complete replacement, is but one of the possible courses of an innovation following its maturity. Several variations to the standard life cycle pattern exist (Figure 3.1), representing the different ways industries may react when faced with saturated markets. From empirical observation, it would appear that very few of the late-19th and early-20th

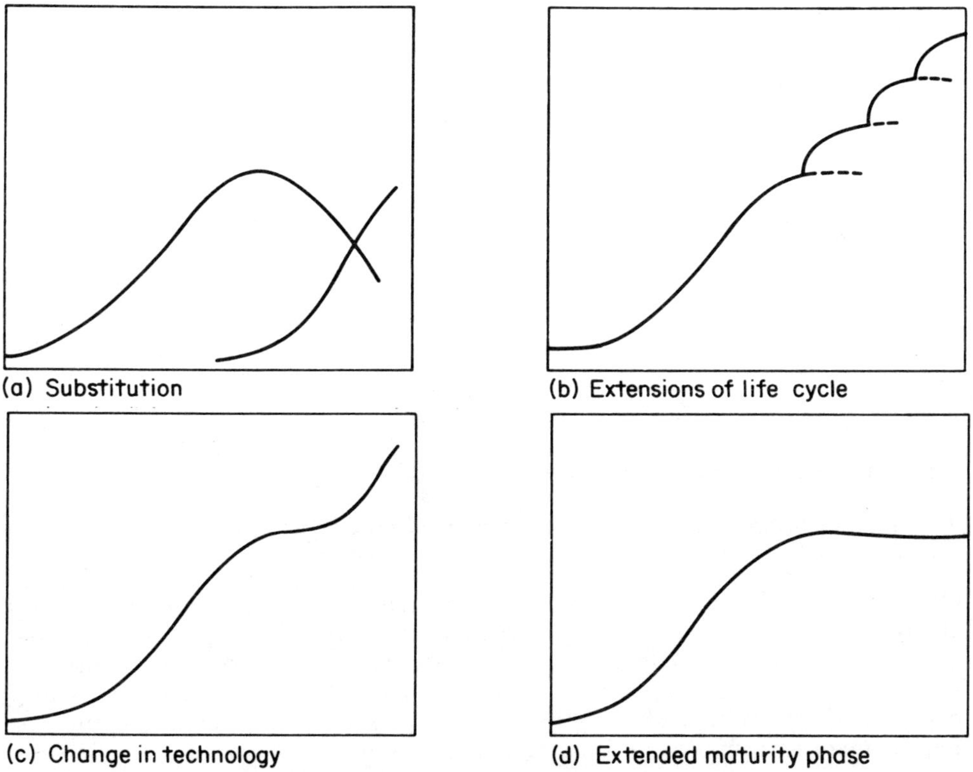

Figure 3.1. Variations to the simple life cycle pattern

Source: J.J. van Duijn, "Another look at industry growth patterns", Faculty Working Papers number 667 (College of Commerce and Business Administration, University of Illinois at Urbana-Champaign, April 1980).

century major innovations have been completely replaced by substitutes. Even the number that faces prolonged decline is rather limited. This may imply something about the length of the life cycles of major innovations, although one should realise that population growth has been an important intervening factor, which has increased the size of markets considerably (even in fairly closed economies such as the US economy).

Little can be said about the length of the various phases of the innovation life cycle. Innovations are too diverse in nature and area of application to allow for any generalisations with respect to the length of their lives. Also, the market conditions at the time and place of introduction will vary considerably. Compare, for instance, the introduction of television in the USA with the introduction of the motor car in that country. The latter innovation needed an introduction phase of approximately 15 years before the growth phase set in; television sales, on the other hand, reached their peak a few years after introduction. After 1950, a slightly decreasing trend has even set in. The quick growth to maturity of television can be attributed to pent-up consumer demand that had no outlet during World War 2.

Kuznets[4] mentions a time span of about 20 years for what roughly corresponds to our introduction phase, and much more than 20 years for the phase of diffusion, so that, even when the obsolescence phase is disregarded, life cycles tend to be half a century or longer. Obviously this matter can only be resolved through detailed studies of particular innovations, and by being very specific about geographical market area (do we study the life cycle of television in the USA, or on a worldwide basis?).

Despite our limited knowledge of the duration of innovation life cycles,the question of life cycle length is very important. For innovations contribute to growth, but the size of that contribution depends, among other things, upon the phase of the life cycle. Insight into the duration of various phases would give us insight into the growth potential of an economy. For the present situation, it is important to know to what extent the transition from rapid growth to maturity (or even to decline) of the post-war leading sectors could offer an explanation for the present economic stagnation.

Leading sectors and macro-economic growth

The view of long-term macro-economic development as a succession of leading sectors is best represented by Rostow:[5]

> Old sectors may be declining, others may be stagnant, others may be moving forward at about the average rate of industrial production as a whole (like raw material production); but there will be one or more leading sectors, reflecting the introduction of major new technologies, moving ahead more rapidly than the average, absorbing a disproportionate volume of current investment and able entrepreneurs, stimulating requirements to sustain it, and, quite often, bringing about accelerated urban growth in new regions.

The essence of leading sectors is that they incorporate major new technologies. Rostow's leading sectors are therefore nothing but our major innovation life cycles in their growth phase. Major innovations create leading sectors. Historically, the classic sequence of leading sectors is: cotton textiles, railways and iron; steel, chemicals, and electricity; and the motor car industry.[6]

These are broad categories, however. Chemicals include fertilisers as well as plastics; electricity includes the incandescent lamp as well as the electronic computer. A more refined industrial classification should shed light on the question of sequential appearance or (partial) coincidence of major innovation life cycles.

Elsewhere, I have argued that the sequence of (clusters of) innovation life cycles could offer an explanation for what I have termed the long-cyclical character of macro-economic growth, ie the observed alternation of growth acceleration and growth retardation.[7] Certainly, fluctuations in capital good production do also affect these long-cyclical developments. Besides theoretical support, there are empirical indications that corroborate the importance of (infrastructure) investment as a causal factor in long waves. The most spectacular evidence of the fluctuating character of capital construction is the appearance of excess capacity in the 1970s in such sectors as steel production, shipping, synthetic fibres, and chemicals. Yet excess capacity is always the result of capital construction on the one hand, and market sales developments on the other hand. Capital good explanations of the long wave, too, have to deal with the role of leading sectors.

TABLE 3.1. MACRO-ECONOMIC CHARACTERISTICS OF LONG-WAVE PHASES

Characteristic	Depression	Recovery	Prosperity	Recession
gross national product	little or no growth	increasing growth rates	strong growth	decreasing growth rates
investment demand	excess capacity; rationalisation	increase in replacement investment	strong expansion of capital stock	scale-increasing investment
consumer demand	for a while continued growth at the expense of savings	purchasing power seeks new outlets	expansion of demand in all sectors	continued growth of new sectors

Following Schumpeter, I have divided up the long wave into four phases: prosperity, recession, depression, and recovery. Prosperity and recession together form the upswing phase; depression and recovery, the downswing phase. Table 3.1 summarises the macro-economic characteristics for each phase, and Table 3.2 presents my long-wave chronology (using Juglar peaks—ie business booms of 7–11 years—as demarcation points). If we were to perceive the long wave as a succession of life cycles of leading sectors, it would be tempting to match long-wave phases with life cycle phases:

- prosperity—growth,
- recession—maturity,
- depression—decline, and
- recovery—introduction.

But even though it cannot be denied that periods of prosperity are carried by the strong growth of one or more leading sectors, phases do not coincide so smoothly. An important deviating pattern can be observed: leading sectors can survive macro-economic depressions (ie continue to grow during depressions, albeit at a reduced rate), and resume expansion again during the next upswing. The US motor car and aircraft industries are cases in point. Both resumed growth after World War 2. Micro-electronics is an innovation-packed sector that is growing through this depression.

Major innovation life cycles can therefore be stretched out over more than one long wave. Yet it should be emphasised that growth rates are not nearly as high 'the second time around'.[8] In the case of the motor car industry, it was initial demand that led to rapid expansion in the 1920s. When this demand was becoming saturated, excess capacity resulted, thus starting a negative multiplier-accelerator effect.[9] Post-war growth was mainly based on replacement demand, and Rostow concludes that:

TABLE 3.2. A LONG-WAVE CHRONOLOGY

1st Kondratiev	2nd Kondratiev	3rd Kondratiev	4th Kondratiev
prosperity 1782–1792	prosperity 1845–1857	prosperity 1892–1903	prosperity 1948–1957
prosperity 1792–1802	prosperity 1857–1866	prosperity 1903–1913	prosperity 1957–1966
(war 1802–1815)		(war 1913–1920)	
recession 1815–1825	recession 1866–1873	recession 1920–1929	recession 1966–1973
depression 1825–1836	depression 1873–1883	depression 1929–1937	
recovery 1836–1845	recovery 1883–1892	recovery 1937–1948	

Source: J.J. van Duijn, Chapter 20, in S.K. Kuipers and G.J. Lanjouw, eds, *Prospects of Economic Growth* (Amsterdam, North-Holland, 1980), pages 223–233.

by the late 1950s, the automobile and durable consumers' goods complexes, while massive elements in the American economy, lacked the momentum to qualify as leading sectors.[10]

So a simple match between long-wave phase and innovation life cycle phase does not exist. Long waves are also caused by fluctuations in industrial infrastructure investment; major innovations, while being diffused at gradually decreasing growth rates, can extend their potential over more than one long-wave expansion phase. Yet every expansion phase has coincided with the introduction of new clusters of innovations. What then determines the timing of these innovations?

Before attempting to answer this question, a distinction should be made between various types of innovation. I will distinguish between:

- major product innovations which create new industries;
- major product innovations in existing industries;
- process innovations in existing industries; and
- process innovations in basic sectors (such as steel or oil refining).

To the first category belong such innovations as the motor car, the airplane, radio, synthetic fibres, plastics, or the electronic computer—to name but a few. Their introduction often gave rise to new firms that eventually became an industry, although in some instances existing firms have been responsible for the main innovations (as in chemicals), with new firms joining in. All these innovations have in common the creation of completely new markets.

The second category contains innovations introduced by firms as an answer to the market saturation of their existing product line. For these firms, the pressure to innovate is strong in the maturity phases, and even more so in the decline phases, of their industry life cycles. Figure 3.1 applies to these industries. Current examples are television production and the gramophone record industry. I assume that, for some industries at least, saturation will coincide with, and partly cause, a long-wave downturn. This would be true at present for not only the two industries just mentioned, but also for such industries as the European car industry, synthetic fibres, and some household durables.

Process innovations that improve the production process in existing industries constitute the third category. Process innovations will most likely be a response to two different forces: demand increases pushing firms to increase labour and capital productivity; and cost increases pressuring firms to reduce their requirements for their most expensive input factors (labour, energy).

The final category is innovations in basic sectors, of which steel and oil have so far been the most important. Here, increased demand will be a strong impetus to innovate.

In Table 3.3 I have hypothesised how strong the propensity to innovate is likely to be, during each of the four phases of the long wave, for each of these four categories of innovations. Major product innovations, which create new industries, are certainly possible during a depression phase. However, it is much more likely that they will be introduced during the recovery phase, when increasing demand for replacement investment will turn the pessimism of the depression phase into a more optimistic economic outlook. Projects which carry

TABLE 3.3. THE PROPENSITY TO INNOVATE DURING DIFFERENT PHASES OF THE LONG WAVE

Type of innovation	Depression	Recovery	Prosperity	Recession
product-innovation (new industries)	+	++++	++	+
product-innovation (existing industries)	+++	+++	+	+
process innovations (existing industries)	+++	+	++	++
process innovations (basic sectors)	+	++	+++	++

major risks are only undertaken in a favourable economic climate. Prospects are even better during the prosperity phase, although the presence of new growth sectors will make the introduction of still more radical innovations less urgent.

In existing industries, the majority of innovations are likely to be introduced during the depression and recovery phases. These industries can respond more quickly to long-wave downturn, provided that this downturn coincides with demand saturation for existing products. One reason for this quicker response is that existing industries are more aware of the life cycle phases in which their own products are; another is that changing the technological base of a product is less risky when it continues to serve the same market (ie to meet the same need). As soon as the new generation of products (eg gramophones, or records) has gained acceptance by the public, the need for further, radical product innovations decreases.

Process innovations in existing industries are seen as the reaction to two kinds of forces: market demand increases, and cost increases. While the effects of process innovations are most noticeable during long-wave downturns (through increases in unemployment), these innovations are implemented just as well during long-wave expansion. Innovations in basic sectors (oxygen steel making, catalytic cracking) usually occur as responses to final demand increases. Demand-induced innovations in producer goods sectors will be mainly introduced during long-wave expansion.

Distinguishing various categories of innovations yields a more detailed view of the process of technological change as it develops over time. Table 3.3 would hypothesise that the overall propensity to innovate is highest during the recovery phase, and lowest during the recession phase. It can also be used to explain the contradiction that is apparent in present discussions on innovation. On the other hand, the lack of innovation is emphasised (as is witnessed by the many government memoranda on the subject); on the other hand, expressions like 'the chip revolution' and the 'information explosion' suggest an abundance of innovations. Table 3.3 shows both: there is indeed a lack of (employment-creating) product innovations in new sectors, but certainly no lack of (labour-saving) process innovations in existing sectors.

Table 3.3 offers no more than a stylised picture of the propensity to innovate as determined by meso- and macro-economic factors. In reality, several other forces affect this propensity. War conditions, for instance, have historically been a major force in bringing about innovations. In confronting the hypotheses of Table 3.3 with empirical evidence, this should be borne in mind.

Major innovations in 13 industrial sectors

This section lists some 80 major innovations that shaped the lives of 13 20th-century growth sectors. Presenting lists of major innovations is a hazardous venture. Every author in the field agrees that it is extremely difficult to measure the contribution of a particular innovation, and this makes every list subject to much criticism.

Considering the lack of any alternative measure of fluctuations in major innovations over time, I feel the best one can do is to reduce the arbitrary character of these lists by, first, relying on the judgements of various experts and, second, by classifying them according to sector, thus combining them into less heterogeneous groups. An additional advantage of this latter procedure is that the timing of innovations can be put in a life cycle perspective.

Space limitations prevent me from commenting extensively on the timing of the innovations listed in Table 3.4. I present brief summary statements instead.

In only one sector—medicine—does there seem to be no relation between major innovations and macro-economic development or industry life cycle. The timing of innovations in synthetic fibres and plastics seems determined more by technology push than by demand pull. These sectors are dominated by large firms capable of financing large and long-lasting research programmes. In other sectors, either the long-wave phase or the life cycle phase offer clues as to when major innovations are to be expected.

We can confront our preliminary findings with the views of Mensch and the

TABLE 3.4. MAJOR INNOVATIONS IN 13 INDUSTRIAL SECTORS

	Innovation	Invention	Innovator
A. Printing	1846 rotary press	1790	Hoe Rotary, USA
	1886 linotype	1884	New York Tribune, USA
	1894 monotype	1887	Sellers & Co, Phil, USA
	1946 phototype	1936	American Intertype Corp, USA
	1950 xerography	1937	Haloid Corp, USA
Note: cost-reducing innovations notably during recovery phases.			
B. Oil refining	1859 drilling for oil	1859	Pennsylvania Rock Oil Co, USA
	1913 thermal cracking	1909	Standard Oil of Indiana, USA
	1920 continuous thermal cracking	1909	Texas Co, USA
	1937 catalytic cracking	1927	Sun Oil, Socony-Mobil, USA
	1942 continuous catalytic cracking	1942	Standard Oil of New Jersey, USA
Note: derived demand (car industry, aircraft, petro-chemicals); innovations as reactions to increasing demand from these sectors.			
C. Resins, etc	1870 celluloid	1865	J.W. Hyatt, USA
	1910 bakelite	1872	Bakelite Corp, USA
	1917 cellophane	1912	La Cellophane, France
	1930 polystyrene	1925	I.G. Farben, Germany
	1932 polyvinylchloride	1931	I.G. Farben, Germany
	1935 plexiglas	1912	Röhm & Haas, USA
	1939 polyethylene	1936	ICI, Britain
	1943 silicones	1904	Dow-Corning, USA
	1959 polyacetates	1924	DuPont, USA
	1960 polycarbonates	1935	Bayer, Germany; General Electric, USA
Note: technology push dominates, but strongest growth is during prosperity phases.			

Table continues

Table 3.4 continued

	Innovation	Invention	Innovator
D. Rayon, synthetic fibres	1890 rayon (nitro-cellulose process)	1884	De Chardonnet, France
	1898 rayon (cuprammonium process)	1890	
	1901 viscose rayon	1892	
	1920 acetate	1902	
	1938 nylon	1934	DuPont, USA; I.G. Farben, Germany
	1948 orlon	1945	DuPont, USA
	1950 terylene	1941	ICI, Britain

Note: technology push character dominates, but strongest growth is during prosperity phases.

E. Medicine	1899 aspirin	1853	Bayer, A.G., Germany
	1923 insulin	1921	Connaught Labs, Toronto, Canada
	1932 antimalaria drugs	1932	Eli Lily Co, USA
	1932 sulphur drugs	1917	I.G. Farben, Germany
	1942 penicillin	1929	Kemball, Bishop & Co, Britain
	1946 streptomycin	1924	Merk & Co, USA
	1948 cortisone	1931	Merk & Co, USA

Note: innovations respond to dominant causes of death and illness; no relation with macro-economic developments.

F. Rubber	1840 vulcanised rubber	1839	Goodyear, USA
	1889 pneumatic tyre	1845	Dunlop Pneumatic Tyre Co, Britain
	1906 chemical accelerator for rubber vulcanisation	1906	Diamond Rubber Co, USA
	1932 synthetic rubber	1882	DuPont, USA
	1950 radial tyre		Michelin, France

Note: derived demand (car industry); innovations as reactions to demand increases in these sectors.

G. Steel	1856 steel making (Bessemer process)	1855	various countries
	1864 steel making (Siemens-Martin process)	1857	various countries
	1879 steel making (Thomas process)	1877	various countries
	1913 stainless steel	1911	various countries
	1923 continuous hot strip rolling	1892	Armco, USA
	1952 continuous casting	1927	Mannesmann, Germany
	1953 oxygen steel-making	1939	Vöest, Austria

Note: derived demand (follows macro-economic developments); innovations respond to increasing macro-economic demand.

H. Illumination	1880 incandescent lamp	1854	Edison Lamp Works, USA
	1938 fluorescent lamp	1896	Westinghouse, General Electric, Sylvania Electric, USA

Note: timing of innovations may have been determined by existing life cycle.

I. Telecommunications and related equipment	1877 telephone	1860	Bell Telephone, USA
	1888 cylindrical record player	1877	Columbia, Edison, USA
	1889 mechanical record player	1887	Kämmerer & Rheinhardt, Germany
	1913 vacuum tube	1904	AT&T, USA
	1920 AM-radio	1900	Westinghouse Co, USA
	1924 dynamic loudspeaker	1906	
	1925 electric record player	1908	Brunswick Co, USA

Table continues

Table 3.4 continued

I. Continued	**Innovation**	**Invention**	**Innovator**
	1935 magnetic tape recorder	1899	Magnetophon (AEG), Germany
	1936 television	1907	Electrical & Musical Industries, Britain
	1936 FM-radio	1902	Telefunken, Germany
	1948 long-playing record	1948	CBS, USA
	1951 transistor	1947	Bell Telephone Labs, USA
	1953 colour television	1925	RCA, USA

Note: timing of innovations determined by existing life cycles and scope for substitution.

J. Electronic computers	1951 transistor	1947	Bell Telephone Labs, USA
	1951 electronic computer	1944	Remington Rand, USA
	1961 integrated circuit	1959	Fairchild; Texas Instruments, USA
	1971 micro-processor	1959	Intel, USA

Note: cost-reducing innovations continue during depression phase.

K. Motor cars	1860 internal combustion engine	1853	Societé des Moteurs Lenoir, France
	1876 four-stroke engine	1862	Gasmotorenfabrik Deutz, Germany
	1888 automobile	1883	Benz, Germany
	1894 motor cycle	1885	Hildebrand & Wolfmüller, Germany
	1895 electric automobile	1874	Acme & Immisch, Britain
	1913 assembly line	1913	Ford Motor Co, USA
	1948 automatic transmission (passenger cars)	1904	Buick, USA
	1951 power steering (passenger cars)	1926	Chrysler, USA
	1967 Wankel-motor	1954	NSU, Germany

Note: strong demand has evoked few major innovations.

L. Aircraft	1910 airplane	1903	Military airplanes, France
	1938 helicopter	1907	Focke-Wulf, Germany
	1942 jet airplane	1930	Messerschmitt, Germany

Note: timing of innovations seems to be determined by existing life cycles and scope for substitution.

M. Photography	1888 portable camera	1881	Eastman Kodak, USA
	1924 Leica	1913	Leitz, Germany
	1935 colour photography	1912	Eastman Kodak, USA
	1948 Polaroid Land camera	1937	Polaroid, USA

Note: timing of innovations seems to be determined by existing life cycles.

Sources: R. Baker, *New and Improved ...* (London, British Museum Publications, 1976); J.L. Enos, "Invention and Innovation in the Petroleum Refining Industry", in *The Rate and Direction of Inventive Activity* (New York, National Bureau of Economic Research, 1962), pages 299–321; C. Freeman, *The Economics of Industrial Innovation* (Penguin, 1974); J. Jewkes, D. Sawers, and R. Stillerman, *The Sources of Invention*, 2nd edition (Macmillan, 1969); B.J.A. van der Kooy, *Microcomputers* (Deventer, Kluwer, 1978); D.S. Landes, *The Unbound Prometheus* (Cambridge University Press, 1969); K.B. Mahdavi, *Technological Innovation* (Stockholm, Beckmans, 1972); W.F. Mueller, "The origins of the basic inventions underlying Du Pont's major product and process innovations, 1920 to 1950", in *The Rate and Direction of Inventive Activity*, pages 323–346; P. Robertson, *The Shell Book of Firsts* (London, Ebury Press, 1974).

empirical results of Schmookler. Mensch argues that basic innovations will be introduced during depression phases ("in der Talsohle der Konjunktur"[11]); Schmookler found that changes in inventive activity tended to lag output changes.[12] While both authors accentuate the induced character of the innovation process, their conclusions on the timing of innovations seem to be essentially different. Mensch sees the depression as a trigger to innovative activity. Schmookler finds that the number of patents awarded in an industry increases only after demand has increased. Considering that patenting often precedes innovation, this would seem to imply that Schmookler's innovations occur much later than Mensch's.

There is a contradiction here, but it can be easily resolved. Both authors would agree on the importance of demand pull in stimulating innovations. Also for those basic innovations which actually occurred, the demand-pull theory holds. Mensch, however, would add that with respect to potential basic innovations, or 'opportunities', the pull-mechanism may fail because of some intervening variables, which give way most often in times of depression.

In addition, it is necessary to distinguish between different types of innovations once again. Mensch's list includes product innovations in new sectors as well as process innovations in basic sectors. I have argued that these two categories will follow different timing patterns. If we lump together all 80 innovations listed in Table 3.4, the time pattern shown in Table 3.5 emerges for the 1920–1960 period.

This time distribution indicates that the 1930s and 1940s were innovation-rich decades—something no one denies—without, however, revealing any tendency for innovations to be clustered in the midst of the Great Depression, rather than during and after the war. One certainly could not argue, therefore, that firms in the 1930s jumped at the opportunities that were offered by the then available innovations, as Mensch suggests. If a specific cause of basic innovations in the 1930s and 1940s has to be singled out, it would have to be rearmament and war-related demand.

Clearly, there is little incentive to embark on risky, innovative ventures during long-wave expansions, when the current crop of growth sectors is still in full bloom. In that respect, the entering of maturity and decline phases of particular industry life cycles can be said to force new product innovations out of necessity. Ultimately, however, innovators need to be convinced that they are tapping new growth markets. That conviction comes easier during recovery than in the midst of a depression.[13]

Then what about Schmookler's findings? At first glance, his empirical work would seem to indicate that innovations are abundant in recovery and prosperity phases. However, one should be careful here. First of all, Schmookler's conclusions are based on empirical evidence for a number of older sectors (railways, agriculture, petroleum, and paper). Secondly, his fluctuations in patent activity are short waves, not long waves. Thirdly, patent statistics do not discriminate between major and minor technological improvements. So Schmookler's empirical findings have to be placed in their proper perspective: within established sectors, technological change will follow increases in market demand; business-cycle expansions will stimulate inventive activity, even though the resulting innovations may be marginal improvements.

TABLE 3.5. MERGING THE 80 INNOVATIONS FROM TABLE 3.4

recession	1921–1929	5 innovations
depression	1930–1937	11 innovations
recovery	1938–1948	15 innovations
prosperity	1949–1957	9 innovations

Schmookler is not concerned with the succession of life cycles and the dating of major innovations. His choice of sectors implies that he was mainly concerned with the economic motives underlying improvement innovations.

My emphasis on the recovery phase as the period during which major product innovations are most likely to be introduced does not conflict with Schmookler's thesis. With Schmookler I agree on favourable market prospects as a necessary condition for innovation. But since he and I focus on different types of innovations, the expected timing will differ.

Notes and references

1. J.J. van Duijn, *De Lange Golf in de Economie* (Assen, Van Gorcum, 1979). English version: *The Long Wave in Economic Life* (London, Allen & Unwin, 1983). See also J.J. van Duijn, "The long wave in economic life", *De Economist*, 1977, *125*, 4, pages 544–576.
2. J.J. van Duijn, *Eb en Vloed: de Lange Golf in het Economische Leven* (Inaugural Lecture, Erasmus University, Rotterdam, 1977) and van Duijn, *De Lange Golf in de Economie* (reference 1), pages 69 ff.
3. S. Kuznets, "Innovations and adjustments in economic growth", *Swedish Journal of Economics*, 1972, *74*, pages 431–451; S. Kuznets, "Technological innovations and economic growth", in P. Kelly and M. Kranzberg, eds, *Technological Innovation: A Critical Review of Current Knowledge* (San Francisco Press, 1978), pages 335–356, reprinted in S. Kuznets, *Growth, Population, and Income Distribution* (New York, W.W. Norton, 1979), pages 56–99; J. Schmookler, "Technological change and the law of industrial growth", in J. Schmookler, *Patents, Invention and Economic Change* (Harvard University Press, 1972), pages 70–84.
4. S. Kuznets, "Technological innovations and economic growth", page 64 (of the 1979 reprint; see reference 3).
5. W.W. Rostow, *The World Economy: History and Prospect* (Macmillan, 1978), pages 104–105. Also by the same author: *The Process of Economic Growth* (Oxford, Clarendon Press, 1960), and *How it All Began* (London, Methuen, 1975).
6. Rostow, *The World Economy* (see reference 5), pages 105–106.
7. Van Duijn, *De Lange Golf in de Economie* (see reference 1), pages 88–89.
8. Compare these annual growth rates for the US motor car industry:

1900–13	43·8%	1929–48	−0·7%
1913–29	15·2%	1948–73	3·7%

   Source: see Figure 3.1.
9. See J.F. Willis and M.L. Primack, *Explorations in Economics* (Boston, Houghton Mifflin, 1977), page 204.
10. Rostow, *The World Economy* (see reference 5), page 263.
11. G. Mensch, *Das technologische Patt* (Frankfurt u/Main, Umschau Verlag, 1975), page 180; see also G. Mensch, *Stalemate in Technology* (Cambridge, MA, Ballinger, 1979).
12. J. Schmookler, *Invention and Economic Growth* (Harvard University Press, 1966).
13. In his 1979 *Stalemate in Technology*, Mensch seems to agree with this timing, when he predicts (page 201): "Only a small number of basic innovations will be achieved prior to 1984 [ie during the depression phase]. Approximately two-thirds of all the technological basic innovations that will be achieved in the second half of the twentieth century will occur in the decade around 1989 [ie during the recovery phase]".

CHAPTER 4

CHANGING CAPITAL VALUES AND THE PROPENSITY TO INNOVATE

Gerhard Mensch, Charles Coutinho and Klaus Kaasch

Basic innovations tend to cluster in periods of discontinuity, when changes in the marginal efficiency of capital (amongst other things) lead to disappointing depreciation in the operative value of installed capital goods in stagnating industries, thus inducing investors to seek alternatives, and so making the economic system ready for new technologies. The conditions encouraging innovation are formulated in a contingency theory of changes in capital values, changes which occur as the economic influence bestowed by owning production facilities moves from one sector to another. This non-equilibrium theory allows short- and long-term forces of balance and dominance in economic exchange to be considered together.

WE SHARE the view of many economists that, at the heart of the troublesome issue of long-term industrial fluctuations, lie some unexpected, yet patterned appreciations and depreciations of certain parts of the capital stock, and subsequent changes in the propensity of capital owners either to invest further in these existing kinds of capital good, or to invest into alternative means of production.

"The weakness of the inducement to invest has been at all times the key to the economic problem", wrote Keynes during the Great Depression, in his *General Theory*.[1] This statement, though probably correct in general, does not penetrate down to the specific causes of temporal weakness of the investment propensity, and of sectoral differences in this weakness at a given time, or to changes over time. It nevertheless offers a very fruitful first approach to the problem. For example, it comes as no surprise that, as expenditure on new plant and equipment in US non-farm business has been doubling from decade to decade (1950, \$27 320m; 1960, \$48 630m; 1970, \$99 520m; and 1978, \$178 080m), real investment has levelled off since 1966, and investors have come to think twice before starting a project. Figure 4.1 depicts the discrepancy between nominal and real investment. The average investor seems to be facing diminishing returns to increasing effort, the marginal efficiency of capital having decreased. This is consistent with both the wave model (Kondratiev, Schumpeter) and Mensch's metamorphosis model,[2] as depicted in Figure 4.2.

The authors are at the Department of Economics, and the Weatherhead School of Management, Case Western Reserve University, Cleveland, Ohio 44106, USA.

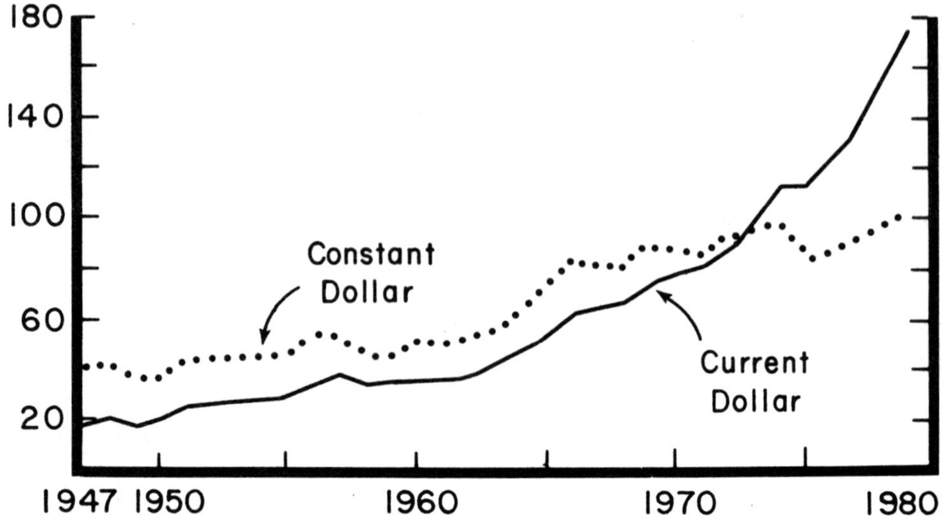

Figure 4.1. Investment expenditure on new plant and equipment, USA. All business; in current and in 1972 dollars; thousand millions.
Source: Thirty-third Annual McGraw-Hill Spring Survey of Business Plans for New Plant and Equipment, 1980–83.

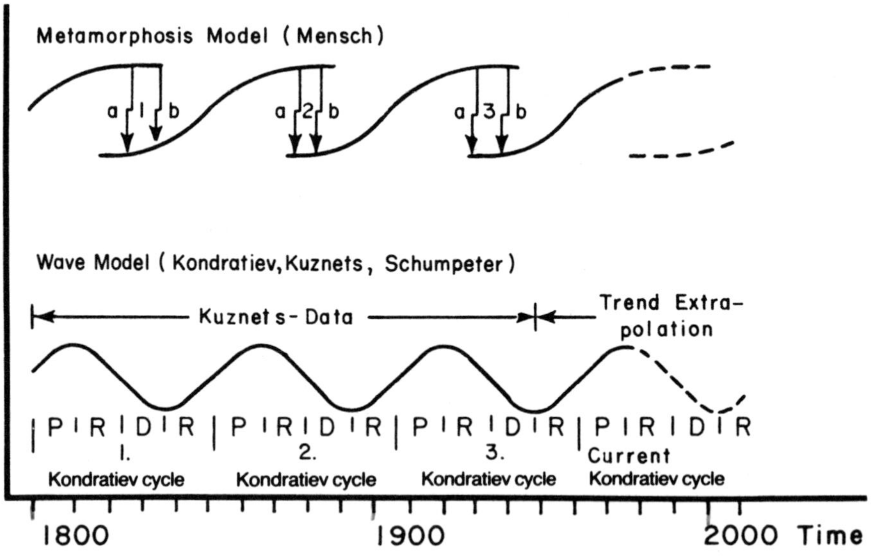

Figure 4.2. The metamorphosis model of long-term industrial fluctuations.
Source: Mensch, reference 2, page 73.

A closer approximation to the point we are trying to make is Keynes' suggested explanation of the Depression:

I suggest that a more typical, and often the predominant, explanation of the crisis is, not primarily a rise in the rate of interest, but a sudden collapse in the marginal efficiency of capital.

Figure 4.3 depicts, in terms of fixed capital investment in US producer durable equipment, what Keynes in the 1930s might have meant by a "sudden collapse". Inspection of the 80 years time series leads to two preliminary conclusions:

- The aggregate data show little resemblance to a pattern approximating a sinusoidal (long-wave) model.
- The aggregate data show much resemblance to a pattern suggested by the metamorphosis model—a succession of long-term, sigmoid trend cycles.

Furthermore, the metamorphosis model offers two additional advantages over the mechanistic wave model. Indeterminism is the first. During the trend period of stagnation on a high plateau, there could be a sudden collapse as referred to by Keynes, or there could be a uniform decline as suggested by Burns,[4] rejected by Gold,[5] and discussed by van Duijn;[6] or there could be a temporary, stationary equilibrium state of retardation, as envisioned by Mensch,[2] where the collapse is avoided and the depression overcome by basic innovation. In fact, all these developments can and do happen simultaneously in different sectors—and this needs explaining.

The second merit concerns structuralism: at any time, the short-term economic performance of a firm, a branch, or an industrial sector, can be studied, compared cross-sectionally, and evaluated longitudinally, all relative to some long-term average or trend. This, of course, can be done without recourse to any long-term theory. However, we firmly believe that long-term theory not only could be much more sophisticated than the mechanistic black-box (eg

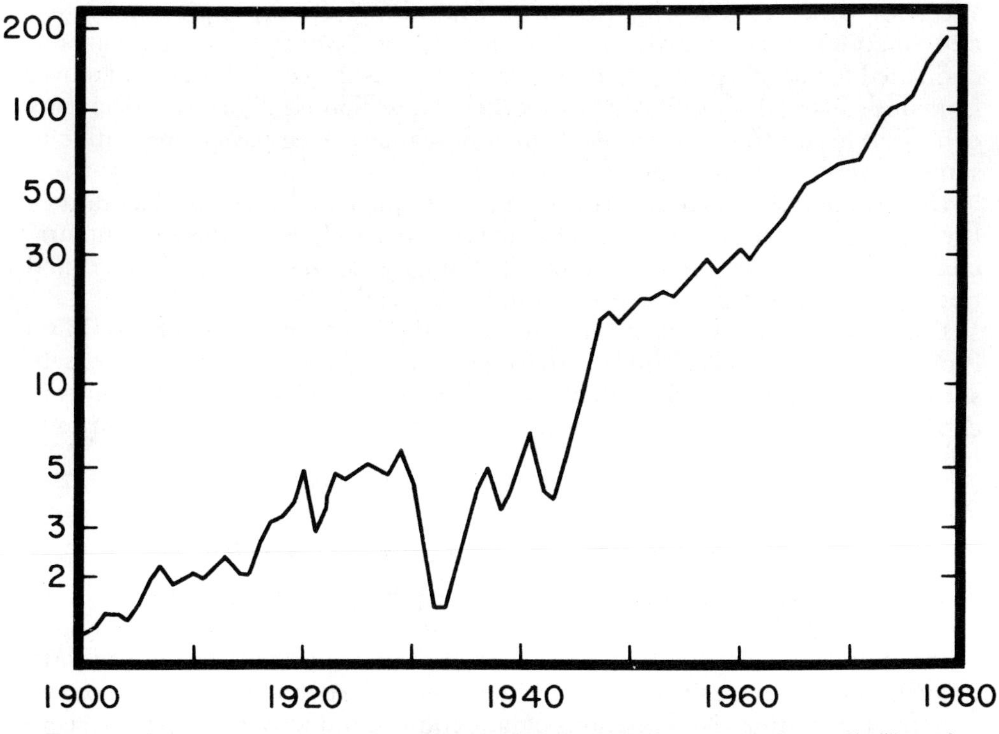

Figure 4.3. Fixed investment in producers' durable equipment, USA, 1900–79. In thousand millions of current dollars.

Kondratiev) model, but that the application of improved long-term theory would pay off handsomely in economic performance, if incorporated into corporate strategies and public policies.

Necessary, but not sufficient, for long-term theory to be useful for decision making is that it be able to explain inter-sectoral and intra-sectoral differences in the marginal efficiency of capital which is invested under the conditions prevailing at that time, relative to the efficiency if it were invested under different conditions. The function of long-term theory is to provide a framework for examining alternative conditions (contingencies), and to provide some predictions concerning the shift from one set of conditions to another (ie phase transitions). The purpose of this paper is to go one step in that direction by facing the fact that, to the best of our knowledge, the economic process is hardly ever in an equilibrium state, but nearly always proceeds from disequilibrium to disequilibrium. In other words, it proceeds by a non-equilibrium motion.

Disequilibrium implies an imbalance in market power among economic agents participating in an exchange system, and non-equilibrium motion implies that the distribution of market power will shift over time, possibly with some degree of regularity. Thus, the study of long-term economic change is by necessity a study of performance and power.

Non-equilibrium movements in industrial structures

Main Proposition

We start with the following premises (Main Proposition): appreciations and depreciations of the operative value of capital goods, which are designed for and dedicated to specific production purposes, occur as the economic system moves from one state of disequilibrium to another. These non-equilibrium movements exhibit at least some observable degrees of regularity (eg pattern repetition in some partially ordered way).

In particular, what has been observed as a sequence of stages of a Kondratiev cycle is a succession of phases, characterised by different types of structural change. If long waves exist, they are changing trends in structural change, and not mere fluctuations in the level of production and prices.

Before we formulate a substantive hypothesis on the basis of our Main Proposition, we should address the questions of built-in equilibrium, self-stabilisation, and causation, that trouble the discussion about long-term concepts. We answer these questions by formulating three negative corollaries to our main proposition.

Three negative corollaries to the Main Proposition

(1). Non-equilibrium movements are neither equilibrating nor disequilibrating

Evidently, we economists are much better equipped to describe a state of the economic system at a point in time than to describe a movement of the system over a stretch of time. We are in possession of a fairly sophisticated body of static theory, where the equilibrium state of the economy is defined as the solution of a system of equations. No equations of non-equilibrium movements in the economy are known to economists. The best approximation we have are models of 'pseudo-motion' from one quasi-equilibrium state to another. This simply is a bad approximation if the actual movement is non-equilibrium—ie proceeds from one disequilibrium state to another, as we are suggesting.

It then follows from the main proposition of non-equilibrium motion that we have to be careful about assuming the motion is only in one direction—as assumed by Schumpeter and Kirzner, with two simple types of motion. The first negative corollary is that non-equilibrium motions tend towards neither equilibrium nor disequilibrium.

It was J.A. Schumpeter, in his *Theory of Economic Development*,[7] who taught that entrepreneurship is "entirely foreign" to equilibrium states, tending to disrupt it whenever it can, and running counter to "the tendency towards equilibrium" assumed by most equilibrium schools (stable and temporal equilibrium schools alike).

As we must disagree with a uni-directional interpretation of the economic process, maintaining that parts of the economy might drift towards some equilibrium while other parts move towards more disequilibrium, we should agree with Kirzner's critique of Schumpeter, on the one hand, but should not follow him when he argues for a converse uni-directional interpretation on the other. Kirzner wrote in 1973,[8]

Schumpeter's entrepreneur acts to disturb an existing equilibrium situation. Entrepreneurial activity disrupts the continuing circular flow. The entrepreneur is pictured as initiating change and as generating new opportunities. Although each burst of entrepreneurial innovation leads eventually to a new equilibrium situation, the entrepreneur is presented as a disequilibrating, rather than an equilibrating force.

By contrast my own treatment of the entrepreneur emphasizes the equilibrating aspects of his role. I see the situation upon which the entrepreneurial role impinges as one of the inherent disequilibrium rather than of equilibrium—as one churning with opportunities for desirable changes rather than as one of placid evenness. Although for me, too, it is only through the entrepreneur that changes can arise, I see these changes as equilibrating changes.

(2). Discontinuity in economic processes

Facing the fact that the economy moves in a non-equilibrium fashion calls for a backing off from the "Principle of Continuity". The extent of this backing-off is determined by the room needed in our models for long-term regularities in non-equilibrium movements. With respect to gaining this room, we have to go back to Marshall[9] whose concept of "continuity" is one of normality. In the preface to the first edition of his *Principles*, he elucidates the "Principle of Continuity" by reference to "the normal willingness to save", "the normal willingness to undergo a certain exertion of a certain pecuniary reward", "the normal alertness to seek the best markets in which to buy and sell", and, "when that is once understood, the theory of normal value".

And as there is no sharp line of diversion between conduct which is normal, and that which has to be *provisionally neglected as abnormal* [emphasis added], so there is none between normal values and "current" or "market" or "occasional" values. The latter are those values in which the accidents of the moment exert a prepondering influence; while normal values are those which would be ultimately attained, if the economic conditions under view had time to work out undisturbed their full effect.

Clearly, the justification for provisionally neglecting the "abnormal cases" has always been derived from a basic supposition of randomness: that the abnormalities are randomly distributed, and occur independently from one another. This supposition rules out

- the tolerance for regularities in the occurrence of "irregular cases", and
- the emphasis on these "abnormal cases" which have recently become much more interesting to economists and innovation researchers.

Today, as in other periods of unsatisfactory economic performance, economists feel uneasy about the implications of the Principle of Continuity, and are attracted to the notion of discontinuity. The strongest attack on the continuity theory was launched by Schumpeter in his *Theoretical Problems of Economic Growth,*[10] because it "comes near to being a general theory of economic growth. It proved extremely durable, for we find it first in Adam Smith and, with minor developments and improvements, in J.S. Mill and Alfred Marshall.

Briefly, it may be stated like this. Taking institutional, political, and natural factors for granted, these economists (and most others) start from the assumption that a social group—we may just as well call it 'nation'—will, unless periodically plundered, experience a certain rate of economic growth that is accounted for by increase in numbers and saving. This induces a 'widening of markets' which, in turn, increases the division of labour and thus 'productivity'. 'Induced inventions' appear as a separate factor of growth, whereas inventions that are not induced by previous increases in production (so-called revolutionary inventions) are outside the theoretical picture—they just happen, as manna might fall from heaven—and break the line of growth, establishing a new one.

In the Smith-Mill-Marshall theory, the economy grows like a tree. The process is no doubt exposed to disturbances by external factors that are not economic or not strictly so. But in itself it proceeds steadily and continuously... I submit that this is not so".

(3). Causation—not heroism but incrementalism

Just as we should not follow Schumpeter in his perception of entrepreneurship as a disequilibrating force (because it could work on a disequilibrium state, and push towards equilibrium), we should not follow Schumpeter all the way into the heroic interpretation of the causes of discontinuity. But first, we have to recall Schumpeter's concept of the Principle of Discontinuity, with which he wanted to replace the Continuity Principle.

The obvious question, invited by the observation of so-called "abnormality" in economic performance, has been asked by Schumpeter in his 1927 article:[11]

> Why is it that industrial and commercial change is not continuously distributed in time, but proceeds by leaps?

In his *Business Cycles,*[12] Schumpeter then formulated an alternative supposition for such "abnormalities" as innovations, thereby rejecting both the randomness and the independence implied in the Principle of Continuity. According to Schumpeter,

Innovations are not at any time distributed over the whole economic system at random, but tend to concentrate in certain sectors and their surroundings.

Innovations do not remain isolated events, and are not evenly distributed in time... on the contrary, they tend to cluster, to come about in bunches, simply because first some and then most firms follow in the wake of successful innovation.

The dividing line between our agreement and our disagreement with Schumpeter's notion of discontinuity is indicated by the words "simply because" in the second paragraph of the above quote. Of course Schumpeter was well aware that entrepreneurial initiative and success depend on certain preconditions. To the dwindling away of these supportive preconditions, he devoted a whole book (*Capitalism, Socialism, and Democracy*).[13] However there is a lack of theory about the origins of innovation, and the tautology 'innovation follows innovations' holds the place of explanation. Furthermore, there is evidence of this gap being intended. Ruttan[14] observed, correctly, that

the business cycle, in Schumpeter's system, is a direct consequence of the appearance of clusters of innovations. But no real explanation is provided as to why innovations appear in clusters, or why the clusters possess the particular type of periodicity which Schumpeter identified in terms of Kitchin (forty months), Juglar (ten years), and Kondratiev (sixty years) cycles. Indeed, in discussing the cyclical pattern of innovation that emerges from this review of economic history, Schumpeter[15] specifically eschewed any theoretical basis for the observed pattern of behaviour.

Both the strong emphasis upon the entrepreneur innovator, and the weak emphasis upon the economic and environmental preconditions for innovation and entrepreneurship, have been the reason why economists have been hesitant to follow Schumpeter's "Heroic Innovator" paradigm. For the purpose of theory formation, it indeed seems to constitute a *cul de sac.*

In our current analytical work we are operating on the basis of another type of discontinuity theory: it is the discontinuity in the preconditions for innovation which change the market opportunities and propensities of entrepreneurs to innovate. This view of the causal linkages is called the Principle of Self-Organisation in modern, non-equilibrium, thermo-dynamics. In eco-dynamics, this principle has a respectable history, going back to Eugene Slutzky and his notion of disarrangement of a regime, and the transition to another regime. One important point is that a change in regime can be the result of incremental endogenous changes, and need not be the result of a strong external shock:[16]

The summation of random causes generates a cyclical series which tends to imitate for a number of cycles a harmonic series of a relatively small number of sine curves. After a more or less considerable number of periods every regime becomes disarranged, the transition to another regime occurring sometimes rather gradually, sometimes more or less abruptly, around certain critical periods.

In our discontinuity theory, a change in regime allows for, and coincides with, a swarm of basic innovations. In Schumpeter's theory, innovation comes from prior innovation. In Mensch's theory, basic innovation comes from prior lack of innovation, and the subsequent devaluation of capital goods dedicated to old lines of business, and a subsequent shift in the propensity to invest in innovation.

Innovation and our central hypothesis

From the main proposition (non-equilibrium motions of the economic process), it follows that a fruitful direction for pursuing an improved long-term theory is characterised by the co-evolution of economic conditions of valuation, and economic values themselves. In equilibrium theory, this can be, but often is not, the case. For example, in the Smith–Mill–Marshall equilibrium theory, "normal values are those which would be ultimately attained, if the economic conditions under view had time to work out undisturbed their full effect". Economic conditions are taken as the cause, economic values as the effect; and it is precisely this continuity in the conditions of the economic valuation process which Schumpeter attacked ("Taking institutional, political, and natural factors for granted, these economists, and most others. . ."). Methodologically, this type of equilibrium theory is characterised by the *ceteris paribus* clause, which is presupposed in all marginal analysis, and which therefore has failed to be of much use in the study of long-term economic development.

In any contingency theory of economic change, the conditions of the economic valuation process, and the economic values themselves, may change *pari passu*. Hence, causes and effects appear intertwined, and one would need cybernetic tools for analysing changes in the economic system. In several substantive areas of economic reasoning (such as in production theory, employment theory, income theory, distribution theory, growth theory), this methodology has found promising application, thus preparing the ground in these areas for incorporation of long-term, non-equilibrium aspects.

In this paper, we are focusing upon the long-term aspects relevant to the substantive theory of investment, especially innovation, as these are particularly interrelated with the micro-economic theory of action, and the macro-economic theory of development.

Using Mensch's metamorphosis model, we define phases of extensive and of recessive structural change in the economy. Extensive structural change is characterised by an expanding economy, which is growing while the technological base is rapidly changing through the diffusion of product and process innovations in sectors that have recently been established or modernised by basic innovations. Structural changes during recessions are characterised by slow-downs in the rate of diffusion of improvement innovations in the economy, and a relative decline in the rate of improvement. Furthermore expansionist (recessional) structural change is a symptom of strengthening (weakening) of the prevailing economic regime.

From this we derive our Central Hypothesis: that, under certain circumstances (which we identify as typical of either expansionist or recessional structural change), entrepreneurs find that their stocks of fixed capital in plant and equipment either enjoy or suffer systematic up- or downgrading in operative value (appreciation or depreciation). These shifts in value of parts of the capital stock either induce capital owners to invest their funds in more of the same type of capital goods, or induce them to invest in alternative types of capital goods. We denote this behavioural theory of the propensity to invest the contingency theory of capital valuation.

To the best of our knowledge, Kaldor came closest to a statement of this contingency theory of switches in the propensity to invest when he was contemplating the amount of capital goods which can be "usefully" employed.[17]

In normal circumstances, stocks of all goods possess a yield... The marginal yield falls sharply with an increase in stock above "requirements" and may rise very sharply with a reduction of stocks below "requirements". When redundant stocks exist, the marginal yield is zero. . . [The] amount of stock which can be thus "useful" is, in given circumstances, strictly limited.

In the theory of business cycles, this contingency theory found an expression in Kaldor's model of a sigmoid investment demand function.[18] According to Kaldor, the propensity to invest, dI/dX, will be small, both for low and for high levels of X, relative to its "normal" level. Kaldor linked this proposition to work of contemporary economists such as Kalecki, Keynes, and Tinbergen. The similarity between Kaldor's model and Mensch's metamorphosis model is acknowledged.

Additional hypotheses with long-term implications

Phase transition and innovation rate and direction

Under the large umbrella of the central hypothesis of capital valuation, there is room for several hypotheses about longitudinal innovative behaviour of macro-economical entities, such as whole sectors or regions. Examples of such hypotheses, which incorporate the distinction between basic innovation and improvement innovation,[19] and product and process innovation, are as follows:

- In periods of prolonged recessional structural change (depressions), the economy becomes structurally ready for basic innovations, and swarms of basic innovations occur.
- After a cluster of basic innovations has occurred, diffusion of improvement innovations (series of product and process innovations) drive the economy into growth cycles (Schumpeter).
- Over time, a substitution of process innovations for product innovations will take place (Utterback and Abernathy[20]).
- As the series of improvement innovations are truncated by mere pseudo-innovations, the growth cycles level off, possibly after investment overshoot in industries that formerly grew disproportionately strong.

Although many complementary hypotheses on innovation behaviour are possible, this set constitutes a line of interlocking statements which, taken together, form a preliminary statement of an innovation-oriented long-term theory of industrial evolution.

How are the many individual actions and incremental changes coordinated at the macro-economic level? The concept of an economic regime appears relevant and fruitful.

Macro-economic coordination in micro-economic chaos

No economic theory of long-term development will be satisfactory if it does not explain how micro-economic interaction is macro-economically ordered, and how the order feeds back on the interaction. The theory also has to explain how both the order and the interactions change—if they change—over time (longitudinally) and in kind (cross-sectionally).

Since we propose that this motion of the economic structure is a non-equilibrium movement, proceeding from one state of disequilibrium to another state of disequilibrium, we need to clarify two issues:

- internal (short-term) consistency of the movement; this is the issue of interrelated cost–benefits, or balance in exchange relationships; and
- external (long-term) consistency of the movement; this is the issue of dependence upon scarce resources, or dominance in exchange relationships.

As we are not intending at this juncture to hypothesise the final word on these issues of balance and dominance, we shall satisfy ourselves with a 'classical' approximation; eg we look at the school of pre-equilibrium economists for concepts of non-equilibrium motion. In particular, we shall recall how the issues of balance and dominance were treated by economists such as Ricardo, Mill, Say, and Jevons, before the aspects of dominance had been dwarfed in the exchange models of Marshall and Walras, who put the question of market dominance on the back burner.

We conjecture that "the one all-ruling law of supply and demand" (Marshall) suffices to explain the balance issue, and is also necessary but not sufficient for understanding the dominance issue in non-equilibrium economic motion. An imbalance in market power or influence of one economic agent over another can have many causes. Nevertheless, a good many of the seemingly non-economic reasons for obtaining or maintaining a dominant position in exchange relationships can be traced either to some scarce resource or to some mechanism for reducing uncertainty (certainty being a scarce resource).

The economically determined part of dominance in exchange can thus be defined as either the influence of cost of production (Ricardo–Mills), or the influence of final utility (Say, Jevons). For the classical economists, this was an either/or question. Ricardo, the proponent of the Cost of Production Principle, wrote in October 1820 to Malthus:[21]

> M. Say has not a correct notion of what is meant by value when he contends that a commodity is valuable in proportion to its utility. This would be true if buyers only regulated the value of commodities; then indeed we might expect that all men would be willing to give a price for things in proportion to the estimation in which they held them; but the fact appears to me to be that the buyers have the least in the world to do in regulating price; it is all done by the competition of the sellers, and, however really willing the buyers might be to give more for iron than for gold, they could not, because the supply would be regulated by cost of production. . . You say demand and supply regulates value; this I think is saying nothing, and for the reason I have given in the beginning of this letter: it is supply which regulates value, and supply is itself controlled by comparative cost of production.

Nearly identical views on the dominance of cost of production factors over the wants and needs of partners in exchange have been expressed by J.S. Mill:[22]

> From this, it appears that demand and supply govern the fluctuations of prices in all cases, and the permanent values of all things of which the supply is determined by any agency other than that of free competition: but that, under the regime of free competition, things are, on the average, exchanged for each other at such values and sold for

such prices as afford equal expectations of advantage to all classes of producers; which can only be when things exchange for one another in the ratio of their cost of production.

This position on producer sovereignty was strongly contested by Jevons, the proponent of the Final Utility Principle:[23]

Repeated reflection and inquiry have led me to the somewhat novel opinion that value depends entirely upon utility. We have only to trace out carefully the natural laws of variation of utility as depending upon the quantity of commodity in our possession, in order to arrive at a satisfactory theory of exchange of which the ordinary laws of supply and demand are a necessary consequence.

Of course, Marshall was quite correct in his way of putting it all in perspective, by concluding from the writings of the antagonising classics that the Cost of Production Principle and the Final Utility Principle[24]

are undoubtedly component parts of the one all-ruling law of supply and demand; each may be compared to one blade of a pair of scissors. When one blade is held still, and the cutting is effected by moving the other, we may say with careless brevity that the cutting is done by the second; but the statement is not done to be made formally, and defended deliberately.

We conjecture that the mechanism of macro-economic coordination of the chaos of micro-economic activities works like the Marshallian scissors. Sometimes it is more the cost of production (determined by scarcities and abilities of producers) that governs the emergence of economic linkages, and the terms of trade; at other times, it is more the final utility of the product (determined by needs and wants of users), and the degree of abundance, that governs the regime of exchange.

Our theory of capital good appreciation and depreciation, and of the resulting change in the propensity to innovate, leads us to hypothesise that the propensity for basic innovations increases if the economic regime changes in a particular way, ie when the producer's sovereignty (in sellers' markets) is eroded in many old industries. The old economic regime, which was once built upon the Principle of Production Cost, becomes structurally unstable as it faces uncertainty caused by a moody final demand in times of affluence. This structural instability translates into hidden devaluations of capital goods, dedicated to producing mass consumption items, and of the means for producing these; it translates into attempts to have others carry the loss of value (stagflation); and into attempts to enter alternative lines of business (innovation).

Empirical analysis of structural readiness for basic innovation

A testable hypothesis

The acceptability of our theory is probably not helped by some of its components (intervening variables and relationships) not being directly observable. Appreciations and depreciations in the operative value of installed capital goods are usually not measurable (the usual method of measuring the exchange value of a capital good when being traded is totally inappropriate for determining the operative value when in use in an ongoing concern). Changes in the usefulness of production goods, and underlying changes in the market position of the producer, can at best be inferred from indirect observation of related changes in dependent variables.

As we propose that the propensity to innovate shifts as a result of prior shifts in the operative value of capital goods, and that these value shifts in turn are symptoms of power shifts in the economic regime, the issue of measurability is indeed a crucial issue, to which F.A. Hayek, in this 1974 Nobel Memorial Prize Lecture, made the following pertinent remarks:[25]

> . . . while in the physical sciences the investigator will be able to measure, on the basis of a *prima facie* theory, what he thinks important, in the social sciences what is treated as important is often that which happens to be accessible to measurement. This is sometimes carried to the point where it is demanded that our theories must be formulated in such terms that they refer only to measurable magnitudes.
>
> It can hardly be denied that such a demand quite arbitrarily limits the facts that are to be admitted as possible causes of the events in the real world. This view, which is often quite naively accepted as required by scientific procedure, has some rather paradoxical consequences. We know, of course, about the market and similar social structures, very many facts that we cannot measure and on which indeed we have only some very imprecise and general information. And because the effects of these facts in any particular instance cannot be confirmed by quantitative evidence, they are simply disregarded by those sworn to admit only what they regard as scientific evidence. And they thereupon happily proceed on the fiction that the factors they can measure are the only relevant ones.

We agree with Hayek that measurability can be distinct from meaningfulness in the formulation of formal theory, but in empirical work, measurability is a necessity for quantitative analysis. If a latent variable can only be inferred from indirect observation, we speak of semi-quantitative analysis, and pattern recognition is one possible objective of such work. It is in this way that we shall proceed here.

A short-hand version of our formal theory goes like this: structural readiness for a new cluster of basic innovations results from, or is identical with, structural instability of the old economic regime. The emergence of structural instability can be inferred by semi-quantitative analysis of data on "Manufacturers' Evaluations of Their Plant and Equipment Facilities", which manufacturers in the USA report to the Bureau of Economic Analysis (BEA) on a quarterly basis. The manufacturers report to the BEA whether:

1 More capacity is needed;
2 Capacity is about adequate; or
3 Capacity exceeds needs.

The Bureau transforms these questionnaire answers into percentage points X_{ij}, where i denotes a sector and j = 1, 2 or 3 (above), such that

$$X_{i1} + X_{i2} + X_{i3} = 100\% \text{ for each sector.} \tag{1}$$

Excess capacity scores X_{i3} are available for the following sectors:

i = 1 All US manufactures.
i = 2 All durable goods.
i = 3 Primary metals.
i = 4 Metal products.

i = 5 All non-durable goods.
i = 6 Food including beverages.
i = 7 Chemicals and petroleum.

We may assume that devaluation of the operative value of capital goods invested in some branch of industry correlates with the extent of excess capacity, and the expectation that this disappointing excess is not just a short-term phenomenon. On the other hand, we may assume that the extent of the excess capacity is a measure of the deviation of corrected expectations from prior expectations. At any rate, this is Pigou's notion of the extent of entrepreneurs' errors during the business cycle. In order to arrive at a testable hypothesis, we define the extent of excess capacity as a measure of the disappointment.

Testing hypothesis. The greater the influence of a manufacturing sector in a given economic regime, the greater the disappointment within that sector if the regime changes. If the disappointment in a sector is indicated by the extent of excess capacity in that sector, the inelasticity of supply (specialisation of production facilities; non-durability of output) will also play a role, but a secondary one, in the analysis of structural change.

Results

According to the time-frame suggested by the metamorphosis model (Figure 4.2), we should apply the testing hypothesis to the period beginning around 1966–67, when real investment levelled off (Figure 4.1). In terms of the proposed shift from the governance of the Principle of Production Cost to the governance of the Principle of Final Utility, we would expect:

(1) cross-sectional differences between sectors, according to their rank as "primary good", "intermediate good", and "final good" producing sectors.

At the time around 1966–67, we would expect excess capacity to have been reported, with orders of magnitude identical to (or at least similar to) the rank order of these sectors in the prevailing regime;

(2) inter-temporal differences between the magnitudes of excess capacity, such that the dynamics of the shift in regime becomes apparent.

In a comparison of excess capacity levels in the period around 1966–67 with a later period, we would expect that, during this time of recessional structural change, the signs of disappointment will have moved from "primary" to "intermediate" goods, or even to some "final" goods, around 1976–77.

In order to obtain significant results from regression analysis, which we employ in the inter-sectoral and inter-temporal comparisons, we have partitioned the period under observation thus:

A = fourth quarter 1963 to third quarter 1980; this is also divided into two sub-periods of 24 quarters each, namely:
B = first quarter 1964 to fourth quarter 1969.
C = first quarter 1974 to fourth quarter 1979.

We have calculated the trend parameters for the seven sectors over these three intervals. These parameters (ALPHA and BETA), and significance statistics, are given in Table 4.1.

In Figures 4.4–4.6, some meaningful results are depicted as straight lines

TABLE 4.1. RESULTS OF THE REGRESSION ANALYSIS

Sector	Period (Quarter/Year) t=1 t=T	Estimates Y1	Estimates YT	Coefficients Alpha +/– Beta	Significance of Beta T–value	Significance of Beta S–level
1 All US manufacturing	A 4/1963 – 3/1980	4·113	8·104	4·05 + .060 t	3·99	99%
	B 1/1964 – 4/1969	4·512	3·838	4·54 – ·029 t	2·11	95%
	C 1/1974 – 4/1979	10·468	5·191	10·70 – ·229 t	2·16	95%
2 Durable goods	A 4/1963 – 3/1980	7·182	10·353	7·14 + ·047 t	2·01	95%
	B 1/1964 – 4/1969	7·442	6·269	7·47 – ·050 t	1·88	90%
	C 1/1974 – 4/1979	13·970	6·340	14·30 – ·332 t	2·29	95%
3 Primary metals	A 4/1963 – 3/1980	17·718	6·688	17·88 – ·165 t	4·76	99%
	B 1/1964 – 4/1969	15·649	16·460	15·61 + ·035 t	·81	n.s.
	C 1/1974 – 4/1979	9·710	6·350	9·85 – ·146 t	·83	n.s.
4 Metal products	A 4/1963 – 3/1980	2·572	11·784	2·43 + ·138 t	4·66	99%
	B 1/1964 – 4/1969	3·189	1·627	3·26 – ·068 t	1·63	(87%)
	C 1/1974 – 4/1979	15·650	6·240	16·06 – ·049 t	2·41	95%
5 Non - durable goods	A 4/1963 – 3/1980	1·414	5·822	1·35 + ·066 t	5·07	99%
	B 1/1964 – 4/1969	2·009	1·525	2·03 – ·021 t	2·03	90%
	C 1/1974 – 4/1979	7·032	4·043	7·16 – ·130 t	1·57	(85%)
6 Food, including beverages	A 4/1963 – 3/1980	3·974	6·588	3·93 + ·039 t	4·04	99%
	B 1/1964 – 4/1969	5·032	3·009	5·12 – ·088 t	3·74	99%
	C 1/1974 – 4/1979	4·794	6·356	4·73 + ·068 t	2·00	95%
7 Chemicals, including petroleum	A 4/1963 – 3/1980	0·068	3·805	0·01 + ·056 t	3·64	99%
	B 1/1964 – 4/1969	0·608	0·576	0·61 – ·001 t	·10	n.s.
	C 1/1974 – 4/1979	5·878	2·272	6·04 – ·157 t	1·54	(86%)

Note: n.s. = not significant.

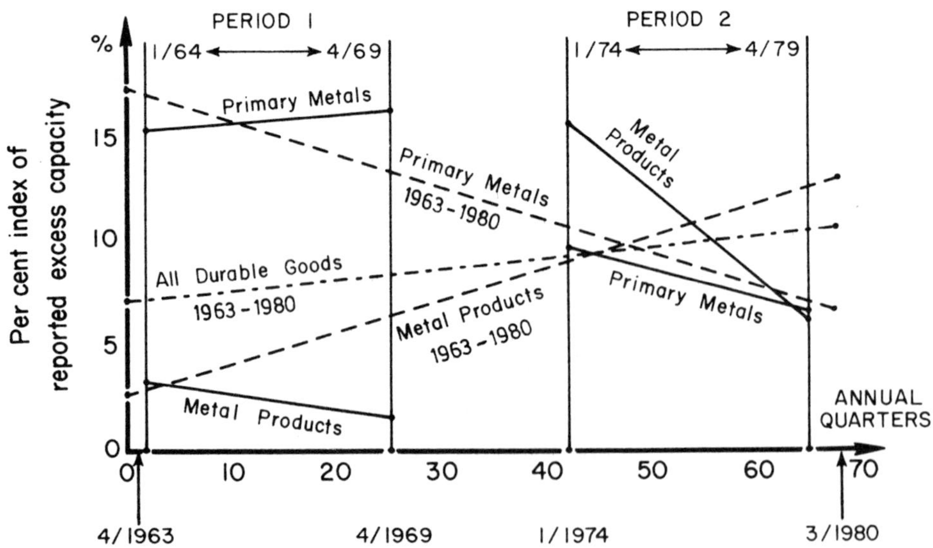

Figure 4.4. Reported excess capacity, USA; see observations 1.1–1.3.

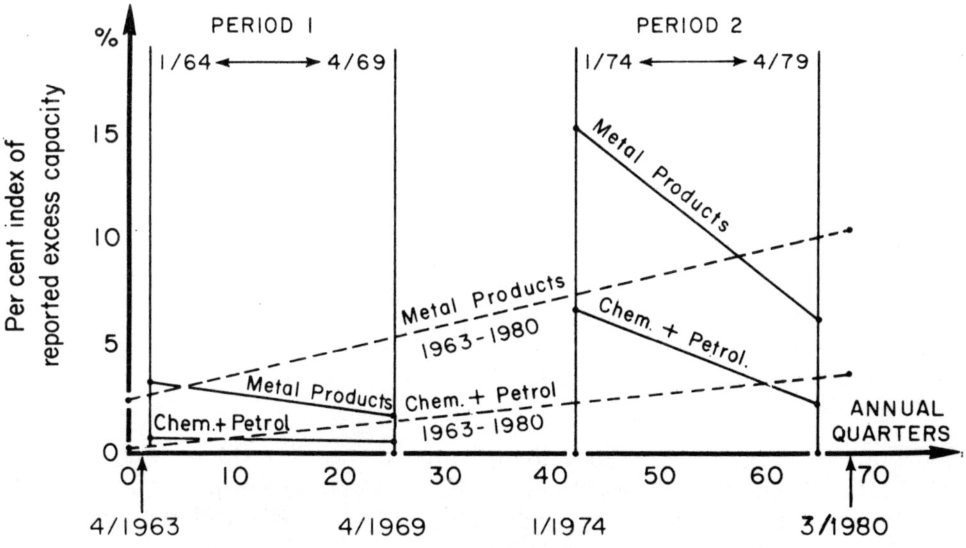

Figure 4.5. Reported excess capacity, USA; see observations 2.1–2.3.

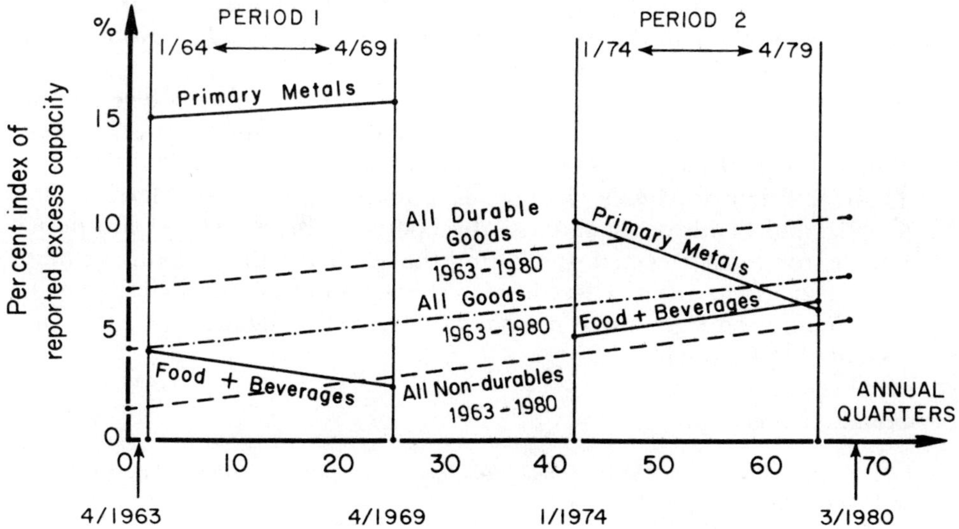

Figure 4.6. Reported excess capacity, USA; see observations 3.1–3.3.

(intercept and trend lines). Inspection of the Figures confirms the testing hypothesis through the following three sets of observations:

1.1 During the period 1964–69, when the recessional structural change had set in, the Primary Metals sector showed much stronger signs of disappointment than any other sector (see left side of Figure 4.4).

1.2 While these signs gradually diminish over the longer run (1966–80) in the Primary Metals sector, they shift to the Metal Products sector (Figure 4.4).

1.3 Within the Durable Goods sector, this shift in market dominance is reflected by the cross-over of the long trend lines of Primary Metals and Metal Products, and is further illustrated by the structural break visible in the short trend shifts in both industries.

2.1 The shift of signs of disappointment from "primary" to "intermediate" products is not only visible in Metal Products (1.2), but also in Chemicals and Petroleum (Figure 4.5).
2.2 Both sectors suffered a considerable increase in the 'normal' level of excess capacity over the period of recessional structural change, and both experience a similar structural break.
2.3 However, Chemicals and Petroleum suffered less on average, owing to the fact that many of these products are closer to final demand, and non-durable.

3.1 The structural transformation in "intermediate" good sectors, such as Metal Products and Chemicals, which gained market power in the first phase of recessional structural change, is markedly different from the structural transformation in "primary" and "final" goods sectors, such as Primary Metals, and Food and Beverages, respectively (Figure 4.6).
3.2 In general, producers of non-durable goods seem less easily disappointed than those who manufacture durable goods. Throughout the long period, the errors in the Durable Goods sector exceeded those in the Non-durable sector.
3.3 With this qualification in mind, it appears that not only the level of error itself, but also the change in that level, indicates a shift of market dominance. In the first sub-period, at the beginning of recessional structural change, Primary Metals changed in accordance with All Manufactured Goods, and Foods changed cross-directionally. In the second sub-period, this feature was reversed, indicating that primary good producers might have learned to adjust effectively to this kind of structural change, while final good producers might not have (fully) comprehended the nature and extent of this change.

Notes and references

1. J.M. Keynes, *The General Theory of Employment, Interest and Money* (New York, Harcourt, Brace and World, 1936), page 347.
2. G. Mensch, *Stalemate in Technology* (Cambridge, MA, Ballinger, 1979) page 73.
3. Keynes, reference 1, page 317.
4. A.F. Burns, *Production Trends in the United States Since 1870* (New York, National Bureau of Economic Research, 1934).
5. B. Gold, "Industry growth patterns", *Journal of Industrial Economics*, 1963, *13*, pages 53–73.
6. J.J. van Duijn, "Another look at industry growth patterns", Faculty Working Papers, College of Commerce and Business Administration, University of Illinois, Urbana, 1980.
7. J.A. Schumpeter, *The Theory of Economic Development* (Harvard University Press, 1934), page 64.
8. I.M. Kirzner, *Competition and Entrepreneurship* (University of Chicago Press, 1973), page 72.
9. A. Marshall, *Principles of Economics*, 8th edition (Macmillan, 1930).
10. J.A. Schumpeter, *Essays* (Cambridge, MA, Addison–Wesley Press, 1951), pages 233–234.
11. J.A. Schumpeter, "The explanation of the business cycle", *Economica*, 1927, pages 286–311; quote from page 297.

12. J.A. Schumpeter, *Business Cycles*, 2 volumes (McGraw-Hill, 1939), page 75.
13. J.A. Schumpeter, *Capitalism, Socialism and Democracy*, 2nd edition (New York, Harpers, 1947).
14. V. Ruttan, Usher, and Schumpeter, "Invention, innovation and technological change", in N. Rosenberg, ed, *The Economics of Technological Change* (Penguin Books, 1971), pages 73–85.
15. Schumpeter, reference 12, page 173.
16. E. Slutzky, "The summation of random causes as the source of cyclic processes", *Econometrica*, 1937, *5*, page 123.
17. N. Kaldor, "Speculation and economic stability", *Review of Economic Studies*, 1939, *VII*, pages 1–27; quote from pages 3–4.
18. N. Kaldor, "A model of the trade cycle", *Economic Journal*, 1940, *L*, pages 78–92; quote from page 81.
19. G. Mensch, "Basisinnovationen und Verbesserungs-innovationen", *Zeitschrift für Betriebswirtschaft*, 1972, *42*, pages 291–297.
20. J.M. Utterback and W.J. Abernathy, "A dynamic model of product and process innovation", *Omega, The International Journal of Management Science*, 1975, *3*, pages 639–656.
21. The citation is taken from Marshall, reference 9; for the complete wording of Ricardo's statement see, for example, P. Sraffa, ed, *David Ricardo, Letters*, Vol VIII (Cambridge University Press, 1952), pages 276–277 (Marshall's quote is from Bonar's edition).
22. J.S. Mill, Book III, Chapter XVI, section 1.
23. W.S. Jevons, *The Theory of Political Economy*, 5th edition (New York, Kelley, 1965), pages 1–2.
24. Marshall, reference 9, Appendix I, page 820.
25. F.A. Hayek, "The Pretence of Knowledge", Nobel Memorial Prize Lecture, 1974.

CHAPTER 5

OBSERVATIONS ON THE SCHUMPETERIAN SWARMING OF INNOVATIONS

Alfred Kleinknecht

The hypothesis that depressions trigger innovations is examined first on theoretical grounds, and then with a random sample of 120 innovations. Radical product innovations, the alleged basis for new 'Schumpeterian' industries, indeed cluster around the 1930s crisis, but are closely associated with the less severely hit modern industries. Process innovations and less radical product changes are associated more with periods of prosperity. The final section considers whether new 'Schumpeterian' industries are a precondition for, or the result of, economic upswings.

IN A RISING number of publications, the Schumpeterian hypothesis on the clustering of innovations is taken as an explanation of longer-term economic fluctuations.[1] There are some authors on long waves who question the relevance of the Schumpeterian innovation hypothesis,[2] but even those who consider the Schumpeterian hypothesis important seem to have considerable differences as to the theoretical concept of innovation clusters.

Clark, Freeman and Soete make a strong criticism of the position of Mensch. Mensch had concluded from his data that major innovations ("basic innovations") have occurred in bunches at times of deep and prolonged crises in the world economy. Mensch argues that there were significant clusters of basic innovations around 1820–30 as well as in the 1880s and 1930s. These innovations contributed to overcoming depression by giving rise to new growth industries.[4]

Clark, Freeman and Soete question the reliability of Mensch's empirical observations (representativeness of his sample) as well as his theoretical explanation. In their view,[5]

> innovations are pushed forward ... in ... expectation of profits associated with expanding market prospects. To suggest that this would be more likely to occur during depression flies in the face of all economic theory of whatever description ... It also contradicts the evidence of case histories of innovation, and the general consensus that high risks are one of the main factors inhibiting innovations.

A. Kleinknecht is from the Research Group on Long Term Fluctuations, Free University of Amsterdam, Koningslaan 31, 1075 AB Amsterdam, The Netherlands.

Consequently, it should be expected that—if there is any bunching of innovations—this would occur not in depression phases but in periods of recovery and boom.[6]

At a first glance, the two positions appear to stand in sharp contrast: do enterprises primarily innovate when favourable market prospects promise high profits and low risks for innovative products, or do they need the experience of "rien ne va plus" in economic crisis as an incentive to go new ways?

Profits, risks, and types of innovation

It may be that the profit maximisation strategy of firms during prosperity periods will change into a strategy of minimising risks, losses and uncertainties during long periods of economic distress. From this point of view, the argument that radical innovation projects are not started in times of crises and recession, when risks and uncertainties are highest, seems plausible. But it may be asked whether this argument does not backfire when looking at the *relative* risks and uncertainties of investment alternatives: innovation processes are *generally* of a risky nature. This risk is certainly lower in prosperous times—but so may be the incentive to innovate. Why should an enterprise deal with the uncertainties, risks and costs of introducing completely new products into the market, as long as its established products can be easily sold? Does not the firm make money much more easily and with less risk by restricting R and D activity to the mere improvement of existing products? The logical conclusion from this is that prosperous times are good times for quality-augmenting and cost-reducing improvement and process innovations, but they are bad times for radical innovators.

Thus not only the risks but also the incentives for radical innovations could considerably increase in periods of distress. There may be decreasing returns to R and D expenditures for further improvements of mature products. Moreover, demand saturation for individual products, as well as a general decline of demand due to the crisis, may lead to overcapacity in existing product lines. In this situation, further expansionary investments become risky, if not disastrous. As to investment and R and D strategy, there remain only two reasonable alternatives: firms may try to cope with sharper price competition for their old products, by means of rationalisation investment; or they may try to open up completely new investment areas by innovating radically new products.

In other words, in times of prolonged crises and sharper competition, investments in established lines of production, as well as in innovation projects, bear higher risks and uncertainties. This explains a general decline of investment expenditures and some shift in favour of rationalisation investment within the remaining investment volume. At the same time, however, vanishing growth prospects might lead to an orientation towards completely new markets, in which the innovating firm has a chance of obtaining a temporary monopoly position and to achieve high Schumpeterian profits. Thus, longer depression periods are favourable times for process innovations (biased towards rationalisation), and for radical product innovations, giving the basis for new growth industries. On the other hand, the rise of new 'Schumpeterian' industries may be accompanied by a large stream of additional innovations (the 'bandwagon effect'), which lead to price reductions and further quality improvements, opening up a larger range of possible applications for the innovation. Clark, Freeman and Soete obviously meant this latter type of innovation when pointing out:[7]

Mensch has been looking at the wrong 'swarms'. The swarms which matter in terms of Schumpeterian analysis are the diffusion swarms *after* the basic innovations and the swarming effects associated with a set of interrelated basic innovations ...

The two positions are not incompatible. The argument that depression is acting as a trigger for major innovations (let me call this the "depression-trigger" hypothesis) does not exclude the existence of a swarm of related innovations which accompany the diffusion of newly introduced products ("demand-pull" innovations). Perhaps the controversy could be reduced to the question of how causality is supposed to run; do the dynamics of the new Schumpeterian industries induce the upswing, as suggested by Mensch ("innovations overcome depression"), or is the emphasis laid on the upswing movement, which after being primarily induced by factors other than innovation, then prolongs itself by facilitating the diffusion of innovations?[8] Here, I cannot deal further with this question, but it certainly deserves more detailed future investigations.

In this paper I shall concentrate on the question of further empirical testing of the depression-trigger hypothesis. The Anglo-American literature on R and D and growth performance,[9] or Schmookler's results on patents and growth,[10] give some general empirical evidence for the existence of demand-pull effects. The depression-trigger hypothesis, as criticised by Clark *et al*, certainly appears to stand on weaker empirical ground.

Some more empirical evidence

If we interpret Mensch's list of basic innovations as a random sample from an unknown population, the criticism of Clark *et al* can be understood as calling into question the representativeness of the sample on grounds of size and selection principles.[11] There are two possibilities of checking whether this criticism is justified: attempts can be made to increase the size of the sample by including further cases of basic innovations from other sources; or we can try to evaluate other random samples from independent sources. The first possibility has been explored by J. van Duijn.[12] In this approach, the advantage of a larger sample must be set against the fact that we must have some reservations as to whether this sample is randomly selected. That is why I will use the other approach. I will try to evaluate a collection of 120 case studies of "important innovations" as given in an extensive study by K.B. Mahdavi.[13]

Mahdavi was not interested in the hypothesis we are investigating, but was looking at the efficiency of R and D processes (unnecessary time lags in the development of innovations between early invention and market introduction; duplication of R and D efforts between firms). Therefore, it can be supposed that, for the purpose of testing our hypotheses, his 120 cases are randomly selected. Furthermore, Mahdavi's sample is not exclusively restricted to basic or major innovations, but contains innovations of various types.

Classification procedure

In Table 5.1 I have listed Mahdavi's 120 cases. They are grouped roughly according to those industrial sectors which are supposed to be primarily concerned with them. I also made a classification into three categories:

SI = Scientific Instruments
IP = Improvement and Process Innovations
PI = Product Innovations

The separation of the category of Scientific Instruments is justified by their origins and purpose; they are primarily developed and used by R and D departments themselves, and can be considered as being directly induced by the direct needs of research processes.[14] Their secular increase over time reflects the augmenting scientific level of R and D work.

The category IP covers cases of (a) more or less radical quality improvements of already existing products; and (b) improved or completely new production processes for existing products. These also include those cases in which significant quality improvements and process innovations are overlapping, as, eg, in the iron and steel industry (airblowing, open hearth, or oxygen process), or in petroleum refining (petroleum cracking). The IP category covers a lot of cases (eg petroleum refining, atomic power) which are certainly not less important for economic development in the 20th century than many of the product innovations.

The category of product innovations (PI) covers only completely new products and materials. It is not identical with, but comes relatively near to, Mensch's or van Duijn's "basic innovations". It has the advantage of being defined independently of its implications for economic performance[15] and so avoids the problem of ranking heroic basic innovations versus minor improvements.

Separation into the two categories PI and IP seems to me to be justified on the following grounds. The IP type innovations concern products that are already in the market, and it seems justifiable to assume that direct market demand may be an important stimulus for innovation. Therefore innovations of the IP type could be expected to occur mainly in times when market demand is the strongest, ie in prosperous times. Completely new products (PI innovations) are introduced into the market when entrepreneurs expect that they will meet a *latent* demand. This speculation on potential demand implies higher risks, and incentives to take such risks are stronger in times of economic distress, when investment alternatives in existing product lines are exhausted. This classification thus provides a test for the hypothesis and explanation outlined above.

Before proceeding to discuss the results of this test of the hypotheses of demand-pull clusters (improvement and process innovations) and depression-trigger clusters (product innovations), it is essential to make some comments on problems of classification.

In some cases, the reader might prefer to classify innovations under other sectors, or to make a finer disaggregation. But this is not important for my further argument. More care is required in those cases in which the assignment to a certain sector is decisive for the classification of an innovation as an IP or PI type. First, there is a problem with capital goods innovations (gas turbine,

TABLE 5.1. 120 INNOVATIONS: THEIR DISTRIBUTION OVER TIME AND INDUSTRIES

	General Classification	WAS[a]
1. Chemical industry		
New and synthetic materials		
1880s Optical glasses (67)	IP	
1906 Chemical accelerator for rubber vulcanisation (19)	IP	
1909 Bakelite (16)	PI	
1920 Rayon (82)	PI	x
1930 Polystyrene (78)	PI	
1930s Oxygen production (71)	IP	
1933 Organic glass (Plexiglass) (68)	PI	
1937 Nylon (Perlon) (64)	PI	
1939 DDT (25)	PI	
1939 Polyethylene (77)	PI	
1940s Silicones (91)	PI	x
1949 Terylene (106)	PI	x
1959 Delrin (polyacetals) (24)	PI	
1960 Polycarbonates (76)	PI	
Pharmaceutical products		
1923 Insulin extraction (50)	PI	
1930s Antimalaria drugs (8)	PI	x
1930s Sulphur drugs (97)	PI	
1934 Vitamin C (117)	PI	
1936 Vitamin B_1 (115)	PI	
1938 Vitamin B_2 (116)	PI	
1942 Penicillin (72)	PI	x
1946 Streptomycin (95)	PI	x
1947 Vitamin A (114)	PI	
1948 Cortisone (23)	PI	x
2. Petroleum production and refining		
1859 Drilling for oil (26)	IP	
1920 Thermal pressure cracking (107)	IP	
1930s Antiknocking gasoline (7)	IP	
1942 Continuous catalytic cracking process (18)	IP	
1950s Shale oil (88)	IP	x
3. Electrical Industry		
Electricity production		
1870 Dynamo (27)	IP	
1884 Steam turbine	IP	
1888 Alternating current system (5)	PI	
1932 Gas turbine (42)	IP	
1954 Atomic energy (11)	IP	x
1955 Solar battery (92)	IP	
1958 Fuel cell (40)	IP	x
Communication, data processing, and final consumer goods apparatus		
1860s Refrigerator (84)	PI	
1866 Submarine telegraph (transatlantic cable) (96)	IP	x
1879 Incandescent lamp (48)	PI	
1881 Telephone (105)	PI	
1892 Automatic switchboard (14)	IP	
1899 Long-distance telephone communication (54)	IP	
1900 Wireless telegraphy (118)	IP	
1920 AM radio (6)	PI	x

Table continues

Table 5.1 continued

	General Classification	WAS[a]
1920s Cinema (20)	PI	
1920s Record player (83)	PI	
1926 Facsimile (35)	IP	
1936 Electronic TV (34)	PI	
1936 FM radio (39)	PI	x
1938 Fluorescent lamp (36)	PI	
1940s Radar (80)	PI	x
1940s Radio astronomy (81)	IP	x
1940s Xerographic machine (119)	PI	x
1947 Tape recorder (104)	PI	x
1948 Long playing record (55)	IP	
1951 Transistor (109)	PI	
1960 Tunnel diode (110)	PI	
1960 Vacuum switch (112)	IP	
1960s On-line computer systems (65)	IP	
1960s Time-sharing computer system (108)	IP	
1965 Lucalox lamp (56)	PI	
4. Transport		
Aircraft		
1920s Airplane (4)	PI	x
1938 Helicopter (45)	PI	x
Automobiles		
1885 Internal combustion engine (51)	PI	
1913 Assembly line (9)	IP	
Ship		
1903 Marine diesel engine (57)	IP	x
1911 Flying boat (37)	IP	
1913 Electric ship propeller (28)	IP	x
1949 Ship stabilizers (89)	IP	
1950s Hydrofoil and hovercraft (47)	IP	x
Railways		
1879 Electric railway (29)	IP	
5. Iron, steel, non-ferrous metals		
1861 Open hearth furnace (66)	IP	
1878 Air-blowing process in steelmaking (2)	IP	
1886 Electrolytic process in aluminium (30)	IP	
1906 Vanadium steel alloy tool (113)	IP	
1909 Age-hardening aluminium alloy (1)	IP	x
1913 Stainless steel (93)	IP	x
1953 Oxygen process in steelmaking (70)	IP	
6. Machine construction		
1955 Numerically controlled milling machine (63)	IP	x
1950s Remote control (85)	IP	x
7. Metal working		
1876 Universal grinding machine (111)	IP	
1877 Automatic gear-cutting machine (13)	IP	
8. Printing		
1886 Automatic composing machine (12)	IP	
1945 Photocomposing machine (74)	IP	

Table continues

Table 5.1 continued

		General Classification	WAS[a]
9. Watch manufacturing			
1930	Automatic watch (15)	IP	
1969	Quartz crystal clock (79)	IP	
10. Scientific instruments			
1937	Electron microscope (31)		
1938	Oscilloscope (69)		
1940	High-energy accelerator (46)		
1943	Infrared spectrometer (49)		x
1943	Mass spectrometer (59)		
1946	Phase microscope (73)		
1953	Nuclear magnetic resonance spectrometer (61)		
1954	Gas chromatograph (41)		
1957	X-ray microscope (120)		
1960s	Electron spin resonance spectrometer (32)		
1960s	Maser (58)		x
1960s	Microwave spectrometer (60)		
1963	Atomic absorption spectrometer (10)		
1963	Nucleopore filter (62)		
1964	Gel permeation chromatograph (43)		
1966	Biolaser (17)		
1966	Laser-hologram-camera system (53)		
1966	Scanning electron microscope (87)		

		Branch	General Classification	WAS
11. Difficult cases				
1869	Synthetic Alizarin (98)	Chem. Ind.	?	
1906	Airconditioning (3)	?	?	
1911	Gyrocompass (44)	?	?	x
1930s	Synthetic detergents (99)	Chem. Ind.	?	x
1931	Freon (38)	Chem. Ind. or Electr. Ind.	?	
1932	Short-wave radio (90)	Electr. Ind.	?	x
1932	Synthetic rubber (103)	Chem. Ind.	?	x
1935	Colour photography (21)	Photo. Ind.	?	
1936	Synthetic oil (102)	Chem. Ind.	?	x
1940	Jet airplane (52)	Aircraft Ind.	?	x
1951	Electronic computers (33)			x
1953	Colour TV (22)	Electr. Ind.	?	
1957	Synthetic diamonds (100)	Metal Ind.	?	x
1958	Pipelines (75)	?	IP	
1960s	Synthetic leather (101)	Chem. Ind.	?	
1962	Satellite communication (86)	Electr. Ind.	?	x

Note: [a]WAS equals war armament strategy.

atomic power, NC-machines, etc) which, from the standpoint of their producers, are new products, but for the final user serve exclusively as process innovations. Irrespective of their classification by producer or user sectors, I classified them generally as IP innovations, according to their final purpose as process innovations.

Classifying is much more problematic in some cases of inter-industry innovations. There is, for example, Freon, innovated in 1931. As a new chemical substance, we could classify it under the chemical industry as a product innovation. But it was innovated through the cooperation of Du Pont and the Frigidaire Corporation. The latter was urgently looking for an improved substance to overcome problems of flammability, corrosion, toxicity, etc. Thus we could just as well classify Freon as a further improvement of the refrigerator, which was innovated in 1863.

Airconditioning could be interpreted as an improvement innovation, because the main impulse for its development came from textile mills, trying to reduce breakages and the consequences of static electricity; in 1906, it was first applied to a lithographic plant. But due to its later diffusion in offices and private households, it could be regarded as a new final product of the electrical industry. Because the gyro-compass originally played an important role in the exploration of the North Pole, we could conceive of it as a new scientific instrument, but its generalised use for ship and aircraft would justify classification to the transport industries. Moreover, it is a matter of subjective judgment whether, in the latter case, we conceive of it as an improvement of its predecessor (the magnetic compass) or as a completely new product, because of its quite different technical functioning.

Similar problems arise in some other cases. Are the jet engine, colour TV, colour photography, and short-wave radio, new products or decisive improvements on already existing basic products (propellor aircraft, black and white TV and photography, long-wave radio)? The same question is relevant with synthetic substitutes for natural materials.

To avoid an arbitrary decision in these cases, I omitted them from Figures 5.1a and 5.1b. They are listed separately in Figure 5.1c and further included in 5.1d and 5.1e. As is well known to innovation researchers, dating of individual innovations is sometimes somewhat difficult.[16] In order to ensure replicability of my testing, I use the dates as given by Mahdavi,[17] even if in some cases other dates might appear more plausible.[18]

In dealing with frequencies of innovations, I counted in 10-year periods, to account for the fact that Mahdavi's data do not allow for a finer periodization. Instead of individual years, Mahdavi often gives rough dates like "during the 1920s", or "early 1960s".

Testing procedure

Looking at PI type innovations, a smaller cumulation with a peak in the 1880s and a larger one with a peak in the 1930s can be discerned (Figure 5.1a); the IP type innovations (Figure 5.1b) show some clustering tendencies which could be interpreted as confirming the prosperity-pull hypothesis of Clark *et al*. If we make a 'soft' definition of PI, including the difficult cases (Figure 5.1c) into PI, the observed cluster in the 1930s becomes even stronger (Figure 5.1d). With a 'hard' definition of PI, classifying all difficult cases under IP, the clustering tendency from 1900 to 1919 and in the 1950s becomes stronger, but there is also some clustering in the 1930s (Figure 5.1e).

Roughly speaking, in terms of long-wave theories, the 1870s and 1880s, as well as the inter-war period, can be considered as periods of relatively unstable and weaker growth (B-periods), whereas the time from the 1890s till World War 1, and from the late 1940s to the late 1960s, are periods of relative prosperity (A-periods).[19] Thus we could interpret the PI data as confirming the depression-trigger hypothesis, and the IP data as giving evidence for the demand-pull hypothesis; but as the clustering of the latter data is less strong, I conclude that either the demand-pull effect for IP innovation is somewhat weaker than depression triggering for PI innovations, or that another effect (the depression-trigger effect?) is acting as a countertendency.

The question must now be asked, are the clusters to be observed in Figure 5.1

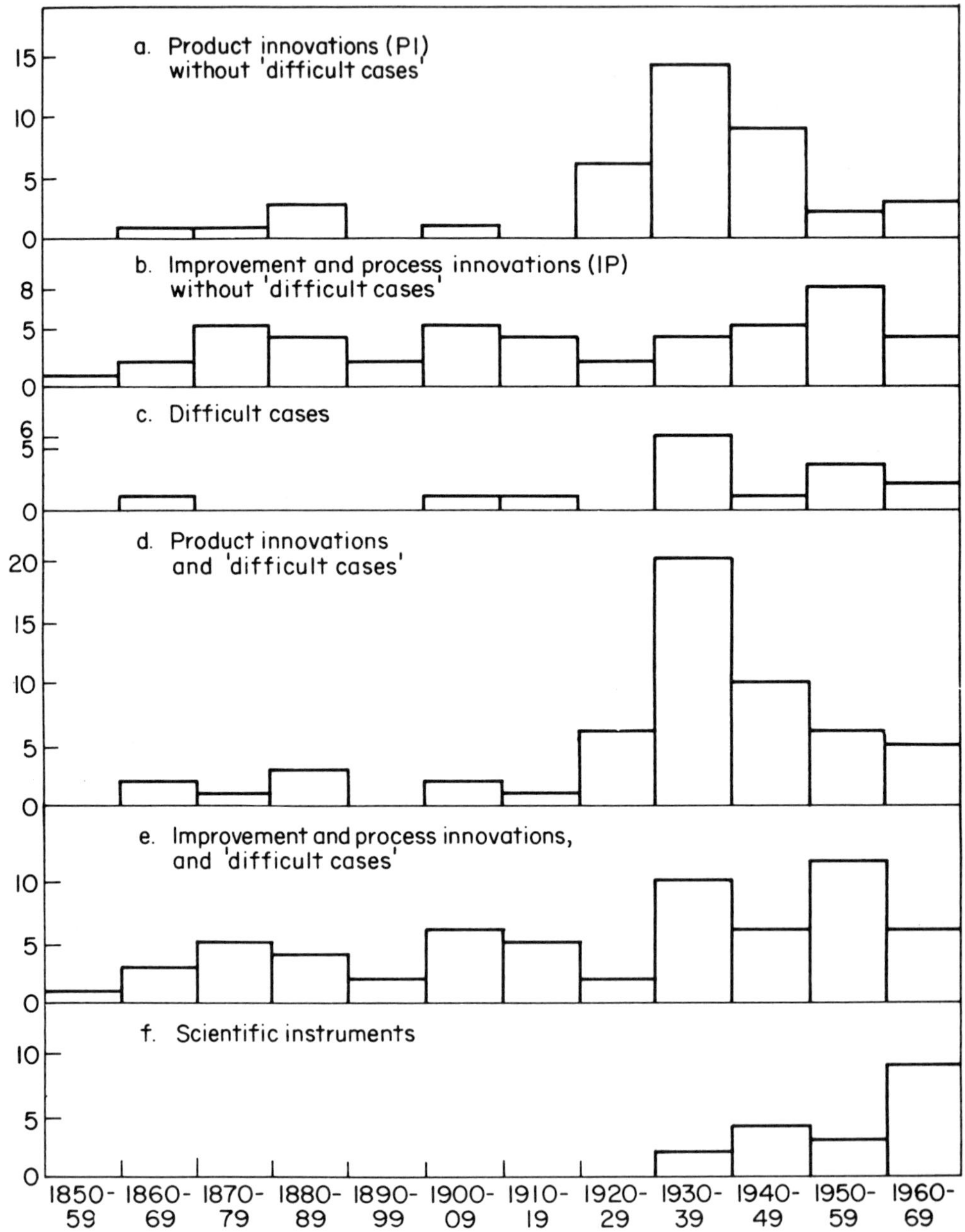

Figure 5.1. Numbers of innovation by type (as in Table 5.1), 10-year periods.

significant, or could they be mere chance variations? This will be answered by means of the nonparametric χ^2 one-sample test[20] on the following null hypothesis (the Random Walk hypothesis):

There is no difference in the number of innovations in different 10-year periods; any observed differences are merely chance variations to be expected in a random sample.

To test the null hypothesis for the PI clusters, I shall choose the sub-periods such that the B-periods in which the clusters occurred (1880s, 1930s) are compared with the preceding and the following A-periods. According to the long-wave framework, the two sub-periods from 1850 to 1909 (or to 1919) and from 1890 to 1969 (or from 1900 to 1969) can be chosen.

Testing the hypothesis for the IP type innovations, several sub-periods from 'trough' to 'trough' in the distribution of innovations (1850s to 1890s, 1890s to 1920s, and 1920s to 1960s) will be taken.

Test results

First, the χ^2 one-sample test reveals that the null hypothesis has to be rejected at the 0·001 level of significance for the product innovations during the 20th century; the clustering of radical product innovations during the 1930s cannot be interpreted as a chance variation. Thus the test gives strong support for the existence of a depression-trigger effect.

This result is robust against changes in periodisation (taking 1890 to 1969 instead of 1900 to 1969), as well as changes in classification. Whether the "difficult cases" (Figure 5.1c) are excluded from the PI category (Figure 5.1a) or included (Figure 5.1d), the χ^2 value is always far above that required for an 0.001 level of probability that such a cumulation could occur under the null hypothesis.

Secondly, the random hypothesis has to be accepted for the PI type innovations of the 19th century, and for all IP type innovation clusters for the above-named sub-periods. There remain probabilities of between 0·3 and 0·5 that such clusters could occur as mere chance variations.

Interpretation

The acceptance of the random walk hypothesis for product innovations during the 1880s is somewhat surprising. It contradicts the results I obtained from applying the same test on Mensch's sample[21] and van Duijn.[22] In both cases, the clusters of "basic innovations" in the 1880s and 1930s prove to be significant on levels between 0·01 and 0·001.[23] From the sheer number of cases, I see no problems about the representativeness of Mahdavi's sample for the 20th century. However, looking at the small number of cases in his sample for the 19th century (see Table 5.1), the confirmation of the random walk hypothesis should be interpreted with the utmost caution: Between 1859 and 1919, Figure 5.1a covers 6 (and Figure 5.1d, 9) cases, 3 of which fall into the 1880s. Even if this clustering has the relatively high probability of about 0·3 to 0·4 of being a chance variation, this can hardly be taken as strongly contradicting the result from the Mensch and van Duijn samples.

As mentioned above, the IP type innovations in Mahdavi's sample come near the diffusion-oriented definition of innovations given by Clark *et al.* Does the acceptance of the null hypothesis justify the assumption that the demand-pull effect can be neglected, or that prosperity has only a minor influence on IP type innovations? I tend more to the conclusion that the observed pattern is the result of two relatively strong factors which have partly contradictory effects. On the one hand, depression is acting as a trigger not only for radical product innovations, but also for some outstanding improvement and process innovations. On the other hand, the favourable market conditions during prosperity periods promote IP innovations in old industries as well as related innovations within the new Schumpeterian industries ('band-wagon').

Finally, in considering possible prosperity-pull effects, an important limitation of the data should not be forgotten. As Mahdavi was looking for "important innovations" (without further defining "important"), it has to be assumed that he selected more or less outstanding breakthrough innovations. It may be that, in addition to spectacular advances, the large stream of smaller steps is a decisive element of the band-wagon. Therefore, the discussion of IP innovations from Mahdavi's sample has to be interpreted in its proper perspective. Considerable doubts about the demand-pull hypothesis are justified insofar as a significant clustering of *outstanding* innovations during prosperity periods is postulated. However, emphasising the large stream of "smaller", less spectacular innovations, the demand-pull hypothesis may still be important. The results of Schmookler on patents and growth are a good example of the relevance of such prosperity-pull effects.

The above result confirms the depression-trigger hypothesis and, at the same time, it does not contradict the prosperity-pull position. As suggested below, however, the controversy between the two positions could still be relevant when we look at another perspective: is it primarily the dynamics of the new Schumpeterian industries, set in motion by trail-blazing innovations, which overcome depression, as suggested in the subtitle of Mensch's book: *innovations overcome depression?* Or is the rapid diffusion of major innovations only possible in a climate of general prosperity, as suggested by Clark *et al*?[24] In the present paper, this aspect of the controversy has not been worked out. However, it should be clear that, because of the quite different policy implications of the two positions, this aspect deserves much fuller investigation.

Concluding remarks

Statistical tests on Mahdavi's data offer weak support for the prosperity-pull hypothesis on important innovations, but they give strong support for the depression-trigger hypothesis: there is a strong clustering of radical product innovations in the period of relatively unstable growth between the two world wars. This appears in sharp contrast to the periods of relative prosperity before the first and after the second world wars, which were obviously poor in initiating (but not in the diffusion of) radical product innovations.

Furthermore, the distribution between sectors of important product innovations in the Mahdavi sample corresponds well with the distribution of "basic innovations" from the Mensch sample.[25] *Product* innovations in the Mahdavi sample did mainly occur in those sectors which can be considered new from the standpoint of the 20th century, such as the electrical and electronics industries, the chemical industry (plastics and resins, drugs) or the aircraft industry. Again judging from the Mahdavi data, innovative activity in old industries (iron and steel, printing, railways, etc) seems to be restricted almost exclusively to improvement and process innovations, often exploiting technical progress achieved in new industries. Such spill-over improvements are, for example, the electric ship (ie submarine) propeller or the electric railway.

At present, I do not feel able to make more than some rough and tentative notes on the question of why product innovations are concentrated precisely within these new industries. One possibility is that part of these industries had already well developed R and D departments, and that depression induced a re-orientation of R and D policy in favour of longer-term projects. It is a general pattern in the Mahdavi case studies that, for many projects which were completed during the 1930s and 1940s, important inventions (or even development steps) had been made during the preceding prosperity periods. However, projects were often dropped when they hit an important technical obstacle which hindered a *prompt* market introduction. During longer periods of relative prosperity, less risky investment alternatives in established sectors, which promised to be profitable within the short term, seemed to leave no chance of continued expenditure on longer lasting innovation projects.

Consequently, it can be expected that, during prosperous times, some half-finished innovation projects will be stored up and, in periods of prolonged crises, it may become interesting to resume work on such projects, because short-term alternatives are unattractive.[26] This argument does not necessarily contradict the argument of Clark *et al*[27] that firms sometimes delay the launch of new products until the market recovers. Under conditions of high risk and uncertainty, firms may be rational in ordering their R and D departments to initiate longer lasting projects, while simultaneously ordering the marketing department to postpone the introduction of finished projects until the deepest point of the recession has passed.

In general, however, we should also guard against overestimating the influence of the *macro*-economic recession on the introduction of new products. The extent to which an individual firm, or an entire industry, will be hit by a macro-economic crisis depends largely on its ability to rejuvenate its production programme by means of innovation. In other words, the influence of innovation on the so-called *gespaltene Konjunktur* (simultaneous depression in one sector and prosperity in another) should not be neglected. Compare for example the current state of the electronics industry with steel or textiles.

A simple notion of depression-triggering would lead one to expect that those industries which are suffering most from economic crisis would be pressed to become most innovative. However, a survey of production growth in the inter-war period, such as that of Wagenfuehr,[28] shows the contrary. New industries were less hit than old by the crisis. The relatively higher profits in the new industries apparently not only made financing of innovation projects easier, but also provided additional incentives for further expansion into new spheres of investment.

A further warning against constructing a simple depression-trigger hypothesis of product innovations comes from Table 5.1. The column 'WAS' (war armament strategy) indicates those cases in which Mahdavi explicitly mentions that a government institution gave substantial support to the innovation process. In all these cases, this support was motivated by the importance of the innovation to the armament industry or the war machinery built up in response to wars, or the supposed possibilities of wars. As such projects are often kept secret, the real influence of this factor may be rather underestimated but, nonetheless, its quantitative weight is impressive. During the 1940s, in 8 out of 9 cases of product innovations (1920s: 3 out of 6; 1930s: 3 out of 14), preparation for war, or demand management through armament contracts, led to substantial government support.

Observing the shortsighted R and D strategies of firms which are obvious in many of Mahdavi's case studies,[29] it seems that the simple myth of the dynamic entrepreneur whose creativity is spurred by economic distress will not yield much of an explanation of the observed clustering tendencies. To work out a satisfactory theory of innovation clusters, the whole complex of social, political, and economic changes during the crisis periods has to be considered. The argument that risk and uncertainty are important obstacles for innovation should be taken seriously. But, finally, most product innovations do occur in times of distress, when risks and uncertainties are at their highest. Under these circumstances, from a macro-economic view or from the standpoint of society as a whole, the question may well be justified as to whether the mechanism of creating innovations works efficiently and rationally. It cannot be ignored that, during the inter-war period, innovative activity was accompanied by stronger tensions in social and political relationships within individual countries and that, in a worsening international climate, strong armament efforts promoted an important proportion of innovation projects. This fact does not confirm that economic conditions and decision mechanisms for achieving innovations in the market economy worked rationally. The coincidence of innovation and war armament efforts with all its consequences is not accidental—and neither are some parallels with the present.

Notes and references

1. J.J. van Duijn, *De lange golf in de Economie* (Assen, van Gorcum, 1979); L. Fontvieille, "Cycles longs et régulation", *Issues*, Cahiers de Recherches de la Revue Economie & Politique, 1979, *4*; C. Freeman, "The Kondratiev Long Waves, Technical Change and Unemployment", in *Structural Determinants of Employment and Unemployment*, papers prepared for OECD Experts Meeting, January 1977 (Paris, OECD, 1979), pages 181–196; A.K. Graham and P.M. Senge, "A long-wave hypothesis of innovation", *Technological Forecasting and Social Change*, 1980, *17*, 4, pages 283–311; G. Ray, "Innovation in the long cycle", *Lloyds Bank Review*, January 1980, *135*.
2. H.H. Glismann, H. Rodemer and, F. Wolter, "Lange Wellen wirtschaftlichen Wachstums, Replik und Weiterführung", Kiel Discussion Papers number 74, December 1980.
3. J. Clark, C. Freeman, and L. Soete, "Long waves and technological developments in the 20th century" in D. Petzina and G. van Roon, eds, *Konjunktur, Krise, Gesellschaft: Wirtschaftliche Wechsellagen und sozialer Entwicklung im 19 und 20 Jahrhundert* (Stuttgart, Klett-Cotta, 1981), pages 132–169.
4. G. Mensch, *Das technologische Patt* (Frankfurt/M, 1975); English translation: *Stalemate in Technology* (Cambridge, MA, Ballinger, 1979).
5. Clark, Freeman, and Soete, reference 3.
6. Clark, Freeman, and Soete, reference 3.

7. Clark, Freeman, and Soete, reference 3.
8. Clark *et al* seem to sympathise with this latter position when pointing out: "If we are thinking of economic impact, then the immediate effect of 10 or 20 basic innovations is negligible in the first few years of their introduction, as well as in the gestation period. The resources committed are absolutely tiny compared with the scale of investment involved in the large-scale follow-through of these basic innovations with the concomitant development of entirely new branches of industry, supporting services, component industries, materials supplies and so forth". [Reference 3]
9. For a survey, see R.R. Nelson and S.G. Winter, "In search of useful theory of innovation", *Research Policy*, January 1977, *6*, 1.
10. J. Schmookler, *Invention and Economic Growth* (Harvard University Press, 1966).
11. Clark *et al*, reference 3.
12. Van Duijn, reference 1, pages 139ff.
13. K.B. Mahdavi, *Technological Innovation. An Efficiency Investigation.* (Stockholm, Beckmans, 1972).
14. Reading the first draft of this paper, Rod Coombs (Manchester) remarked that there are some cases of scientific instruments which also have some importance for industrial production (especially quality control). Such cases are: infrared spectrometer (1943), mass spectrometer (1943), gas chromatograph (1954), maser (1960s), or atomic absorption spectrometer (1963). Grouping these 5 cases with the IP innovations did not change the results significantly (there still remains a probability of between 0·1 and 0·2 that the resulting cluster could be a mere chance variation).
15. To test the influence of innovations on economic growth, we need an independent definition of 'innovation'. Defining 'basic innovations' as giving the base for new markets and growth industries (Mensch, van Duijn) is somewhat unsatisfactory, because it bears the danger of tautological testing, besides being less precise.
16. Historians may argue for a long time about when to date 'railways' as a 'basic innovation'; whether it was 1817, 1825, 1830—or much earlier because of the railways in the mines, or later with first networks. (See Clark *et al*, reference 3.)
17. In two cases, Mahdavi gives two dates: pipelines—1865 and 1958; quartz crystal clock—1927 and 1969. Obviously, in these cases the first innovation was not successful. That is why I decided to take the dates of the re-innovation.
18. In the case of insulin, Mahdavi takes 1923 (first successful mass extraction of insulin). One year earlier, the first human patient recovered from his diabetes after the injection of insulin. But we could as well take 1969, when the first synthetic insulin was produced in a laboratory (Mahdavi, reference 13, page 138).
19. I tried to give some empirical evidence for this in my paper, "Innovation, Akkumulation und Krise", *PROKLA*, *35* (Berlin, Rotbuch Verlag) 1979.
20. S. Siegel, *Nonparametric Statistics for the Behavioral Sciences*, International Student Edition (McGraw-Hill, 1956), pages 42ff.
21. Mensch, reference 4, pages 136f.
22. Van Duijn, reference 1, page 139f.
23. Only the cluster presented by Mensch for 1820–30 (which does not appear in van Duijn's list) does not stand the χ^2 test: There remains a probability of between 0·1 and 0·2 that such a distribution could occur under the null hypothesis.
24. See reference 8.
25. An allocation of the 42 cases of Mensch's 20th century sample to industrial sectors has been made in my paper, "Basisinnovationen und Wachstumsschübe: das Beispiel der westdeutschen Industrie", *Konjunkturpolitik* (Berlin, Duncker and Humblot), 1979, 5/6. When comparing this result with the Mahdavi data, only one correction seems necessary: the category of scientific instruments, which becomes particularly important during the 1960s in Mahdavi's sample, is insufficiently represented in the Mensch sample.
26. This possibility is explicitly mentioned by Mahdavi in the case of polyethylene (polythene): "In 1932, some scientists of Imperial Chemical Industries Limited in England proposed to the Board of Directors that high-pressure chemistry, as a long-run research project, be undertaken ... Economically, the Great Depression of 1929–1932 had eliminated the possibility of introducing a new profitable commodity immediately. Thus the object was a long-run program". [Mahdavi, reference 13, page 161.]

27. Clark *et al*, reference 3.
28. R. Wagenfuehr, ed, *Die Industriewirtschaft. Entwicklungstendenzen der deutschen und internationalen Industrieproduktion 1860 bis 1932* (Berlin, Vierteljahreshefte zur Konjunkturforschung, Sonderheft 31, 1933 edited by Institut für Konjunkturforschung).
29. It was none other than the famous Thomas A. Edison who fought with various kinds of psychological warfare against the introduction of the alternating current system. The Edison Electric Light Company was already financially committed to DC systems (Mahdavi, reference 13, page 30ff). In another case, even Prussian bureaucracy proved to be much more flexible than American entrepreneurs: "The innovation of the automatic switchboard was not at first welcomed by the telephone companies. They had already invested a relatively large amount of money in the mechanical system. The application of the automatic switchboard was, therefore, restricted to small towns for some time. Germany, with her public telephone ownership, was the world's first major centre to use an automatic switchboard, in 1909." (*Ibid*, page 103).

CHAPTER 6

LONG WAVES, INVENTIONS, AND INNOVATIONS

John Clark, Christopher Freeman and Luc Soete

It is argued that aggregate R and D and patenting activity, while generally less volatile than short-term economic activity, is more closely related to the latter in the longer term. Second, 'fundamental' or 'basic' inventions exhibit a clustering behaviour, although there appears to be no clear-cut relationship with the overall level of economic activity. Third, we find no support for Mensch's argument that lead times between invention and innovation contracted in the 1930s. Fourth, we find some evidence for a bunching of *innovations* in the 1930s, but this does not appear due to depression-induced acceleration. Fifth, in the particular case of plastics, where the 1930s clustering is particularly clear, the forces at work were primarily related to 'science push' and the particular requirements of the German economy.

Long-term trends in R and D and patenting

DURING the post-war period, it has often been assumed that scientific and inventive activities were less vulnerable to the downswings of short- and medium-term business cycles than investment outlays. Generally speaking, R and D expenditures for the OECD countries[1] showed a strong and steady upward trend in the 1950s and 1960s, apparently little affected by short-term fluctuations. Whilst in the late 1960s there was a slowing-down of the previously very high rate of growth of these indicators, they still appeared to be little affected by business cyclical patterns. This was usually explained in terms of a recognition by firms of the essentially long-term nature of R and D, and the need for stability in this sort of programmed activity. Whereas outlays for new machinery, vehicles and plant could be and were postponed or brought forward in response to changing business expectations, it was believed to be less sensible and less necessary to expand or contract R and D in response to the same pressures. Consequently, whereas investment behaviour occupies a central place in almost all theories of the business cycle, and indeed is the principal determinant of the cycle itself in many such theories, R and D behaviour is usually ignored. This must be qualified by the observation that R and D

The authors are with the Science Policy Research Unit, University of Sussex, Falmer, Brighton, Sussex, BN1 9RF, UK. They wish to thank Joe Townsend for helpful discussions regarding data sources.

statistics only recently became generally available and that even now they are not always published annually. The relative stability of R and D outlays during the 'recessions' of the 1950s and 1960s, moreover, does not mean that they are unaffected by deeper disturbances. Although we do not have systematic comparable annual series for the pre-war period, there is very strong evidence that company outlays for R and D were significantly reduced during the depth of the Depression, from 1931 to 1934, in the leading industrial countries. Terleckyj[2] reports a fall of 10% in US R and D in industry over this period. There is now additional evidence that, in the more serious recessions of the 1970s and early 1980s, R and D behaviour may be more affected and may itself have greater effects on other behaviour than in the previous 20 years.

In this article, we note initially that the possibility of a long-wave pattern of R and D cannot be excluded. But in any case, R and D expenditures are only a measure of some of the *inputs* into the innovative process and, in principle, it would be quite possible for inventions or innovations themselves to follow a cyclical pattern, even if R and D did not do so. In an extremely thorough study of long-term trends in patenting in the USA, Schmookler[3] advanced the view that patenting follows closely the fluctuations in fixed investment in plant and machinery in each industry, with essentially the same pattern of upswings and downswings. This has formed the basis for some 'demand-led' theories of innovation which have, however, been heavily criticised by Mowery and Rosenberg[4] and others.

One of the main points of criticism relates to the relative importance of individual inventions and innovations. Some of Schmookler's critics point out that the large numbers of patents included in his analysis relate mainly to incremental or improvement inventions, which might be expected to respond to changes in demand after a short time lag. As Schmookler himself accepts, this would not rule out the possibility that the more important radical or fundamental inventions and innovations might precede, or indeed induce, changes in both investment behaviour and demand, which is one of the central points in Mensch's[5] neo-Schumpeterian theory of innovations and long waves. The argument is difficult to verify and it centres, like much of the innovation–long-wave debate, around the definition of 'radical', 'basic', or 'fundamental' inventions and innovations.

We have drawn, in Figure 6.1, the total number of patents applied for and granted in the USA from 1840 (year of the Patent Law) to 1979. The diagram suggests that variations in the number of patent *applications* have tended to coincide with overall economic fluctuations. With an average of a four-year time lag—up to the 1960s, the average time needed for the Patent Office to examine and issue a patent was about four years—the evolution of the number of patents *granted* shows the same kind of behaviour.[6] The steep fall in the early 1930s in patent applications coincides most clearly with the Great Depression. The effects of both World War 1 and World War 2 are also obvious. Overall, and despite the relatively weaker evidence for the 1875–1884 depression, Figure 6.1 tends to support Schmookler's hypothesis. It does not, of course, provide an answer to the question of the possible fluctuations in *radical* or *fundamental* inventions.

The advantage of patent statistics is that they provide readily available

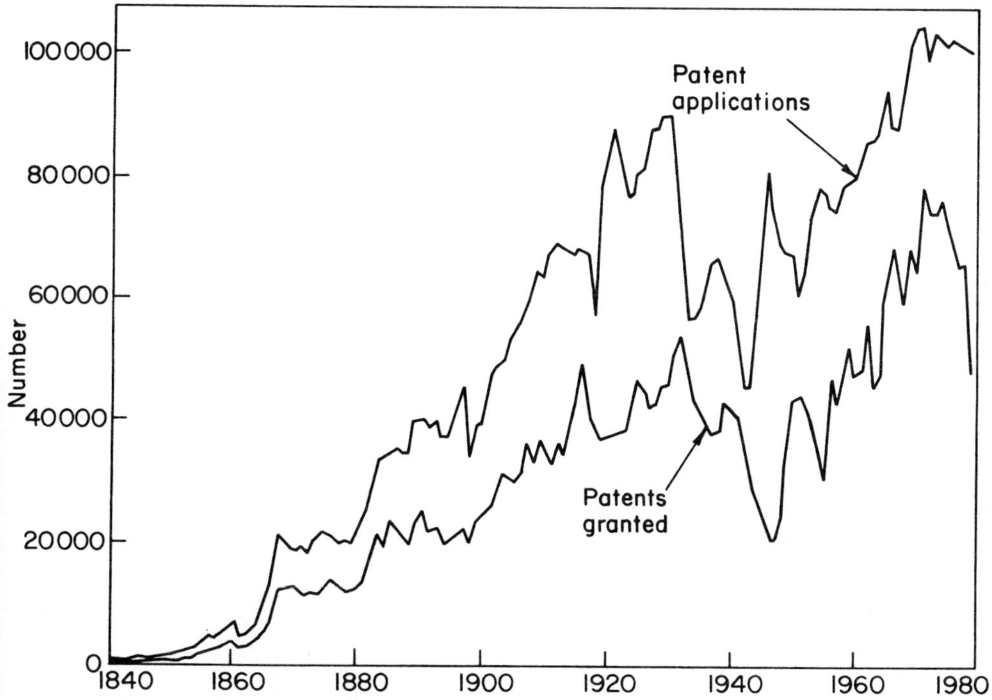

Figure 6.1. US patenting activity, 1840–1979.

information over a rather long period which can be easily classified by year, and which is not affected by changes in relative prices. A more important advantage is that they have been collected and examined over all these years by the same official agency, generally speaking a Patent Office. Moreover, all patents issued in the United Kingdom from 1734 onwards have a specification describing the invention in full. It is on the basis of these specifications that an attempt was made by R. Baker of the British Library to select 'significant' inventions.[7]

Baker's method consisted of identifying first 'significant' subjects, for which the most important inventions, in terms of the major patents, could then be identified. These ranged from the Addressograph to Zip fasteners, and constitute (in our view) a rather coherent and comprehensive sample of the major inventions of the last two centuries. For a number of subjects, Baker also identified some specific subclasses; for example, in the case of radio valves and microwave tubes, he considered the diode, triode, pentode, variable mu, klystron, cavity magnetron, and travelling wave tube subclasses. In total, 400 different major subjects and subclasses were identified for which the available information included the exact year of the first patent on the subject, and what we would call the *master* and the *key* patents. The master patent is defined as "the first to be economically viable",[8] while key patents refer to the most important patents in relation to each specific subject. The total number of patents covered by Baker's analysis was about 1000, and covered the period 1691 to 1971.

On the basis of this information, 10-year moving averages of the master patent data and the key patent data for the period 1775 to 1965 are plotted in

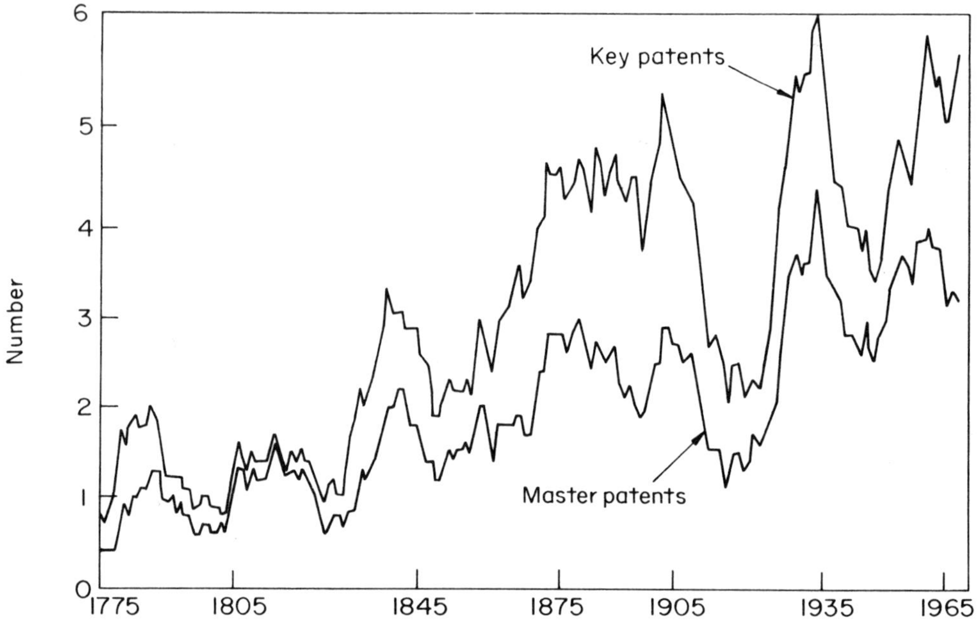

Figure 6.2. Significant inventions, 1775–1965 (10-year moving averages).

Figure 6.2. The evolution of the number of first patents was practically identical to the master patent evolution (r=99·7%), and has not been drawn here (more detailed information can be obtained from the authors).

A number of points emerge from Figure 6.2. First, the evolution of the number of key patents is very similar to the evolution of the number of master patents (r=97·7%). The major difference relates to the most recent period, for which it is rather difficult to judge what will become master patents. The close correlation between the three patent concepts suggests, however, that Baker's criteria for selecting key patents are relatively reliable

Second, despite the steady increase in the number of inventions over time, the data support the view that clustering of basic inventions takes place, and that the evolution over time of such inventions (as measured by numbers of key or master patents) is not characterised by a linear trend.

Third, it seems possible to identify a clustering of major inventions in various phases of the long wave, including depressions (clusters in periods 1874–89 and 1928–36), prosperity phases (clusters in periods 1897–1903 and 1956–1961), and war (1806–1815). There does not therefore appear to be clear *prima facie* evidence that the observed clustering is unambiguously related to particular economic circumstances, whether favourable or adverse. Further examination of actual cases is needed to determine whether particular categories of invention (eg radical/incremental, or product/process) are associated with general economic trends. While these data are highly imperfect, they probably are among the best available, and the trends they suggest are *grosso modo* inconsistent with the overall patent evolution as illustrated in Figure 6.1. This suggests the importance of examining data on significant inventions rather than merely aggregate data.

Long-term trends in innovation

When we turn to innovations, one of the major difficulties is the absence of any satisfactory long-term series of statistics. While we have R and D statistics for the post-war period, and patent statistics for centuries—despite all their imperfections—we do not have any comparable data for innovations. Some people think that we shall never have such data, but we do not subscribe to this counsel of despair. On the contrary, we heartily welcome the recent initiative of the OECD Directorate for Science, Technology and Industry in organising an international conference on output indicators,[9] and the valuable stimulus this gave to the embryonic attempts to measure innovations in West Germany, France, UK, and other countries. We are ourselves involved in a project for the collection and analysis of a data-bank on long-term trends in innovation. This work, however, is necessarily of a long-term nature and, in the meantime, we have to do the best we can with the existing very imperfect sources of information, and to attempt elucidation of the theoretical problems.

We are therefore very sympathetic to the original and difficult work of Mensch,[10] in which he attempts to make use of various lists of basic innovations. It has been a stimulus to our own research and we believe the same is true of much other research and debate, such as that represented in this collection of papers. However, despite this belief, we have found ourselves in disagreement with some of the basic ideas which Mensch advances.[11] In the following sections, we deal with two issues in particular: the general idea that innovations are bunched in periods of deep depression and that, in such periods, time lags between invention and innovation are compressed owing to the economic climate. We make these criticisms as part of the current debate just because we think it is so important for research in this field to develop from Mensch's starting point by achieving the greatest possible clarification and verification of the available statistical evidence.

We first look at the statistical evidence put forward by Mensch to support his hypothesis that the number of innovations increases during a period of chronic depression, primarily as a result of a reduction in the lead time between invention and innovation in such periods (rather than simply because of an 'echo' effect following an earlier bunching of inventions). The following is a brief discussion of the empirical evidence on which the hypotheses of 'bunching' and 'acceleration' are based with particular reference to the first half of the twentieth century.

For this period, Mensch relies heavily on *The Sources of Invention* by Jewkes, Sawers and Stillerman, in which 61 inventions are listed (or 71 in the second edition).[12] Of these, 50 (or 56) are discussed in some detail as case histories. Mensch derives invention and innovation dates for 41 cases, and uses them in an attempt to demonstrate bunching and acceleration in the Great Depression of the 1930s.

The persuasiveness or otherwise of Mensch's argument depends on the adequacy of the data sources used, and the way in which they have been applied. On the first point, it is important to realise that Jewkes *et al* never claimed that their selection of inventions was in any sense complete or statistically reliable. Moreover, there are several obvious sources of bias for the purposes in which Mensch is interested. He relies on the first edition and, since this was compiled in the mid-50s, it is obvious that it could not do justice to the inventions and innovations of the 1950s.

As regards the application of the data, some 22 inventions from the list in the first edition of Jewkes *et al* are not included by Mensch; assuming that Mensch is relying only on the case studies, this number falls to 11. (We understand that the chief criterion for exclusion is that the innovations concerned have an estimated 'lead time' between invention and innovation of ten years or less,[13] although it is not at all clear to us why an innovation with a relatively short lead time should be considered less important for Mensch's purposes.)

A second point with respect to interpretation is that there is a high degree of ambiguity surrounding any estimate of invention and innovation dates, and differences of opinion are likely even when a single source of information (such as Jewkes *et al*) is used. The three of us have independently estimated the dates of invention and innovation for all 60 cases covered in the case histories sections of the two editions of Jewkes *et al* and have found that, in a number of them, we disagreed between ourselves and with Mensch. We have, however, arrived at a joint consensus estimate in each case, and compared our results with those of Mensch. We do not maintain that our estimates are better or more realistic, but merely that they are equally plausible alternatives. These are shown in Table 6.1.

TABLE 6.1. LIST OF INNOVATIONS AS REPORTED IN JEWKES, SAWERS AND STILLERMAN

Estimated invention and innovation years; where these differed from Mensch, his estimates are placed in brackets.

	Invention	Innovation
A. Cases considered by Mensch		
Automatic drive [vehicles]	1904	1939
Ballpoint pen	1888	1946 (1938)
Catalytic cracking of petroleum	1915	1935
Cinerama	1937	1953
Continuous steelcasting	1927	1952 (1948)
Continuous hotstrip	1920 (1892)	1923
Cotton picker (Campbell)	1920	1942
Cotton picker (Rust)	1924	1941
Crease-resisting fabric	1926 (1906)	1932
Diesel locomotive	1895	1913 (1934)
Fluorescent lighting	1901 (1852)	1938 (1934)
Gyro-compass	1900 (1827)	1909
Helicopter	1904	1936
Hydraulic clutch	1924 (1904)	1946 (1937)
Insulin	1920 (1889)	1927 (1922)
Jet engine	1928	1941
Kodachrome	1921 (1910)	1935
Magnetic tape-recording	1898	1937
Neoprene	1906	1932
No-knock gasoline	1912	1935
Nylon, perlon	1927	1938
Penicillin	1928 (1922)	1943 (1941)
Plexiglass	1912 (1877)	1935
Polyethylene	1933	1937 (1953)
Power steering	1925 (1900)	1930
Radar	1925 (1887)	1934
Radio	1900 (1887)	1918 (1922)
Rockets	1923–9 (1903)	1942–4 (1935)
Silicones	1910 (1904)	1943 (1946)
Streptomycin	1942–3 (1921)	1944
Sulzer Loom	1928	1945

Table continues

Table 6.1 continued

	Invention	Innovation
Synthetic detergents	1886	1928
Synthetic light polarizer	1928 (1857)	1932
Television	1923 (1907)	1936
'Terylene' polyester fibre	1941	1955
Titanium	1937 (1885)	1944 (1937)
Transistor	1948 (1940)	1950
Tungsten carbide	1900	1926
Watertight cellophane	1912 (1900)	1926
Xerography	1937 (1934)	1950
Zipper	1891	1923
B. Cases from Jewkes et al, excluded in Mensch, ref 5[a]		
Bakelite	1904	1910
Cyclotron	1929	1937
DDT	1874 (1939)	1942
Electric precipitation	1884	1909
Freon refrigerants	1930	1931
Hardening of liquid fats	1900	1909
Long-playing record	1945	1948
Safety razor	1895	1904
Self-winding wrist-watch	1922	1928
Shell moulding	1941	1948
Stainless steels	1904	1912
C. Additional cases from Jewkes et al, 2nd Edition (estimates by current authors)		
Chlordane, Aldrin, Dieldrin	1944	1946/47
Electronic digital computers	1939	1943
Float glass	1902	1957
Hovercraft	1955	1968
Moulton bicycle	1959	1963
Oxygen steel making	1949	1952
Photo typesetting	1936	1954
Prevention of rhesus haemolytic disease	1961	1967
Semi-synthetic penicillins	1957	1959
Wankel rotary piston engine	1957	1967

Note: [a] Dates for these inventions are given in Mensch[14] or, for Bakelite and the long-playing record, in Mensch[15]. For DDT, our own dating of the invention differs substantially from Mensch's; the latter estimate is in brackets.

Source for Mensch's dates: Mensch, reference 5.

Figure 6.3, reproduced from Mensch, shows the invention year plotted against the lead time from invention to innovation for the list he distilled from Jewkes *et al.* All these cases are estimated by Mensch to have become innovations between 1909 and 1955 inclusive. From the way the diagram is drawn, a point representing an innovation made in any particular year must lie somewhere on a line with a negative 45° slope drawn from that year on the horizontal axis; hence, by definition, all the points representing innovations between 1909 and 1955 *must* lie between the dashed lines we have added to the diagram (the fact that two points lie outside these limits is apparently due to an arithmetic error in calculating the lead times for two types of cotton-picker).

In Figure 6.4, we have constructed a similar diagram on the basis of our own estimates of the invention/innovation dates of the inventions discussed by Jewkes *et al.* This differs from Figure 6.3 in that, in some cases, our estimates diverge from those of Mensch; these differences are listed in Table 6.1A; we have

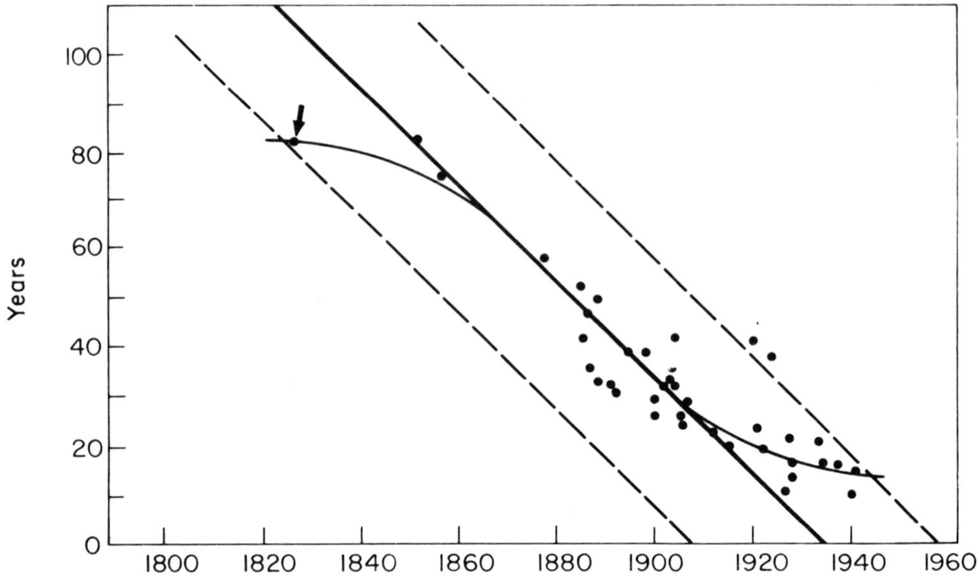

Figure 6.3. Lead times from invention to innovation—Mensch's estimates.
Source: Mensch, reference 5; reprinted with permission from *Stalemate in Technology: Innovations Overcome Depression*, © 1979, Ballinger Publishing Co., Table 4.4, page 127.

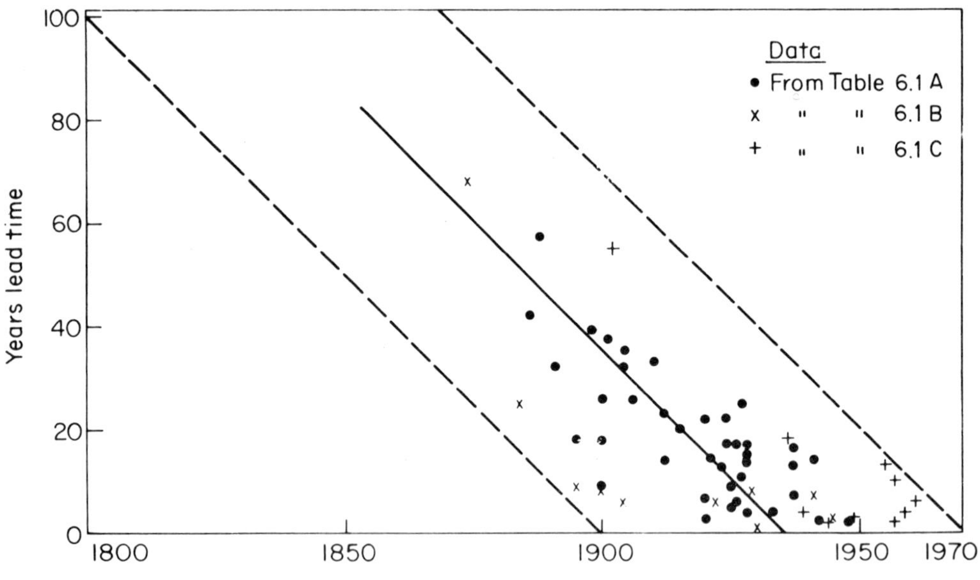

Figure 6.4. Lead times from invention to innovation—authors' estimates.

included those innovations excluded in Mensch (Table 6.1B); and we have also included additional case histories covered in the second edition of Jewkes *et al* (Table 6.1C).

With regard to the bunching hypothesis, it is clear that a disproportionately large number of innovations fall in the 1930s (ie lie between lines with negative 45° slopes drawn from 1930 and 1940). However, we would make three points in this connection. First, the unsuitability of the Jewkes *et al* list for the purpose in hand must cast some doubt on the conclusion. Second, most of the 1930s innovations fall within the second half of that decade, which is arguably a period of recovery rather than a period of depression; and third, as we shall argue later, there is little evidence that those which did fall within the period of depression were in fact induced by adverse economic circumstances.

The acceleration hypothesis, ie the notion that lead times are reduced in periods of depression, would be reflected in these diagrams by a preponderance of points to the left of the solid diagonal lines in the upper parts of the diagrams, and a similar clustering to the right of this line in the lower parts. This would mean, for example, that in the earliest part of the 20th century, a disproportionate number of innovations had a comparatively long lead time whereas, on average, this lead time declined as the century wore on. It is clear that this hypothesis is very poorly supported by Figure 6.4; in Figure 6.3, its support depends to a great extent on the single point arrowed, representing the gyro-compass which, we believe, has been allocated an unrealistically early invention date by Mensch.

Thus we regard the data of Jewkes *et al* as providing no evidence of the acceleration hypothesis and, while superficially supporting the bunching hypothesis, as unsuitable as evidence for the conclusions that Mensch draws. We would certainly not exclude the possibility that there are some bunching effects in relation to the flow of basic innovations over time—indeed, we would be surprised if it was a smooth process—but we cannot regard Mensch's empirical evidence as adequate.

But even if the data were more reliable, simply to demonstrate that a bunch of basic innovations precedes or follows an upswing or a downswing in a long wave (or any other fluctuations) in the economy does not in itself elucidate any causal mechanism which may be at work, even if (as Mensch maintains) the statistical probability of some association is extremely strong. To establish that depressions induce basic innovations, or that booms may lead to a "crowding out" of basic innovations or vice-versa, or to demonstrate that a bunch of basic innovations could be one of the factors stimulating a long-wave upswing, or to validate any other hypothesis about the reciprocal relationship between economic fluctuations and innovations, it would in our view not only be necessary to collect good statistics and use them well, but also to show by the evidence of case studies that business (or government) behaviour was indeed influential in the manner postulated by the theory.

To go through all the case histories of basic innovations of the past 60 years, and check them for the operation of a "depression-induced accelerator" mechanism, would be a very time-consuming exercise although, in our view, a worth-while one. However, a preliminary test can be made from the Jewkes case studies from which Mensch drew up his list. None of the case studies mentions that the Depression of the 1930s influenced the decision-making of innovators in a positive way, whether by the initiation of a new development project, by the acceleration of an existing project, by the earlier-than-expected launch of an innovation, or by taking a completed or nearly-completed project "off the shelf". On the other hand, no fewer than ten of the case studies (penicillin, hotstrip rolling, jet engine, radar, rockets, shell moulding, silicones, titanium, tungsten carbide, and computers) suggest positive effects on decision-taking associated with strong demand in the recovery and prosperity phases, mainly rearmament and war but also the vehicle boom of the 1920s. Two of the case studies mention the negative effects of depression on decision-making (cotton picker and power steering). Ten of these twelve cases were used by Mensch in compiling his list of 41 innovations derived from Jewkes *et al.*

We would not suggest that this preliminary screening of the Jewkes *et al* case studies is a conclusive test or refutation of the hypothesis. The studies are highly condensed and more concerned with the personal experiences of the inventor and the technical problems of the innovation than with the influence of general economic conditions. However, we do believe that, if depression is as strong an accelerator mechanism on innovation as Mensch suggests, then at least some pale reflection of this mechanism would have been present in these summary case studies, instead of the opposite. The fact that ten of the studies (eight of which Mensch includes in his list) mention in one way or another the positive effects of strong demand during rearmament, war-time, and/or the prosperity phase, suggests at least a *prima facie* case for the alternative hypothesis that this was a more important inducement mechanism than depression.

We are not suggesting that only "carrots" are effective in relation to basic innovations, and that "sticks" play no part. Actual or expected shortages have been demonstrably important inducement mechanisms for basic innovations in materials and energy, and these may be construed as both stick and carrot. It can also be argued plausibly (as for example by Downie[16] in a totally different context) that the stick of failure or prospective failure in existing product lines should induce firms to attempt innovations. Since this stick will beat more people during a deep depression than at any other time, it could conceivably operate in the way which Mensch suggests. However, as against this the risks of radical innovative behaviour must appear even greater during a deep slump. The balance cannot be assessed on a purely theoretical basis, and we would like to see some evidence from actual case histories of major innovations that the stick mechanism really did work in the manner suggested.

A thorough test would require meticulous reconstruction of a fairly large number of case histories, and is beyond our resources. However, an alternative approach (in addition to the brief test already described) is to take one group of basic innovations which have been fairly well enumerated and described in the literature and see if the hypothesis is valid for that group. The group which we have chosen for this test—synthetic materials—would appear at first sight to offer support to the Mensch bunching hypothesis, and it was chosen partly for this reason. Hufbauer did a great deal of careful research for his PhD to date all the major innovations in all the leading industrial countries, and the results are summarised in Table C3 of the book based on his thesis.[17] We have used this table as our source for dating the innovations, as Hufbauer's definition is close to that of Mensch. We have made only two changes, both indicated by Hufbauer himself and favourable to Mensch's theory: we have placed two synthetic rubber innovations in the 1930s rather than later, because the first production in Germany is relevant to our purpose, and Hufbauer's reason for dating the innovations as the later US launching dates was peculiar to his own research purpose (international trade).

TABLE 6.2. NUMBER OF MAJOR INNOVATIONS

Before 1900	4	1930–39	18
1900–09	4	1940–49	15
1910–19	3	1950–59	10
1920–29	5	After 1959	1

Source: Hufbauer, reference 17.

Table 6.2 summarises the numbers of innovations in each decade. As we have already said, this listing does appear at first sight to support the hypothesis of bunching in the 1930s, and we do not seek to deny this. What we are contesting is rather the explanation of this bunching in terms of depression-induced acceleration. One of us was involved in an earlier study of innovation in the plastics industry and from this evidence,[18] as well as from Hufbauer's account, that of ter Meer[19] (one of the principal participants), and other case studies, we simply have not found support for the depression mechanism suggested by Mensch. This is of course a negative finding but, on the other hand, virtually all these sources agree in suggesting two alternative (and to us more plausible) mechanisms which are sufficient to explain the bunching. One of these is the advance in fundamental science associated with the work of Staudinger on the structure of long chain molecules, and the other was the autarchic pressures from the demand side associated with German rearmament and the German war economy affecting especially the synthetic rubbers.

On the influence of basic science and technology, it is worth quoting Hufbauer's comment in full:[20]

The similarity of nomenclature between various plastics, rubbers and man-made fibres indicates another important feature of the industry: materials with diverse economic uses are sometimes related chemically. It is this characteristic which justifies the designation of a synthetic materials "industry". ... investigation of one high polymer often borders very closely on another. W.H. Carothers' discovery of nylon (both a plastic and a fibre) laid the groundwork for the discovery of polyester fibre by J.R. Winfield and J.T. Dickson in Britain. Likewise the IG Farben discovery of polystyrene plastic led naturally to the invention of styrene rubber and IG Farben's research on polymethyl methacrylate plastic opened the way to nitrile rubber which in turn led to acrylic fibre. ......

The technical bonds joining plastics, synthetic rubbers and man-made fibres are to some extent reflected in the concentration of innovation found among synthetic materials. Due to the chemical similarities of synthetics, one innovation by a given firm almost certainly increases the likelihood of the next. Aside from generating a pertinent body of high polymer research findings, a tradition of innovation attracts the brains and produces the financial successes requisite for its own propagation.

We agree completely with Hufbauer in this emphasis on common "technical bonds" linking the various innovations, and we think it highly probable that this factor also helps to explain some earlier clusters and bunches of innovations, such as those associated with the introduction of electric power in the 1880s. This means that we disagree with Mensch in his rejection of breakthroughs in fundamental science and inventions as one possible explanation of bunching.

The importance of these "technical bonds" can also be illustrated by looking at the evolution over time of *inventions*. In Figure 6.5, we have plotted, as in Figure

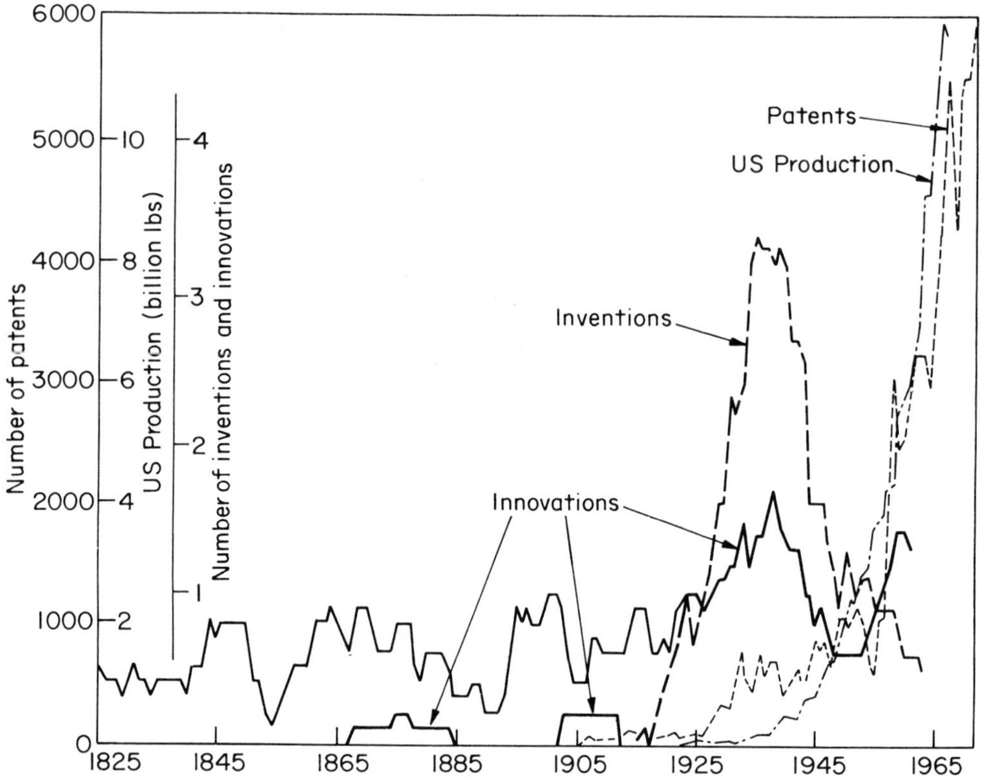

Figure 6.5. The plastics industry, 1825–1975.

6.2, the 10-year moving averages of the number of inventions and innovations in the plastics industry over the period 1825 to 1963. These data were compiled by J.H. Dubois from the *SPE Journal* and *Plastics World* staff.[21] To complete the picture, we added data in relation to the number of patents and total US production of plastics (in volumes) as given by Walsh *et al.*[22]

Despite the difficulties of uniquely categorising the advances listed by Dubois into inventions and innovations, it appears that these data do clearly show a clustering or bunching of inventions and innovations and, with a clear time lag, its impact on overall plastics patenting and actual production. The data are suggestive of a 'trajectory' of technological advance, in which a bunching of inventions has been followed closely by a bunching of innovations, although other interpretations are possible; further work on actual cases is needed here also.

The work of Walsh *et al* is particularly interesting since it shows that, during the growth of new sectors of the chemical industry, there is clear evidence of a tendency for the basic science indicators to lead the indicators of invention, and for both to lead investment and production (ie a "counter-Schmookler" pattern). At a later stage, as the new sectors take off into rapid growth, all the indicators tend to surge strongly together and later a "Schmookler-type" pattern may emerge. During these phases, Walsh and Townsend show that the share of process patents in relation to product patents increases strongly. All of this supports the notion of bunches of basic inventions and innovations leading to the take-off of new industries and a further set of basic and secondary inventions and innovations, especially in the process field as the industries grow rapidly. It does not, however, demonstrate any direct connection between this process and the 'trigger' of depression.

Thus, whereas in the case of synthetic materials the bunching of basic innovations came in the 1930s and 1940s, in the case of drugs it was about a decade later, peaking strongly in the 1950s. In this case too, advances in basic science, together with the 'pull' of war-time demand, played a very important role. The role of basic science as a 'trigger' for a cluster of basic innovations is even more obvious in the more recent case of 'bio-technology'.

We are certainly not advocating a pure 'science-push' theory of innovation. In our view, most case study work on the history of innovations, and especially Project Sappho,[23] point to the interdependence of 'supply' and 'demand' factors and to the importance of a favourable combination of circumstances. Therefore we would not underestimate the significance of demand, whether actual or anticipated, or that of entrepreneurial attitudes and behaviour. The German chemical trust IG Farben accounted for an extraordinarily high proportion of the world's synthetic materials innovations and, in our view, its success was related both to the strength of its R and D, including its strong links with basic research through its consultancy arrangements with Staudinger and other outstanding chemists, and to the peculiar nature of German demand, strongly influenced by aspirations to overcome dependence on imported natural materials. That both these factors were important is indicated by the relative failure of the attempts to develop synthetic rubber and other materials during World War 1. It was only after the advances in basic science in the 1920s that the way was open for much greater success. The continuing importance of intense war-time demand pressures is suggested by the fact that innovations were bunched not only in the 1930s, but also in the 1940s. Nowhere however did we find evidence that it was depression which stimulated these bunches of innovations. Ter Meer[24] reported that an intense joint effort to develop new materials was one of the motives for founding the chemical trust itself, but that was in the 1920s; and he also reports that development expenditures were cut back in the worst years of the Depression, despite IG Farben's extremely strong commitment to R and D.

Concluding remarks

Our own 'neo-Schumpeterian' theory of innovation and long-term fluctuations in the economy[25] has much in common with that of Mensch.[26] Even in terms of other (non-Schumpeterian) paradigms in economics, there is actually little reason to expect a smooth adjustment process to a high and steady equilibrium growth rate. Among the many sources of disequilibrium we attach particular importance to the rise of new industries and the decline of old ones, together with the often long time lags involved in the adjustment of capital stock and the skills of the labour force. Schumpeter himself insisted that the opportunities for technical innovations were very unevenly spread across different sectors of the economy and that they were not continuous over time, being subject to explosive bursts as imaginative entrepreneurs realised the possibility of exceptional growth arising from new combinations of technical and organisational change. In our view, the 'swarming' of innovative behaviour, which gives rise to economic fluctuations in Schumpeter's model, arises from the imitation and diffusion process and from the bunching of technically related families of innovations and inventions, rather than from a depression-induced bunching of a set of individual basic innovations. We agree strongly with Rosenberg's point[27] that the "diffusion" process is not just one of "carbon-copy" replication, but involves a whole series of further innovations, some of which may be extraordinarily important, as the case of micro-electronics amply demonstrates. Sahal[28] has given interesting examples of these clusters and relates them to "technology cycles" and long waves.

Thus we certainly do not rule out the possibility of some form of bunching of basic innovations or inventions; indeed, we would be surprised if it was a smooth long-term process, and Figures 6.2 and 6.5 support our expectations. But we think it rather more likely that any such bunching is related to breakthroughs in fundamental science and technology, to bursts of invention, and to periods of very strong demand (including booms and wars).

Insofar as deep depressions do affect basic innovations, we would think it rather more likely that they would delay rather than accelerate their introduction. A rather more plausible form of the theory which seeks to relate bunches of innovations to the influence of deep depressions is the "indirect" theory, which postulates that the social tensions and conflicts engendered by such depressions, and especially by mass unemployment, may lead to radical changes of various kinds, which in turn may provide a more hospitable environment for basic innovations. This could encompass the rise of the Hitler regime and German rearmament.

In this case too, any bunching would tend to occur in the recovery phase and subsequent war or boom, but these hypotheses require further research. For further elaboration of aspects of this theme, see reference 29.

Notes and references

1. OECD, *International Statistical Year for R and D* (Paris, OECD, various years in 1960s and 1970s).
2. Nestor E. Terleckyj, *Research and Development: Its Growth and Composition*, Studies in Business Economics, no 82 (New York, National Industrial Conference Board, 1963), Table A2, page 100.
3. Jacob Schmookler, *Invention and Economic Growth* (Harvard University Press, 1966).
4. D. Mowery and N. Rosenberg, "The influence of market demand upon innovation", *Research Policy*, 1979, *8*, pages 102–153.
5. Gerhard Mensch, *Das technologische Patt* (Frankfurt, Umschau Verlag, 1975); English edition: *Stalemate in Technology* (Cambridge, Mass, Ballinger, 1979).
6. The sudden decline in 1979 in the number of patents granted does not signal any dramatic fall-off in technological activity, but merely a shortage of resources for printing in the USA patent office. The general trend in the number of patents applied for and granted in the USA is, however, somewhat inflated for the most recent period, in particular from 1960 onwards, because of the very rapid growth of foreign patenting in the USA, primarily as a result of the increased international diffusion of technology.
7. R. Baker, *New and Improved—Inventors and Inventions that have changed the Modern World* (London, British Museum Publications Limited, 1976).

8. Baker, reference 7, page 15.
9. OECD, Directorate for Science, Technology and Industry, Conference on Science and Technology Indicators, 1980 (various papers).
10. Mensch, reference 5.
11. J. Clark, C. Freeman and L. Soete, "Long waves and technological developments in the 20th century", in D. Petzina and G. van Roon, eds, *Konjunktur, Krise, Gesellschaft: Wirtschaftliche Wechsellagen und sozialer Entwicklung im 19 und 20 Jahrhundert* (Stuttgart, Klett-Cotta, 1981) pages 132–169.
12. J. Jewkes, D. Sawers, and R. Stillerman, *The Sources of Invention* (Macmillan: first edition, 1958, pages 72–73; second edition, 1969, page 66).
13. See Gerhard Mensch, "Zur Dynamik des technischen Fortschritts", *Zeitschrift für Betriebswirtschaft*, 1971, *41*, 5, pages 295–314. Problems in estimating dates with respect to these innovations are mentioned in Gerhard Mensch, "Long waves and technological developments in the 20th century: comment", in Petzina and van Roon, reference 11 (a comment on Clark *et al*, reference 11.)
14. Mensch, "Zur Dynamik . . .", reference 13.
15. Mensch, "Long waves . . .", reference 13.
16. J. Downie, *The Competitive Process* (London, Duckworth, 1958).
17. Gary Hufbauer, *Synthetic Materials and the Theory of International Trade* (London, Duckworth, 1966).
18. C. Freeman, J.K. Fuller, and A. Young, "The plastics industry: a comparative study of research and innovation", *National Institute Economic Review*, November 1963, No 26, pages 22–62.
19. F. ter Meer, *Die I.G. Farben* (Düsseldorf, 1953).
20. Hufbauer, reference 17.
21. J.H. Dubois, "Plastics industry", *SPE Journal*, June 1967.
22. V. Walsh, J. Townsend, B.G. Achilladelis, and C. Freeman, "Trends in Invention and Innovation in the Chemical Industry", report to UK Social Science Research Council (Science Policy Research Unit (mimeo), 1979).
23. Science Policy Research Unit, *Success and Failure in Industrial Innovation: Report on Project Sappho* (London, Centre for the Study of Industrial Innovation, 1972).
24. Ter Meer, reference 19.
25. Clark *et al*, reference 11.
26. Mensch, reference 5.
27. N. Rosenberg, *Perspectives on Technology* (Cambridge University Press, 1976), Chapter 8.
28. D. Sahal, "The nature and significance of technological cycles", *International Journal of Systems Science*, 1980, *11*, 8, pages 985–1000.
29. C. Freeman, J.A. Clark and L.L.G. Soete, *Unemployment and Technical Innovation: a Study of Long Waves and Economic Development* (London, Frances Pinter, 1982).

CHAPTER 7

TECHNOLOGICAL PARADIGMS AND TECHNOLOGICAL TRAJECTORIES

The determinants and directions of technical change and the transformation of the economy

Giovanni Dosi

The procedures and the nature of technologies are suggested to be broadly similar to those that characterise sciences. In particular, there appear to be technological paradigms (or research programmes) performing a similar role to scientific paradigms (or research programmes). The model tries to account for both continuous changes and discontinuities in technological innovation. Continuous changes are often related to progress along a technological trajectory defined by a technological paradigm, while discontinuities are associated with the emergence of a new paradigm. The nature of technological progress is likely to generate long-run fluctuations in economic activity and employment. Technology alone, however, is not able to explain entirely these long waves: technology is not the only regulating variable of the system and other "macro" and social variables have to be introduced into the model.

TECHNICAL CHANGE is one of the crucial dynamic factors in modern economic systems. The strict relationship between economic growth and change, on the one hand, and technical progress on the other is a rather evident and well recognised "fact" in economic thought. The nature of the relationship between the two, however, has been a much more controversial issue of economic theory. The theoretical problem concerns the direction of causal relationship, the degree of independence of technical change *vis-à-vis* endogenous market mechanisms (both in the short and long run), the role played by institutional factors, the determinants of the "rate and direction" of innovative activity.

Here, we shall try to overcome the difficulties of the dominant approaches to the process of technical change and we shall attempt an interpretation of

Giovanni Dosi is a Fellow of the Sussex European Research Centre and Science Policy Research Unit, University of Sussex. I am grateful to R. Nelson, W. Walker, D. Jones, M. Salvati, A. Merin and L. Bucciarelli for their comments and criticisms on previous drafts. The responsibility for this draft is obviously mine. A version of this research, focusing more on the policy implications of the model, appeared in *Research Policy*, 1982, *11*, 3, June.

the process of innovative activity, suggesting that there are strong similarities between the nature and the procedures of science—as defined by modern epistemology—and those of technology. The parallel is still rather impressionistic, but leads to the definition of technological paradigms (or technological research programmes) with many features in common with scientific paradigms (or scientific research programmes).

Moreover, we shall analyse the role played by economic and institutional factors in the selection and establishment of those technological paradigms and the interplay between endogenous economic mechanisms and technological innovations, once a technological paradigm has been established.

We shall then consider some implications of the model with respect to industrial structures. In particular, we shall try to translate the logical distinction between the process of search for new technological patterns and its establishment into an historical distinction through the development of an industry, between a Schumpeterian phase of emergence of that industry and its "maturity". We do not provide in this work any empirical backing (or very little). An application of the model to the semiconductor industry can be found in another work by the author.[1]

Finally, we shall consider the implications of our model in terms of long-run patterns of change in the economy. If paradigms and trajectories present strong technological and economic interdependencies between different industrial sectors, a correspondence between discontinuities in the emergence of new technologies and discontinuities in patterns of economic growth becomes quite plausible.

This paper does not have any ambition of providing a general theory of technical change. It simply attempts to focus on questions like "why did certain technological developments emerge instead of others?" "are there regularities in the process of generation of new technologies and in technical progress thereafter?" "is there any regularity in the functional relationship between the vast number of economic, social, institutional, scientific factors which are likely to influence the innovative process?" Our answers to these questions are necessarily tentative. In some ways our model could be considered in itself as an "outlook", an interpretative grid, focusing on questions often neglected by orthodox economic theory, which is mainly concerned with questions of instantaneous adjustments instead of problems of long-run transformation of the economic and institutional environment.

Demand-pull theories: technology as a reactive mechanism

Although everyone recognises that there can be—and generally are—different and contextual origins of inventive activity, in the economic literature one used to define two basic different approaches, the first pointing to market forces as the main determinants of technical change (demand-pull theories) and the second defining technology as an autonomous or quasi-autonomous factor, at least in the short run (technology-push theories). Such a clear-cut distinction is of course hard to make in practice but remains useful for the sake of exposition: there is indeed a fundamental distinction between the two approaches in the role attributed to market signals in directing innovative activities and technical changes. It seems to us that the role attributed to market signals, although overlapping a great deal with the distinction between "demand-pull" and "technology-push" theories, is the main core of the discussion.

Let us consider first a "pure" demand-pull theory. As discussed exhaustively in a comprehensive and critical paper by Mowery and Rosenberg,[2] the causal prime mover in those theories is some supposed "recognition of needs" by the productive units in the market, which leads to their efforts to fulfil those needs through their technological efforts. This "pure" market-pull theory would run more or less as follows (both causally and chronologically).

- At a given time there exists a set of consumption and intermediate goods on the market that embody different "needs" by the purchasers. In passing, one must notice—as we shall recall below—that the definition of "needs" is quite ambiguous: at the one extreme one may define them in very general "anthropological" terms (the needs to eat, have shelter, communicate, etc.) in which case they express a total indifference to the way they are satisfied and do not have any economic relevance; at the other extreme, "needs" are expressed in relation to the specific means of their satisfaction, but then each "need" cannot emerge before the basic invention to which it is related.[2]
- Consumers (or users) express their preferences about the features of the goods they desire (ie the features that fulfil their *needs* the most) through their patterns of demand. This is another way of saying that demand functions are determined by the existence and the forms of utility functions. We may assume now that the demand function shifts upward or downward, or just that (which is basically the same), in a growing economy, given the relative prices of the considered commodities, the income elasticities of demand of the latter are different.
- The theory would argue that, with a growing income relaxing the budget constraint of the consumers/users, the latter demand proportionally more of the goods that embody some relatively preferred characteristics (ie those that more adequately satisfy their needs).
- At this point the producers enter into the picture, realising—through the movements in demand and prices—the revealed needs of the consumers/users: some "utility dimensions" have a higher weight (there is more *need* for them).
- Here the proper innovative process begins, and the successful firms will at the end bring to the market their new/improved goods, letting again the market (as above defined) monitor their increased capability to fulfil consumers' needs.

Of course not even the most extremist demand-pull theorist would support entirely this crude view.[4] The basic argument however maintains that there generally exists a possibility of knowing *a priori* (before the invention process takes place) the direction in which the market is pulling the inventive activity of producers, and furthermore that an important part of the signalling process operates through movements in relative prices and quantities. Thus, in this perspective, the innovative process can be placed—although with consistent difficulty—inside the neoclassical framework. With respect to producers, this viewpoint implies that the "choice sets are given and the outcomes of any choice known".[5] The assumption of known outcomes could perhaps be relaxed to introduce risk and stochastic variables, but the first assumption has to be maintained (given a finite sets of choices).

Some significant problems throw doubt on the entire adequacy of demand-based theories of innovations. First, a theory of innovation is supposed to explain not only (and not even primarily) incremental technical progress on existing products/processes, but above all it is meant to interpret major and minor technological breakthroughs. As far as the latter are concerned, the range of potential needs is nearly infinite and it is difficult to argue that these would-be demands can explain why, at a particular point in time, an invention/innovation occurs.[6] Secondly, even after allowing *a priori* recognition of a need, it is difficult to explain with this approach what happens between that recognition by producers and the final outcome of a new product. Either we have to assume that a set of technological possibilities already exists (but then we must wonder why those possibilities have not been exploited before[7]) or we must assume a limited time lag between research and the outcomes of that research. The concept of technology (and, at least indirectly, of science) underlying this approach is of a very versatile and responsive mechanism which can be directed with limited effort and cost in one direction or another. To avoid a crude conception of technology as a "freely available black box", there have been some efforts in the theory to consider information as an expensive commodity.[8] Those attempts, while representing a big advance in that they account for the micro-economic aspects of technological effort (which has a cost and an expected return for each single firm) and also in that they somehow account for the interrelation between science and technology and production, do not seem to be able to consider the entire complexity of scientific and technological procedures and in particular the timing and the direction of technical change.

To summarise, there appear to be three basic weaknesses in the "strong" version of demand-pull approaches: first, a concept of passive and mechanical *reactiveness* of technological changes *vis-à-vis* market conditions; second, the incapability of defining the *why* and *when* of certain technological developments instead of others and of a certain timing instead of others; third, the neglect of changes over time in the inventive capability which do not bear any *direct* relationship to changing market conditions.

The theoretical ambiguities of demand-pull theories seem inevitably reflected in the empirical studies on the determinants of innovation.[9] Not surprisingly, most of the studies find that the "market is important in determining successful innovations". We find ourselves in agreement with Mowery and Rosenberg though, in that most of the studies with a demand-pull approach fail to produce sufficient evidence that needs expressed through market signalling are the prime movers of innovative activity.[10] And this is precisely the question at stake. Other important empirical works, on the contrary, point to multivariable explanations of innovative activity[11] and to the role of science and technology in fostering innovation along a path leading from initial scientific advances to the final innovative product/

process.[12] Some aspects of the innovative process can, in our view, be considered fairly well established. Among them:

- The increasing role (at least in this century) of scientific inputs in the innovative process.
- The increased complexity of R and D activities, which makes the innovative process a matter of long-run planning for the firms (and not only for them) and is evidence against a hypothesis of prompt innovative response by producers *vis-à-vis* changes in market conditions.
- A significant correlation between R and D efforts (as proxy of the inputs in the innovative process) and innovative output (as measured, for example, by patent activity) in several industrial sectors and the absence, in *cross-country* comparisons, of evident correlations between market and demand patterns, on the one hand, and innovative output, on the other hand.
- The intrinsically *uncertain* nature of inventive activity, which goes against a hypothesis of limited and known sets of choices and outcomes.

Technology push: where does technology come from?

The difficulties incurred by strong versions of "technology-push" theories are in some respects opposite to those discussed above. There, it was the difficulty of taking into account the complexity, the relative autonomy and the uncertainty associated with technological change and innovation. Here, the problem arises in relation to the obvious fact that economic factors are indeed important in shaping the direction of the innovative process. The process of growth and economic change, variations in the distribution of national income and in relative prices all affect the direction of innovative activity and one feels quite uneasy about accepting a view of technical progress—paraphrasing Joan Robinson—as "given by God, scientists and engineers". The main theoretical task with respect to supply-side approaches is the avoidance of a uni-directional conception "science–technology–production" in which the first would represent a sort of exogenous and neutral *deus ex machina*. One realises that, in actual fact, there is a complex structure of feed-backs between the economic environment and the directions of technological changes. A tentative theory of technical change should define—in a form as general as possible—the nature of these interactive mechanisms. In different ways, demand-pull and technology-push theories appear to fail to do so. In the former, technical change and innovation are a basically *reactive* mechanism, which certainly shows some consistency with the traditional assumptions of neoclassical economics (consumer sovereignty, optimising behaviours, general equilibrium, etc) but also presents unavoidable logical and empirical difficulties. On the other hand, if supply-side factors present some independence—at least in the short run— from market changes, it must be possible to show how they are affected in the long run by economic transformation.

Technological paradigms and technological trajectories

Economic theory usually represents *technology* as the combination of a given set of factors, defined (qualitatively and quantitatively) in relation to certain outputs. Technical progress is generally defined in terms of a moving production possibilities curve, and/or in terms of the increasing number of produceable goods. The definition we suggest here is, on the contrary, much broader. Let us define technology as a set of pieces of knowledge, both directly "practical" (related to concrete problems and devices) and "theoretical" (but practically applicable although not necessarily already applied), know-how, methods, procedures, experience of successes and failures and also, of course, physical devices and equipment. Existing physical devices embody—so to speak—the achievements in the development of a technology in a defined problem-solving activity. At the same time, a "disembodied" part of the technology consists of particular expertise, experience of past attempts and past technological solutions, together with the knowledge and the achievements of the state of the art. Technology, in this view, includes the perception of a limited set of possible technological alternatives and of notional future developments. This definition of technology is very impressionistic, but seems useful in exploring the patterns of technical change. One can see that the conceptual distance between this definition and the attributes of science—as suggested by modern epistemology—is not so great.

We shall push the parallel further and suggest that, in analogy with scientific paradigms (or scientific research programmes), there are technological paradigms (or technological research programmes).[13]

A scientific paradigm could be approximately defined as an "outlook" that defines the relevant problems, a "model" and a "pattern" of enquiry.

> The success of a paradigm . . . is at the start largely a promise of success discoverable in selected and still incomplete examples. Normal science consists in the actualization of that promise, an actualization achieved by extending the knowledge of those facts that the paradigm displays as particularly revealing, by increasing the extent of match between those facts and the paradigm's predictions, and by further articulation of the paradigm itself.[14]

In broad analogy with the Kuhnian definition of a scientific paradigm, we shall define a technological paradigm as a model and a pattern of solution of *selected* technological problems, based on *selected* principles derived from natural sciences and on *selected* material technologies.

First of all, the similarities relate to the mechanism and procedures of science, on the one hand, and those of technology, on the other hand.[15] As a scientific paradigm determines the field of enquiry, the problems, the procedures, the tasks (the "puzzles", in Kuhn's words), so does technology in the sense defined above (it would perhaps be better to talk of "cluster of technologies", eg nuclear technologies, semiconductor technologies, organic chemistry technologies, etc). As "normal science" is the "actualization of a promise" contained in a scientific paradigm, so is technical progress defined by a certain technological paradigm. We shall define a *technological trajectory* as the pattern of "normal" problem-solving activity (ie of "progress") on the ground of a technological paradigm.

More precisely, if the hypothesis of a technological paradigm is to be of some use, one must be able to assess also in the field of technology the existence of something similar to a "positive heuristic" and a "negative heuristic".[16] In other words a technological paradigm (or research programme)[17] embodies strong prescriptions on the *directions* of technical change to pursue and those to neglect. Given some generic technological tasks (one could call them generic "needs") such as, for example, those of transporting commodities and passengers, producing chemical compounds with certain properties or switching and amplifying electrical signals, certain specific technologies emerged, with their own solutions to those problems to the exclusion of other notionally possible ones: in our three examples, historically these technologies were the internal combustion engine, petro-chemical processes and semiconductors, respectively. Technological paradigms have a powerful *exclusion effect*: the efforts and the technological imagination of engineers and of the organisations in which they work are focused in rather precise directions, while they are, so to speak, "blind" with respect to other technological possibilities. At the same time, technological paradigms also define some idea of "progress". Again in analogy with science, this can hardly be an absolute measure but has some precise meaning within a certain technology. The identification of a technological paradigm relates to the generic task to which it is applied (eg amplifying and switching electrical signals), to the material technology it selects (eg semiconductors and more specifically silicon), to the physical/chemical properties it exploits (eg the "transistor effect" and "field effect" of semiconductor materials), to the technological and economic dimensions and trade-offs it focuses upon (eg density of the circuits, speed, noise-immunity, dispersion, frequency range, unit costs, etc). Once given these technological and economic dimensions, it is also possible to obtain, broadly speaking, an idea of progress as the improvement of the trade-offs related to those dimensions.

The broad analogy between science and technology we have been drawing should clearly not be taken as an identity. In addition to the obvious difference related to the different nature of the problem-solving activity, technological "knowledge" is much less well articulated than is scientific knowledge—much of it is not written down and is implicit in "experience", skills, etc. This implies also that the definition of a technological paradigm is bound to be much looser, while the distinction between normal activity and problem shifts is likely to be hard to make in practice. The same idea of a technological paradigm should be taken as an approximation, adequate in some cases but less so in others. In our view, however, the analogy keeps its validity in that both scientific and technological activities represent strongly selective *gestalten* embodying powerful heuristics.

A crucial question relates to how an established technological paradigm emerged in the first place and how it was preferred to other possible ones. Let us consider the sequence science–technology–production, remembering that it is meant to be just a *logical* simplification which neglects the crucial long-run influence of the economic and technological environments upon science itself.

Even within science, the problems and the "puzzles" *actually* tackled (and those solved) are of course much more limited in number than the total number of problems and puzzles that the scientific theories potentially allow;

even more limited are the pieces of theory, puzzles, possibilities of development, "passed on" from scientific theory to applied science and to technology (the last two, at least, being significantly overlapping). Leaving aside temporarily the problems of feed-backs, the hypothesis is that along the stream science–technology–production, economic forces (which we shall define below), together with institutional and social factors, operate as a *selective device* (the "focussing device" of Rosenberg[18]). Within a large set of *possibilities* of directions of development notionally allowed by science, a first level of selection (at least in the overwhelming majority of research activity in the enterprise sector) operates on the ground of rather general questions like: "is any practical application conceivable?"; "is there some possibility of the hypothesised application being marketable?", etc. Along the downstream from science to production (on a path which is much easier to conceive as a continuum instead of a strictly defined discrete set of steps), the *determinateness* of the selection increases: at one end we have the "puzzle-solving activity"[19] defined by scientific paradigms *stricto sensu*; at the other end we have a technology totally embodied in devices and equipment. In between, in a field that we must already call technology because it is directed towards specific sectors or problems of the economy, the activities aimed at technical progress still have many procedures and features similar to science, namely the problem-solving activity along lines defined by the nature of the paradigm.

The economic criteria acting as selectors define more and more precisely the *actual* paths followed inside a much bigger set of possible ones. On the other hand, once a path has been selected and established, it shows a momentum of its own,[20] which contributes to defining the directions in which the problem-solving activity moves: these are what Nelson and Winter define as "natural trajectories" of technical progress.[21] A technological trajectory (ie, to repeat, the "normal" problem-solving activity determined by a paradigm) can be represented by the movement of multidimensional trade-offs among the technological variables that the paradigm defines as relevant. Progress can be defined as the improvement of these trade-offs.[22] One could imagine the trajectory as a "cylinder" in the multidimensional space defined by these technological and economic variables. (Thus, a technological trajectory is a cluster of possible technological directions whose outer boundaries are defined by the nature of the paradigm itself.) Some features of these technological trajectories, defined on the ground of technological paradigms, are worth considering:

- There might be more general or more circumscribed as well as more powerful or less powerful[23] trajectories.
- There generally are *complementarities* among trajectories (ie, out of the metaphor, there are strong complementarities between different forms of knowledge, experience, skills, etc).[24] Furthermore, development or lack of development in one technology might foster or prevent developments in other technologies.
- In terms of our model one can define the "technological frontier" as the highest level reached on a technological path with respect to the relevant technological and economic dimensions.[25]

- Progress along a technological trajectory is likely to retain some cumulative features: the probability of future advances is in this case also related to the position that one (a firm or a country) already occupies *vis-à-vis* the existing technological frontier. This is strictly consistent with Nelson and Winter's representation of technical progress at firm and industry levels, with Markovian chains.[26]
- Especially when a trajectory is very "powerful", it might be difficult to switch from one trajectory to an alternative one. Moreover, when some comparability is possible between the two (ie when they have some "dimensions" in common), the frontier on the alternative ("new") trajectory might be far behind that on the old one with respect to some or all of the common dimensions. In other words, whenever the technological paradigm changes, one has to start (almost) from the beginning in the problem-solving activity.
- It is doubtful whether it is possible *a priori* to compare and assess the superiority of one technological path over another. There might indeed be some objective criteria, once some indicators have been chosen, but only *ex post*.[27] This is one of the reasons behind the inherently uncertain nature of research activity (even leaving aside market evaluations of the results and just considering purely technological indicators).

Economics and Technology

The role of economic, institutional and social factors must be considered in greater detail. A first crucial role—as already mentioned—is the *selection* at each level, from research to production-related technological efforts, between possible paths on the basis of some rather obvious and broad criteria such as feasibility, marketability, profitability. On this very general ground, there might still be many possible technological paradigms that could be chosen. Given the intrinsic uncertainty associated with their outcomes, in terms of both technological and economic success, it is hardly possible to compare and rank them *ex ante*.[28] Other more specific variables are likely to come into play such as (i) the economic interests of the organisations involved in R and D in these new technological areas, (ii) their technological history, the fields of their expertise, etc, (iii) institutional variables *stricto sensu* such as public agencies, the military, etc. All these factors are likely to operate as focusing forces upon defined directions of technological development. In particular one must stress the role often played in the establishment of a particular technological trajectory by public ("political") forces. An obvious example is electronics, especially in the fields of semiconductors and computers during the first two decades of the post-war period. Military and space programmes operated then as a powerful focusing mechanism in defining technological targets, while at the same time providing financial support to R and D and guaranteeing public procurement.[29] Other similar cases can be found throughout the modern history of technology: for example, the emergence of synthetic chemistry in Germany bears a close relationship with the "political" drive of that country towards self-sufficiency after the Bismarck period.[30]

These kinds of institutional effect on the emergence of new technologies are not a general rule; the point we want to stress, however, is the general weakness of market mechanisms in the *ex ante* selection of technological directions, especially at the initial stage of the history of an industry. This is, incidentally, one of the reasons that militates for the existence of "bridging institutions" between "pure" science and applied R and D.[31] Even when a significant institutional focusing occurs, there are likely to be different technological possibilities, an uncertain process of search, and different organisations, firms or individuals "betting" on different technological solutions. Proceeding in our parallel with epistemology, this resembles a world *à la* Feyerabend[32] with different competing technological paradigms: competition occurs not only between the "new" technology and the "old" ones that it tends to substitute but also among alternative "new" technological approaches.

We did not say very much about *positive ex ante* criteria of selection among potential technological paradigms apart from rather general ones such as marketability or potential profitability. Another powerful selecting criterion in capitalist economies is likely to be the cost-saving capability of the new technology and in particular its labour-saving potential: this is obviously consistent with Nelson and Winter's suggestion of "natural trajectories" toward mechanisation and exploitation of economies of scale. Certainly in societies where industrial conflict and conflict over income distribution are structural features, substitution of machines for labour must be a powerful determinant in the search process for new technologies.

More generally, within the process of selection of new technological paradigms, the patterns of industrial and social conflict are likely to operate both as negative criteria (which possible developments to exclude) and as positive criteria (which technologies to select). In this respect, one might be able to define some long-run relationship between patterns of social development and actually chosen technological paradigms (one quite clear example could be the association between industrial relations at the turn of the last century and the selection and development of "Tayloristic" patterns of technical change in mechanical engineering).

Let us consider the final stage of this logical sequence from science to production, when—in cases of product innovations—a commodity is produced and sold: at this final stage markets operate again as a selective environment.[33] It must be noted that this final selection has a different nature from the previous stages. In the choices of technological path, some kinds of economic indicator were operating as *a priori* directing devices among a large number of possible and wide technological choices. Here the market operates *ex post* as a selecting device, generally among a range of products already determined by the broad technology patterns chosen *on the supply side*. To further clarify the distinction, in his comments on a previous draft of this paper, R. Nelson suggested a biological analogy. The final market selection may be equated with the selection of mutations (Nelson and Winter models mainly describe this "evolutionary" mechanism within the economic environment). The discussion above relates, on the contrary, to the selection of "mutation generating" mechanisms. Thus the economic and social environment affects technological development in two ways, first selecting the "direction of mutation" (ie selecting the technological paradigm) and then selecting among mutations, in a more Darwinian manner (ie the *ex post* selection among Schumpeterian trials and errors).

At times when new technologies are emerging, one can often observe new (Schumpeterian) companies trying to exploit different technological innovations. Markets perform as a system of rewards and penalties, thus checking and selecting amongst different alternatives. In this respect, the existence of a multiplicity of risk-taking actors, in non-planned economies, is crucial to the trial-and-error procedures associated with the search for new technological paths. These actors take risks, of course, because there are markets which allow high rewards (ie profits) in the event of commercial success.

Incidentally, one must note that, if our interpretation of the process of technical change is correct, the emergence of new technological paradigms is *contextual* to the explicit emergence of economically defined "needs". In other words, the supply side determines the "universe" of possible modalities through which generic "needs" or productive requirements (which as such do not have any direct economic significance) are satisfied. (In this, one can see the element of truth contained in those sociologically based theories suggesting that needs are "induced" by corporate strategies.)

Changing economic conditions clearly interact with the process of selection of new technologies, with their development and finally with their obsolescence and substitution. One has therefore to analyse the feed-back mechanisms "upward" from the economic environment to the technology. (One should also consider the long-run influence of economic and technological factors upon scientific change; this is, however, well beyond the scope of this article.) Changing relative prices and distribution of national income are bound to affect demand for the various commodities and the relative profitabilities in manufacturing them. Producers certainly react to these signals from the economic environment, trying to respond through technical advances. However, this often occurs within the boundaries of a given technological trajectory, which might either be conducive to or place increasing constraints on any development consistent with the signals the economic environment is delivering.[34] Difficulties and unsolved technological puzzles and problems, to use Kuhnian language again, operate upward as focusing devices, sometimes put pressure on other technological fields to go further in their problem solving, and finally facilitate or hinder the switch to other technological trajectories. It must be stressed, however, that unsolved technological difficulties do not imply automatically a change to another path.[35] Of course, changes in market conditions and opportunities (among which changes in demand patterns, in the relative distribution of national income, in costs of production, etc, are very important) continuously put pressure "upward", at various levels, on technological trajectories, and on the same selection criteria on the basis of which those trajectories are chosen. But this does not by any means imply an assumption of malleable "ready-to-use" alternative technological paths, or still less of instantaneous technological responses to changes in market conditions. Furthermore, an implicit consequence of what has just been said is that the upward impact of changing economic conditions on technological research patterns seems directly proportional to the *technological determinateness* of the economic stimuli themselves.[36] So one would generally expect this determinateness to increase coming from consumer goods to investment goods and to other kinds of output that cannot strictly be regarded as market goods (such as military equipment).

Note that changes in the economic environment are a permanent feature of the system: these changes often simply stimulate technical progress (as defined above) *along* one technological trajectory. Again in parallel with epistemology, we can call it *normal* technological activity. *Extraordinary* technological attempts (related to the search for new technological directions) emerge either in relation to new opportunities opened up by scientific developments or to the increasing difficulty in going forward in a given technological direction (for technological or economic reasons) or both.[37]

Technical change and industrial structures: from a Schumpeterian phase to industrial maturity

We tried to make a logical distinction above between the process of search and selection of new technological paradigms and technical progress along a defined path. New technologies are selected through a complex interaction between some fundamental economic factors (search for new profit opportunities and for new markets, tendency toward cost saving and automation, etc), together with powerful institutional factors (the interests and structure of existing firms, the effects of government agencies, etc). Technical change along established technological paths, on the other hand, becomes more endogenous to the "normal" economic mechanism. This distinction between two technological phases is likely to correspond historically to two different sets of features of an industry, related to its emergence and its maturity. In the phase of economic *trial and error*, primary importance must be attributed to (a) the *institutions* which produce and direct the accumulation of knowledge, experience, etc, and (b) the existence of a multiplicity of risk-taking actors, ready to try different technical and commercial solutions. Schumpeterian features properly refer to this second aspect.[38] Note that breakthroughs and innovations, in this phase, need not be developed by Schumpeterian companies themselves. There is evidence, on the contrary, that often in this century the production of major technological advances has been the result of organised R and D efforts as opposed to the "inventiveness" of individuals.[39] What matters are the attempts (by either new companies or old ones), in the first phase, to implement and commercially exploit extraordinary technology, driven by the search for new profit and market opportunities. Often this period of emergence of new technologies is actually characterised by new emerging firms, even in cases where the major technological advances were originally produced in established firms and institutions (an example is semiconductor technology).[40]

In the second phase, which may often correspond to an *oligopolistic maturity*, the production, exploitation and commercial diffusion of innovations are much less divorced, and technical change often becomes itself part of the pattern of oligopolistic competition. The more a fundamental technological pattern becomes established, the more the mechanism of generation of innovations and of technological advances appears to become endogenous to the normal economic mechanism. In this respect, the possibility of enjoying temporary monopolistic (and long-run oligopolistic) positions on new products and processes appears to act as a powerful incentive to innovative activity, improvement of existing products, etc. The prospective differential advantages accruing to successful technological and market leaders, in our view, are likely to influence and stimulate the process of innovation much more than the *ex ante* market structure as such.[41] The process of innovation itself is, of course, bound to affect the industrial structure and shape its transformation.

The establishment of a defined technological paradigm is likely to be paralleled by a process of "internalisation" within companies of the so-called "externalities" related to the innovative activity, capitalising on the previous experience of attempts, successes and failures, etc: within an established technological paradigm the fluid market structure characterised by the "heroic entrepreneurship" often described in the literature on new industries is likely to disappear.

Both phases are likely to show some "oligopolistic power",[42] although its sources significantly differ: whereas in the first one oligopolistic positions mainly relate to *dynamic* economies ("learning curve", etc.) and temporary asymmetries in relation to the capability of successfully innovating, in the second stage the origins of oligopolistic structures would relate not only to the technological progressiveness of firms but also to some *static* entry barriers (economies of scale, etc).

Continuity and discontinuity in technical change

We should stress, first of all, the limitations of the suggested model: the analogy between science and technology is, in some respects, impressionistic and the parallel should not be pushed too far without reaching decreasing returns. This notwithstanding, the model might prove useful in interpreting some important questions related to the process of technical change. First, it can explain—in rather general terms—the role of *continuity* vs. *discontinuity* in technical change. Incremental innovation vs. radical innovations can be reinterpreted in terms of normal technical progress as opposed to new emerging technological paradigms. The distinction might still be difficult to draw in practice, but nonetheless can account for the conditions which allow either normal progress or extraordinary innovative effort to take place.

Second, it can throw some light on the *procedures* through which technical change occurs. The search for new products or processes is never a random process on the entire set of notional technological opportunities. Paradigms are also an "outlook" which focuses the eye and the efforts of technologists and engineers in defined directions. (This, incidentally, might have interesting implications in terms of the sociology of the firm and it would be worth studying the origins and the backgrounds of "revolutionary" engineers as compared to "normal" ones.)

Third, the idea of paradigms and trajectories can account for the often observable phenomenon of *cumulativeness* of technical advances (within an established trajectory). At the same time, the intrinsic *uncertainty* associated with technological shifts can be clearly appreciated. Within one technological path, the idea of technical progress might be rigorously defined (as improvement of the trade-offs between the technological and economic dimensions it embodies), but it might prove impossible to compare *ex ante* two different technological paradigms; even *ex post* there might be overwhelming difficulties in doing it solely on technological grounds.[43]

We tried to suggest some interacting mechanisms between technology factors and economic factors, the latter performing as selective criteria, as final ("market") checking and as a continuous form of incentives, constraints and feed-back stimuli. The evidence on market-induced innovative activity[44] that survives a closer scrutiny of its empirical foundations[45] may indeed be consistent with our model: economic growth and transformation of the economy involve a constant reallocation of resources as well as of research efforts among different sectors, and it is plausible to assume that a greater effort will be put into those areas which offer relatively higher growth and profit opportunities (although the two might not necessarily coincide). This process, however, relates much more to *normal* technology than to discontinuous technological advances. In other words: suppose there are two sectors, both defined by fairly stable technological trajectories, which allow broadly similar possibilities of technological advances,[46] but one experiences higher rates of growth of demand than the other. It is plausible that a firm will put greater research effort into the first rather than the second sector. Moreover, if there is some relationship between research input and innovative output, one may find a higher number of technical innovations (as measured, say, by patents) in the former sector. This induced effect, however, does *not* explain the emergence of significantly radical innovations, which is precisely what we tried to do above. This is not to say that the emergence of new technological paradigms is independent from the evolution and the changes of the social system (of which the economic structure is a crucial component). A reconstruction of the history of technology and history of science would almost certainly show the long-run influences of the evolution of the social and economic structure upon the emergence of new scientific and technological opportunities. Simply, what we want to stress is their relative autonomy *vis-à-vis* short-run adjustment and inducement mechanisms of the economic system (changes in prices, quantities, profitabilities, etc).

In the revival of interest in the Schumpeterian *problematique* of the long-run relationship between technical change and economic growth, various hypotheses on the determinants and directions of technical change have been proposed during the past two decades. (One should of course also mention Marx as the other classical economist who focused on the issue.) It is worth mentioning these models and hypotheses, not only to acknowledge our theoretical debts, but also briefly to discuss their reciprocal consistency. We refer in particular to Freeman,[47] Nelson and Winter,[48] Rosenberg,[49] Abernathy and Utterback,[50] and Sahal.[51] In different ways, and with different analytical aims, one may consider these contributions as part of a painstaking attempt to construct a non-neoclassical theory of technical change capable of giving a satisfactory account of (a) the relationship between economic forces and the relatively autonomous momentum that technical progress appears to maintain, (b) the role of supply-side factors, (c) the role and effects of technical change in oligopolistic environments, (d) its relationship with company behaviour and organisational structures, (e) the relevance of non-market organisations and in particular of public institutions in shaping the patterns of technical change.

Broadly speaking, the interpretation of the procedures, tendencies and shifts in the innovative process proposed here is, in our view, consistent with the approach of the above cited works. A few features should be mentioned. In particular, the continuity (and partial overlapping) between our hypothesis and Nelson and Winter's models should be clear enough. The existence of technological paradigms, with the attributes we have tried to describe, supports the existence of "natural trajectories" of technical change suggested by Nelson and Winter. Their models focus primarily on the *endogenous dynamics* of technical progress in oligopolistic environments (and differential innovative success is, in their simulations of the model, one of the main driving forces toward oligopolistic structures). Translated into the definitions used above, their model[52] gives us a fascinating account of the transition to oligopolistic maturity, rich in terms of firms' behavioural variables, and of the subsequent technology-based oligopolistic competition upon a given technological path.

The model suggested here contributes—we hope—to distinguishing between those aspects of technical progress that are essentially endogenous in the economic system (what we have called normal technical progress along a defined technological trajectory) and those extraordinary technological changes that relate to radically new technologies (the new technological paradigms). Differential perspectives of demand growth are likely to affect the *rate* of normal technical progress. Changes in income distribution, in input prices and, more generally, in relative prices are likely to affect the *directions* of normal technical progress *within* the broad boundaries determined by the technological paradigm. We maintain, however, that the emergence of a new technology cannot be adequately explained only in terms of responsiveness to changes in the economic environment, especially in so far as the development of new technologies is linked to scientific advances. Even if scientific changes are influenced, in the long run, by the social and economic structure and problems of a society, the timing of scientific advances and major technological breakthroughs is not likely to coincide with the timing of economic changes. This is simply another way of saying (as should be clear after our discussion) that the scientific–technological system is not univocally defined at each moment in time by the economic system. These points can be illustrated with a rather obvious example: at least since the beginning of capitalist society the dream of any entrepreneur has been of factories full of cheap and docile robots. The fact that, first, we have not got there yet, and, second, that we are heading in that direction, illustrates both the points made above on the power of the economic system to shape the long-run trends in science and technology, on the one hand, and on the relative degree of autonomy and lack of synchrony between the two systems, on the other hand. Before discussing the relevance of the idea of technological paradigms and trajectories for the analysis of the patterns of growth, however, let us make a brief digression on some other contributions in this volume.

Technical change and changes in the economic system: complete self-regulation?

Modern economic systems undergo deep qualitative transformations and at the same time show different degrees of self-adjustment to change. Technical change is both a factor of transformation and a factor of adjustment. In other words, technical change *induces* socioeconomic transformation (we discussed above the effect of technical progress upon the evolution of industrial structures, but one could consider also the effects upon real incomes, income distribution, international competitiveness, skill structures, employment, ways of life, structure of consumption, etc) and at the same time is an *adjustment mechanism* to change (both in terms of short-run effects of changing socioeconomic environments upon normal technical change and in terms of the long-run search for new scientific and technological paradigms). An implication of the above discussion, however, is that the two mechanisms are *not* completely symmetric and are *not* synchronic.

If the two mechanisms of change-inducement and change-adjustment were timed together or at least showed a regular cycle, with one or the other alternately predominating,[53] we would have a dynamic variable (technology) which could generate fairly regular change in the economic system. One can easily see how this can be linked to a theory of *endogenous* long-run cycles, whereby technology generates both change and adaptation to change. On the other hand, if the two mechanisms are not completely symmetrical, the investigation of (i) what makes the economic system change and (ii) what keeps it relatively stable becomes more complex.

We may try to reinterpret parts of the revived discussion on Kondratiev's long waves in terms of this framework. Mensch's contribution,[54] in particular (see chapter 4 in this book by Mensch, Coutinho and Kaasch), focuses attention on these problems related to qualitative (in addition to quantitative) transformations in the economic system, and on the relationship between factors of stability and factors of change (what is described as "internal consistency" and "external consistency" of the movement). The account of these crucial questions however is—in our view—unsatisfactory, both in terms of the theoretical instruments utilised and in terms of actual answers. It is impossible to provide here a thorough critique of the interpretation of the forces leading to stability ("internal consistency of the movement") and to dynamic changes ("external consistency of the movement"). We can only mention that the attribution of the law of supply and demand together with the so-called Utility Principle to the former set of forces and of the Ricardian Cost of Production Principle, with "supply dominance", to the latter set is deeply misleading. Both principles (supply and demand, on one side, and cost of production theory of value, on the other side) are meant to explain *equilibrium* values and thus should both relate to the first set of questions (ie what are the stabilising mechanisms in the system that operate also in static conditions?) Not only that, but they are incompatible with each other, despite Marshallian attempts to reconcile one of the cornerstones of the neoclassical school (the Utility Principle) with the classical theory of value.[55]

For the purposes of this discussion, however, it is more important to notice that the substantive hypothesis in Mensch's approach is the reactiveness of basic innovation to depression periods. In contrast, as a consequence of our above argument, the relative autonomy in the timing of the emergence of new technological paradigms implies that such an automatic anti-cyclical variable cannot be so easily assumed to exist in the economic system. This is what we understand also from the reply of Clark, Freeman and Soete (chapter 6) to Mensch's theory: if our interpretation is correct, their aim is not to oppose some growth-pull theory of basic innovations to a "depression-trigger" theory (as Kleinknecht seems to suggest in chapter 5) but to stress the lack of a necessary correspondence between state variables of the economy—such as rates of activity, income distribution, profit rates, patterns of industrial structure—and emergence of basic innovations (in the language introduced here, the emergence of new technological paradigms). New technological paradigms might emerge during either depression or prosperity periods simply because—as we tried to show—their rules of generation cannot be defined in terms solely of economic inducements.

The thermostat and the engine: what keeps the system moving?

The relationship between change-inducing and change-adjusting effects of technical progress actually hints at a question that discriminates between the various visions of our socioeconomic system, namely: are capitalist societies strongly homeostatic, so that they are able to guarantee both dynamism and stability? The semi-neoclassical hypothesis of technical change entirely "induced" by the economic system could clearly be a great help in obtaining a positive answer. It is somehow paradoxical that part of the Marxist tradition, obsessed with the dominance of economic structures over everything else, also reaches similar conclusions on technology as a variable passively dependent on capitalist requirements.

The semi-neoclassical hypothesis is then able to define technology as both the thermostat and the engine of economic growth, which—so to speak—cools it down but also makes it start again with relative regularity. According to our argument, on the contrary, the technology-related tendencies to self-regulation in the system are relatively weaker.

As we shall discuss below, our hypothesis concerning some degrees of freedom in the emergence of new technological paradigms actually undermines the role of technology as a variable *sufficient* to generate relatively regular waves in economic activity, although it probably remains a necessary one. In order to develop our argument, however, we must analyse at greater length the role of both normal and extraordinary technical progress in shaping economic growth. It was mentioned above that the emergence of new technological paradigms often involves powerful complementarities between different technological developments and interdependences between industries.[56] One can see the relationship between these complementarities and the clustering of innovations and their diffusions suggested by Clark, Freeman and Soete. The development of new technological paradigms is likely to induce self-reinforcing mechanisms: Coombs in this volume (chapter 9) offers a convincing model of virtuous circles between different industries.

This notion of industrial interrelation and reciprocal stimulation to both growth and technical progress may be captured through the concept—which can be found in French economic literature—of *filière*.[57] A filière is a cluster of sectors or activities which are inter-linked by both input–output relationships and technological interdependences.[58] Almost certainly the precise identification of each filière requires a lot of *a priori* qualitative judgements. A useful theoretical device, however, is that of *vertically integrated sectors*.[59] We can thus re-state our argument as follows: the emergence of new technological paradigms activates the expansion of one or—more likely—several vertically integrated sectors which represent, or are linked to, the new technologies and products. In macro-economic terms, the emergence of new sectors and industrial activities together with radically new technologies for making old things can be seen as an *autonomous expansion of effective demand* (both through *ex novo* demand for new consumer products and through "autonomous" investment, ie investment not induced via acceleration or stock adjustment mechanisms).

We have been arguing that the *emergence* of new technological paradigms cannot be directly imputed to direct economic inducement. On the other hand, the economic take-off and the expansion of new filières find in new emerging technological paradigms only a set of necessary conditions but, to repeat, not sufficient ones. What are often also needed are institutional, social and economic conditions. This point can be illustrated with an example. We often refer to the post World War 2 period as a Kondratiev upswing based on electronics, synthetic chemicals, etc. Let us consider the former filière. The economic take-off of electronic industries has been allowed by the emergence of a new technology (micro-electronics), which is in many ways the necessary condition, but contextually also by institutional factors (eg American military and space policies) and macro-economic and social conditions (eg a public commitment to relatively high rates of growth and relatively "democratic" income distributions which guaranteed a fast-expanding demand for electronic consumer goods, good industrial relations, high rates of profit, etc). Virtuous circles are often as difficult to disentangle as vicious circles. We tend to believe that technological conditions alone are *not* likely to set the economic engine in motion again after a major stoppage. However, the same technological conditions (eg the growth of new filières) are likely to *sustain* rapid growth whenever technological, social and economic conditions together have set the system in motion.[60] Certainly this view is less straightforward than a closed model based on a two-way loop between economic conditions and technology, but it appears to us that it is more adequate to the complexity of the dynamics of our economic systems (see the contribution by Salvati in chapter 15 on the macro-economic and social aspects of long waves).

Let us now briefly examine the technology-related "cooling" factors in the system. Normal technical progress along a technology trajectory is, so to speak, constrained within the technological boundaries defined by a basic technological paradigm, which also, as already discussed, defines the "scope of progress" and the trade-offs implied in the path of technological advance. This is a first factor that can slow down the growth impulse associated with the emergence of a new technology: physical laws can increasingly represent an asymptotic limit for any further technological advance (Metcalfe in chapter 8 in this volume refers to Wolff's Law[61]).

Moreover, if profitability of the new activities starts declining after a certain period and if further innovations, diffusion and investments are, *ceteris paribus*, a function of profitability, then a second retardation factor is at work. (To us it is not clear how much of a possible decrease of the rate of profit can be related to micro-economic factors such as the competing away of "super-normal" profits and the increase in input costs and how much, on the contrary, is due to macro-economic conditions such as changes in income distribution[62].)

Third, as Freeman's work has suggested, after the establishment of a normal trajectory of technical progress, the balance between labour-displacing and labour-creating effects of technical progress might change in favour of the former.[63] We may subdivide the argument in three parts.

First, it may well be that for *given* levels of macro-economic activity the balance between productivity growth (and thus labour displacement) in *existing* vertically integrated sectors and the creation of *new* vertically integrated sectors, corresponding to new products, tends to shift in favour of the former. This should not be surprising: in the stage of selection of new technological trajectories (which correspond also to new products) it is difficult to exploit all the productivity potential of mechanisation, automation, economies of scale, etc, whereas these may come into full effect when a technology is fully established.

Second, the rate of growth of demand (ie the rate of expansion of the physical output of each vertically integrated sector) is likely to slow down above a certain level. This clearly hints at the question of saturation of demand. The underestimation of this factor in the literature (and the corresponding reliance on a significant "compensation effect", through relative prices upon demand) is perhaps due to the neoclassical utility approach to consumption structure which suggests widespread substitution. Substitution, on the contrary, might not be so important, and the basket of consumption may be determined by anthropological and historical rankings of priorities that have little to do with the structure of relative prices. No matter how much the real price of television sets may fall, it is hardly conceivable that people would buy, say, 20 television sets for each family. In a similar way, no matter how much the relative price of plastics falls *vis-à-vis* steel, after a certain point demand will be limited by the final demand for products embodying plastics. Major shifts in the aggregate structure of consumption are likely to be associated with historical changes in social structure (and to some degree in income distribution).

Third, we suggested above that within the boundaries of defined technological trajectories there is a responsiveness of the rate and direction of technical progress to changes in the economic environment. In particular, it is likely that, after a sustained period of growth, near-full employment rates of activities and, probably, rising wage demands and increasing industrial conflict, the labour-saving pressure upon technological change increases. In other words, on top of the "normal" labour-saving effect associated with process innovations and product innovations in the capital-producing sector, an *acceleration* of these trends (to repeat, within the fundamental limits determined by the technological paradigms) is very likely.

Again, these technology-related effects may interact with macro-economic and social factors which bring about recessive pressures on the system (see Salvati's contribution).

Some conclusions

The metaphor of technological paradigms and trajectories may, we hope, highlight the role of continuity and discontinuity in technological progress. The relationship between the two has clear implications for the discussion on long waves in economic activity. The nature of technical progress is likely to generate long-run changes in economic activity and employment. Technology alone, however, does not appear entirely to determine the long-run fluctuations in economic activity. It is not possible to define a self-regulating system whose core adjusting variable is technology. A realistic account of long waves has almost certainly to account for the interaction between technology and "macro" variables, especially those related to income distribution and the patterns of social conflict. Our argument does not support a vision of the economic system as characterised by strongly homeostatic features: certainly it generates pressures toward both transformation and adjustment; the two, however, cannot be said *a priori* to relate to each other in such a way as to ensure a regular and relatively smooth cycle.

Notes and references

1. See G. Dosi, Technical change and industrial transformation. The theory and application to the semiconductor industry" (mimeo), SPRU, University of Sussex, 1982. To be published by Macmillan, London, 1983. A summary of the research, together with a fairly detailed discussion of the policy implications, is in G. Dosi, *Technical Change and Survival: Europe's Semiconductor Industry* (SERC, Sussex European Papers, 1981).
2. D. Mowery and N. Rosenberg, "The influence of market demand upon innovation: a critical review of some recent empirical studies", *Research Policy*, 1979, *8*.
3. In other words, according to the first definition, the "need" to move around can equally be satisfied through either a horse or a space-shuttle. According to the second definition, the "need" for a space-shuttle, obviously, cannot emerge before the space-shuttle itself is conceived.
4. A uni-directional determination of innovative activity, assumed to be stemming from "the market", is quite common. See, among others, S. Myers and D.G. Marquis, *Successful Industrial Innovation* (Washington, National Science Foundation, 1969).
5. R. Nelson and S. Winter, "Dynamic competition and technical progress", in B. Belassa and R. Nelson, eds, *Economic Progress, Private Values and Public Policies: Essays in Honour of William Fellner* (Amsterdam, North Holland, 1977).
6. See Mowery|and\Rosenberg, reference 2, and N. Rosenberg, *Perspectives on Technology* (Cambridge, Cambridge University Press, 1976).
7. Except cases where an already existing *invention* becomes a marketable *innovation* at a certain point in time owing to changes in relative prices and income distribution. Strictly speaking, demand-pull theories appear to apply only to these cases.

8. Generally with particular features such as limited appropriability, indivisibility, etc. See K. Arrow, "Economic welfare and the allocation of resources for invention", in NBER, *The Rate and Direction of Inventive Activity* (Princeton, Princeton University Press, 1962).
9. A critical review of these studies is in Mowery and Rosenberg, reference 2.
10. *Ibid.*
11. See Science Policy Research Unit (SPRU), *Report on Project Sappho* (SPRU, University of Sussex, 1971); M. Teubal, N. Arnon, M. Trachtenberg, "The performance of the Israeli electronics industry: a case study of biomedical instrumentation", *Research Policy*, 1976; and M. Teubal, *On Users Needs and Need Determination: Aspects of the Theory of Technological Innovation*, Maurice Falk Institute for Economic Research in Israel, Discussion Paper No. 774 (1977). Those studies, and especially the first, are concerned primarily with determinants of success and failure in industrial innovations and not so much with the determinants of the direction of innovative activity as such.
12. See Illinois Institute of Technology, *Report on Project TRACES* (Washington, National Science Foundation, 1969).
13. On scientific paradigms, see T. Kuhn, *The Structure of Scientific Revolutions* (Chicago, Chicago University Press, 1962); and on scientific research programmes, I. Lakatos, *The Methodology of Scientific Research Programmes* (Cambridge, Cambridge University Press, 1978). For a thorough discussion, A. Musgrave and I. Lakatos, eds, *Criticism and Growth of Knowledge* (Cambridge, Cambridge University Press, 1973). One does not have any ambition here to argue "what science is" or to tackle the epistemological disputes on the differences between the Kuhnian approach and Lakatos' one. For our purposes the degree of overlapping between the two approaches is great enough to borrow from them a few basic definitions of science which they have in common. The first person to draw the analogy between the dynamics of technology and scientific procedures was C. Freeman (see, for example, "The determinants of innovation", *Futures*, 1979).
14. Kuhn, reference 13, pages 23–24.
15. A very stimulating paper by A. Bonfiglioli, "Universal science, appropriate technology and underdevelopment: a reprise of the Latin-American case" (mimeo), SPRU, University of Sussex, 1979, defines "science" as a "particular technology". Although the aims of that paper are different from ours here, there is in common the strict similarity and overlapping between "science" and "technology" and the role of institutional factors in determining the direction of both. See below.
16. ". . . The continuity evolves from a genuine research programme adumbrated at the start. The programme consists of methodological rules: some tell us what paths of research to avoid (*negative heuristic*) and others what paths to pursue (*positive heuristic*)" [Lakatos, reference 13, page 47]
17. Note that here one is impressionistically using the two concepts as equivalent.
18. Rosenberg, reference 6.
19. Kuhn, reference 13.
20. See Nelson and Winter, reference 5; and Rosenberg, reference 6.
21. R. Nelson and S. Winter, "In search of useful theory of innovation", *Research Policy*, 1977, *6*, 1, pages 36–76. They suggest two general dimensions of these "natural trajectories"—toward progressive exploitation of latent economies of scale and toward increasing mechanisation of operations, quoting as supporting evidence (among others) the studies by Hughes on electric power equipment, Levin on various petrochemical processes and Rosenberg (see reference 6).
22. To take obvious examples, the trade-offs between energy consumption and horsepower in internal combustion engines or between speed and density of the circuits in semiconductors (this refers to the comparison between bipolar and MOS technologies). A definition of technical progress in terms of multidimensional trade-offs is sometimes used in technological forecasting models. D. Sahal, "On the conception and measurement of trade-offs in engineering systems", *Technological Forecasting and Social Change*, 1974, and *Law-like Aspects of Technological Development* (Berlin, International Institute of Management, 1978), utilises a similar definition of technology and technical progress, applied to individual industries and products. After the first draft of this chapter was completed, an important article by D. Sahal ("Alternative conceptions of technology", *Research Policy*, 1981) was published. He

suggests a "system approach" to technology and technical change, seeing it as an evolutionary and continuum process. Moreover he suggests the existence of "technological guide-posts". One can easily see the consistency of his thesis with what is argued here. We hope, in this paper, to throw some light also on the definition, emergence and selection of his "technological guide-posts" and on the implications in terms of evolution of industrial structures.

23. Again one uses the term by analogy with epistemology: in our case a trajectory is more powerful the bigger the set of technologies which it excludes. For instance it seems that the technological path defined by nuclear or oil power-generation equipment is very powerful, meaning that many other sources of energy (many other technologies) are excluded.
24. See Rosenberg, reference 6, and N. Rosenberg, "Technological interdependence in the American economy", *Technology and Culture*, 1978.
25. One may represent that "frontier" as a set of points in a multidimensional space.
26. Nelson and Winter, reference 5. Markovian chains, in their model, link the probability of future innovation/imitation, for each firm and for given R and D expenditures, to the innovative *levels* already achieved. In other words, innovative probabilities are serially correlated.
27. For some examples on semiconductors see Dosi, reference 1. An important attempt to define some precise criteria of "progress" is in Sahal, *Law-Like Aspects . . .*, reference 22. As it should be clear from the discussion above, a univocal criterion can be easily identified only *within* a technological paradigm (ie *along* a technological trajectory). Comparisons, even *ex post*, between different trajectories might sometimes, although not always, yield ambiguous results. In other words, the "new" technology might be "better" than the "old" one in several chosen dimensions, but it might still be "worse" in some others. One can see here a very loose analogy with the epistemological discussion between an "extreme" Kuhnian approach claiming strict incomparability and a Popper-like approach suggesting some progressive continuity.
28. For a discussion of uncertainty in the evaluation of R and D projects, C. Freeman, *The Economics of Industrial Innovation*, 2nd edition (London, Frances Pinter, 1982), chapter 7.
29. A more detailed discussion is in Dosi, reference 1.
30. See Freeman, reference 28, and V.M. Walsh, J.F. Townsend, B.G. Achilladelis and C. Freeman, "Trends in invention and innovation in the chemical industry", Report to SSRC (mimeo), SPRU, University of Sussex, 1979.
31. A convincing and thorough discussion is in Freeman, reference 28.
32. P.K. Feyerabend, *Against Method* (London, New Left Books, 1975).
33. Nelson and Winter, reference 5.
34. Take the example of the oil-powered internal combustion engine. Changing oil prices put increasing pressure on oil substitution and energy saving. The scope for substitution however is limited by the technology, which itself defines the range of possible technological advances.
35. Precisely as unsolved puzzles (or "falsifications") in a scientific paradigm do not imply an alternative paradigm.
36. This broadly corresponds to Teubal's concept of "market determinateness" (see Teubal, reference 11).
37. It can be (and has been) reasonably argued that scientific developments themselves are fostered in the long run by the technological and economic "focuses" of attentions and that they are somewhat directed by the *weltanschauungen* that economic systems provide. This wide issue concerns fields like epistemology, sociology of knowledge, etc, and it is not possible to discuss this here.
38. Here one refers to the "first" Schumpeter of the *Theory of Economic Development* (New York, OUP, 1961; first edition 1919).
39. A review of the discussion on the subject is in Freeman, reference 28. Some, however, still hold the opposite view (J. Jewkes, *et al*, *The Sources of Invention*, London, Macmillan, 1958). The history of chemical innovations is analysed by Walsh *et al*, reference 30. On the role of established firms in semiconductors, see Dosi, reference 1.
40. We try to analyse the factors which allow it, related to different attitudes toward risk, constraints limiting a quick diffusion of innovations by existing firms, and taxation regimes in Dosi, reference 1.

41. The relationship between market structure and the incentive to innovate has produced significant discussion in the literature. See among others, Arrow, reference 8, R. Scherer, "Firm size, market structure, opportunity and the output of patented inventions", *American Economic Review*, 1965; D. Needham, "Market structure and firms' R & D behaviour", *Journal of Industrial Economics*, 1975; M. Salvati, *Monopolio Sviluppo e Distribuzione* (Rome, Edizioni dell'Ateneo, 1971). Salvati shows, under rather general assumptions, that the incentive to *introduce innovations* is not generally lower under oligopolistic conditions than in competitive ones. Arrow, in his seminal article, states the opposite view (at least as far as process innovations are concerned). Two implicit and rather questionable assumptions are, however, crucial to his argument. First, one must assume that there are no economies of scale, no minimum threshold in R and D activities, no cumulativeness of technical progress, or, alternatively, that market mechanisms induce an allocation of innovative activities amongst competitive actors *as if* they were a simple monopolist. Second, one must assume that the "degree of private appropriability" of the innovations is the same under competitive and oligopolistic conditions. Needham offers a thorough and rigorous treatment of R and D and innovation decisions under neoclassical behavioural assumptions on firms' conduct. He shows that, depending on elasticities and expected rivals' reactions, more or less everything may be expected to happen (ie that structural variables like concentration, demand elasticities, etc, may have effects of opposite signs upon firms' propensity to innovate).

    One can find in L. Soete ("Firm size and inventive activity: the evidence reconsidered", *European Economic Review*, 1979, *12*) a critical analysis of the available evidence on the subject and strong support for a Schumpeterian view. Nelson and Winter (reference 5) interpret, in a genuinely dynamic framework, the process of innovation under oligopolistic conditions and market structure itself as endogenous variables.
42. We try to assess the existence and the determinants of oligopolistic margins in the semiconductor industry in Dosi, reference 1.
43. Another example from the semiconductor industry: how could it have been possible to compare in the 1950s the thermionic valve technology and the emerging semiconductor technology? Even *ex post* (ie now), when most of the common dimensions (eg size and density, speed, costs, energy consumption, etc) show a striking superiority of semiconductor technology, valves still maintain their advantage in some narrow technological dimensions. Note that we took here one of the most extreme examples of a new clear-cut "superior" technology; in many other cases even an *ex post* comparison between the different technologies may prove rather difficult.
44. See the work by J. Schmookler, *Invention and Economic Growth* (Cambridge, Mass., Harvard University Press, 1966).
45. Walsh *et al* (reference 30) examine Schmookler's hypothesis on the dependence of innovative activity upon market growth and in the case of chemical innovations find abundant falsifying evidence.
46. Note that within stable technologies the possibilities of advance (that is, the potential rate of technical progress compatible with that technology) might radically differ. A low possibility of further advances and unsolved (or "badly" solved) technical problems might indeed be a stimulus to the search for a new technological paradigm.
47. Freeman, reference 28.
48. Nelson and Winter, references 5 and 21.
49. Rosenberg, reference 6.
50. J.M. Utterback and W.J. Abernathy, "A dynamic model of product and process innovation", *Omega*, 1975, *3*, pages 639–656; W.J. Abernathy and J.M. Utterback, "Patterns of industrial innovation", *Technology Review*, June–July 1978, pages 41–47.
51. Sahal, reference 22.
52. See, amongst others, the model from reference 5.
53. One can take an example from physical systems wherein a state variable (say temperature) is controlled by a mechanism (say a thermostat) which "overshoots"; the time profile of the state variable then shows a cycle, which—for some appropriate values—may be stable (ie neither explosive nor asymptotic to a straight line).
54. G. Mensch, *Das technologische Patt* (Frankfurt, Umschau Verlag, 1975); English edition: *Stalemate in Technology* (Cambridge, Mass., Ballinger, 1979).

55. Both theories are meant to determine the structure of relative prices. One must thus choose between two alternative theories:
    - In a "classical" Ricardo–Marx–Sraffa approach:
      1. commodities are freely reproducible
      2. production technologies are characterised by non-decreasing returns to scale.
    - In the neoclassical view:
      1. production technologies show decreasing returns to scale and/or the system travels at full employment of the factors of production.
      2. patterns of demand show a wide possibility of substitution according to relative prices.

    In the first case the "centre of gravity" for relative prices will be defined by costs of production and in the second by the so-called "law of supply and demand" (ie in actual fact there will be no centre of gravity because if the latter keeps moving it is no longer a centre of gravity, by definition).
56. Rosenberg, reference 24.
57. See, amongst others, F. Perroux, "L'Effet d'entrainement: de l'analyse au repérage quantitatif", *Economie appliquée*, 1973; D. Pastré and J. Toledano, *Filières d'entrainement et effets externes: le développement de la filière "Composants Electroniques" et ses effets sur l'emploi*, ADEFI, Chantilly, 1975; G. Lafay, A. Brender and A. Chevallier, "Trois expériences de la spécialisation internationale: France, Allemagne Fédérale, Japon", *Statistiques et Etudes Financières*, 1977.
58. One can see, for example, a "vertical" filière effect between electronic components, computers, telecommunication, etc, or a "horizontal" one between different chemical technologies—see G.C. Hufbauer, *Synthetic Materials and the Theory of International Trade* (London, Duckworth, 1966), discussed also in C. Freeman, J. Clark and L. Soete, *Unemployment and Technical Innovation: a Study of Long Waves and Economic Development* (London, Frances Pinter, 1982).
59. L. L. Pasinetti, *Structural Change and Economic Growth* (Cambridge, Cambridge University Press, 1981). F. Momigliano and D. Siniscalco are undertaking very stimulating research trying to assess the total amount of employment created in each vertically integrated sector (for a work-in-progress report: "Terziario Totale e Terziario per il Sistema Produttivo", *Economia e Politica Industriale*, 1981).
60. This appears also to be the implicit outcome of the Clark, Freeman and Soete model, which is not "closed" because it does not determine technology in an entirely endogenous way. Thus, it is not able precisely to forecast the switching points and switching mechanisms in the cycles. This is, in our view, an *advantage* because it allows degrees of freedom to other important economic and social variables. In a closed system, wars, for example, can only be either an entirely dependent variable or an accident.
61. See S. Kuznets, *Secular Movements in Production and Prices* (Boston, Houghton Miflin, 1930).
62. In our view, Metcalfe's seminal model (chapter 8 in this volume) is more convincing in the second interpretation than the first. In other words, in the example of electronics, the rate of profit falls—if it does—not because of the rise in the price of silicon and/or of electronics-related labour but because of the *general* rise in wages and possibly other inputs associated with high levels of growth and generally to macro-economic and social factors (see Salvati's paper). As regards the micro-economic level, we try to show in Dosi, reference 1, that throughout the history of semiconductors the degree of oligopoly did not undergo any impressive change and neither did the rate of profit, despite the fact that the sources of this oligopolistic power have been changing. On the role of profitability in long waves see F. Chesnais, "Schumpeterian recovery and Schumpeterian perspective—some unsettled issues and alternative interpretations", in H. Giersch (ed.), *Proceedings of Conference on Emerging Technology: Consequences for Economic Growth, Structural Change and Employment in Advanced Open Economies* (Tübingen, J.C.B. Mohr, 1982).
63. See C. Freeman, "The Kondratiev long waves, technical change and unemployment", in *Structural Determinants of Employment and Unemployment* (Paris, OECD, 1977). A thorough and convincing discussion is in Pasinetti, reference 59.

CHAPTER 8

IMPULSE AND DIFFUSION IN THE STUDY OF TECHNICAL CHANGE

J.S. Metcalfe

A common model of innovation diffusion is extended to incorporate the interaction of demand growth and capacity growth, building in part on work by Kuznets and Burns on industrial growth and retardation. It is shown how price and production cost evolve during a diffusion process, and how Schumpeter's theory of the transient nature of innovator's profits has a natural place within the analysis of diffusion. The framework is intended to assist future analysis of the links between the diffusion and the evolution of a given technology, taking account of, eg, learning by doing, production bottlenecks and the inducement to innovation, and producer/adopter interaction to improve technological performance.

IT WILL not be doubted that some of the more reliable facts concerning technical change relate to the diffusion of innovation, the process by which new technologies displace, or substitute for, existing technologies. A wealth of evidence has accumulated to indicate substantial differences in the rates at which the same innovation is diffused in different industries, and in the rates at which different innovations are diffused in the same industry. Important contributions have been made[1] which successfully link observed diffusion patterns to the few key variables which economic theory predicts are relevant to the process of technology displacement. Despite this undoubted success, an air of unease pervades much recent writing on diffusion,[2] manifesting itself in the dilemma over whether diffusion curves are best viewed as sequences of moving equilibrium, or as a disequilibrium adjustment to a given long-run position.

I shall argue that this is related to the narrow focus of past diffusion research, and that advances are now to be made by recognising a broader scope to, and relevance of, diffusion phenomena. Diffusion of innovation is one dimension of that vitally important but not well understood problem of the transition between different economic equilibria. It represents the study in the small of disequilibrium situations, of Schumpeter's process of creative destruction. From this perspective, compelling links can be made between the micro-study

J.S. Metcalfe is professor of economics at the Department of Economics, University of Manchester, M13 9PL, UK. Thanks are due to R. Coombs and P. Saviotti for helpful comments on a previous version of this paper.

of diffusion phenomena and the general process of modern industrial growth, in which the impulses associated with technological innovations are absorbed into the economic structure.[3]

As one explores these links, it becomes clear that current diffusion theory is dominated by a concern with demand-related phenomena. The role of supply factors is almost entirely neglected[4] and, consequently, important interactions between the growth of demand and the growth of productive potential are ignored. I shall suggest that these interactions are vital to a proper understanding of diffusion phenomena, and that they should play an important part in the relation between diffusion and induced innovation. The third part of this paper will develop a simple framework for analysing the interaction of demand and supply during diffusion. This is preceded by a general discussion of the theme of impulse and transition in industrial growth, notably drawn from the writings of Burns, Kuznets, and Hicks. We start with a brief outline of the standard diffusion model and its more important defects.

The standard diffusion model

Though economists have brought a variety of different approaches to the analysis of diffusion, this section concentrates upon the summary statements of diffusion provided by measures of diffusion speed. If, for example, we focus on inter-firm diffusion, measures of diffusion speed are provided by the parameters of the best-fitting time trend to observations on the number of firms utilising an innovation. Often this best-fitting trend turns out to be a sigmoid (eg logistic) curve, and this is associated with the following rationalisation of the diffusion process.

Imagine that, at a particular point in time, an innovation is 'introduced'; say, a new production process embodied in durable capital equipment. This innovation offers some economic advantage over prevailing methods, and so disturbs the constellation of forces defining the existing equilibrium. The innovation creates a potential for change, and we wish to know how quickly this impulse will be exhausted by a process of diffusion. Part of the answer is provided by an analysis of the speed at which firms in the relevant industry decide to adopt the innovation.[5] We know that firms will not adopt the innovation immediately, even when it appears to be economically rational so to do. Why not? The standard diffusion model argues as follows.

An analogy is drawn between the process of inter-firm diffusion and the spread of an epidemic in a homogeneous, randomly mixing population of fixed size. The spreading of the epidemic is akin to the spreading of information about the innovation among a population of potential adopters, each being uncertain of the precise relevance of the innovation to its own operations. When information is costly to acquire and the innovation typically has several technical attributes, firms can resort to an indirect form of 'learning by doing'—learning by observing the experience of existing adopters. At any point in time, firms view differently the advantages offered by the innovation to their own operations but, over time, the process of adoption involves a gradual convergence of viewpoint as all adopting firms acquire a common perception of the innovation's worth.

Simple though this analogy is, it is a useful starting point. Information imperfections and uncertainty are highlighted and it suggests, for example, that the technical complexity or novelty of an innovation may be an important factor in its diffusion. Most important of all, the epidemic analogy provides an underpinning for the logistic curves which often provide an adequate summary of diffusion trends. Denoting the number of adopters at time t by $x(t)$, the epidemic model leads to the following differential equation to govern the diffusion process

$$\dot{x}(t) = \beta\, x(t)\, [n - x(t)] \qquad (1)$$
$$x(0) = z$$

where, n is the equilibrium number of potential adopters, z is the initial, unexplained number of 'innovating' adopters, and β is the adoption coefficient. The magnitude of β reflects the speed at which information is spread between firms and the amount of information required to make a positive adoption decision. The properties of this logistic process are well known, and we simply list the more relevant ones as follows:[6]

- The impulse provided by the innovation creates a potential for future adoption measured by $n-z$. From the initial level z, adoption approaches its equilibrium level n, monotonically along a logistic curve, with inflexion point at $n/2$.[7]
- There is a continuous retardation in the proportionate growth rate of adoption, which approaches zero as diffusion approaches its equilibrium level.
- The time taken for the number of adopting firms to pass between two arbitrary values is inversely proportional to the product $n\beta$. We call this product the diffusion coefficient, and for any given innovation it is usually estimated as one of the parameters of the logistic time trend fitted to the adoption data.[8]

The attraction of this standard model is that it is possible to relate estimated values of the diffusion coefficient to predictable economic characteristics of the innovation and the adoption environment. The diffusion coefficient is found to be greater, the greater the profitability of the innovation and the smaller the capital investment required to adopt it.[9] There is also evidence that adoption proceeds more quickly in science-based, R and D performing industries, and that such factors as the age and qualifications of senior management, and the size distribution of adopting firms, are important in explaining the pace of diffusion.[10]

Despite the obvious success of this approach, there are powerful reasons for believing it diverts attention away from important aspects of the diffusion process. There are two objections I wish to emphasise. The first relates to the static nature of the standard diffusion model—a given innovation is diffused within an unchanging adoption environment, although there are well-documented reasons for expecting both innovation and environment to change as diffusion proceeds. Improvements to the innovation, general economic growth, changes in relative commodity and input prices, and other complementary or competing innovations, can all be expected to occur during diffusion. Instead of a single diffusion curve, we have an envelope of successive diffusion curves, each appropriate to a given set of innovation and adoption environment characteristics, each with its own values of n and β. While any given set of characteristics generates a logistic process, the envelope need not conform to the logistic pattern, and its exact shape will depend on the temporal incidence of the changes in characteristics.[11] Gold[12] and Davies[13], for example, have argued that observed diffusion paths primarily reflect changes in the innovation and adoption environment, rather than a process of learning within a static situation. As Gold observes, the standard diffusion model

rests on the implicit static assumption that the diffusion levels reached in later years also represent active adoption prospects during earlier years.[14]

In his important study, which skilfully blends the learning aspects of the standard model with changes in the equilibrium adoption level, Davies has shown how non-logistic, but sigmoid, diffusion curves can be generated, the shape of which depends on the pattern of post-innovation improvement in the new technology.[15]

This brings us to the second main objection. We have so far treated changes in the innovation and adoption environment as exogenous to the diffusion process. However, an important category of changes will be endogenous and foremost among these, it seems, will be the profitability of adoption. As soon as profitability is considered, several awkward questions surface. The emphasis in diffusion research is upon the profitability of using an innovation as seen by potential adopters, but what of profitability as perceived by the producers of the innovation? Surely this is of equal importance, since the innovation can only be diffused if it is produced. Moreover, high profitability for the producers would seem to imply low profitability for the adopters. This suggests a crucial weakness in the standard diffusion model: it focuses attention entirely on the demand for an innovation by potential adopters. The supply side is ignored,[16] and with it the question of how the profitability of adopting and producing the innovation is determined. Profitability influences the pace of diffusion but, equally, the pace of diffusion will influence profitability. On this crucial matter, the standard model is silent.

Before we turn to a more detailed treatment of profitability and diffusion, we briefly examine three studies of industrial growth by Burns, Kuznets, and Hicks which implicitly treat demand and supply aspects of diffusion.[17] These studies provide useful perspectives on the diffusion process.

Laws of industrial growth

Their central theme is the close connection between economic growth and qualitative, structural change induced by technological innovation. The following quotations, one from Schumpeter[18] and the other from Kuznets,[19] provide a clear statement of this theme.

industrial change is never harmonious advance with all elements of the system actually moving, or tending to move in step. At any given time, some industries move on and others stay behind; and the discrepancies arising from this are an essential element in the situations that develop.

As we observe the various industries within a given national system, we see that the lead in development shifts from one branch to another. The main reason for this shift seems to be that a rapidly developing industry does not continue its vigorous growth indefinitely, but slackens its pace after a time, and is overtaken by industries whose period of rapid development comes later.

Replace 'industries' by 'innovations', and we see perfect examples of the theme of impulse and diffusion in the study of technical change. The connections with diffusion are further reinforced when we find that both Kuznets and Burns were concerned to explain the retardation of industrial growth.[20] As noted above, this is a corollary of a sigmoid diffusion process.

Leaving aside differences of detail and the general factors of population growth and international competition, the retardation-generating elements which they emphasise are three-fold: inter-commodity competition, inelasticity in supplies of productive inputs, and post-innovation patterns of technical progress.

On the first, innovations may stand in a competitive or complementary relation with existing lines of production, restricting the growth of the former and accelerating the growth of the latter. For each innovation, limits exist on the growth of its market demand. The equilibrium market is constrained by a demand curve, but this equilibrium constraint is only binding after a period of transition to overcome market resistance.[21]

The second set of retarding forces relates to the supply side of industrial growth. Inelasticities in the supply of labour, materials, and machinery imply that as the production of an innovation grows they can raise the supply price of inputs and, *ceteris paribus*, reduce profitability in production. In turn, we can expect that the level of profitability will influence the supply of finance available to invest in the innovation, and thus the pace of transition to the new equilibrium. Of course, finance and machines can only provide a temporary bottleneck, but inelasticities in primary input supplies of labour and materials can permanently constrain the equilibrium utilisation of the innovation.[22]

Hicks has recently incorporated similar considerations in his more macroeconomic discussions of the connection between economic growth and science-based technical progress embodied in durable capital equipment.[23] He discusses the path of transition following an innovation-related impulse to growth, and argues that the bottlenecks experienced during the transition will depend on the characteristics of the innovation and the diffusion environment, though there is no explicit treatment of relevant technical dimensions or their relation to various bottlenecks.

The final factor considered by all three authors is the pattern of post-innovation improvement in a technology. A major growth impulse is provided by any innovation which can generate scope for a sequence of secondary improvement innovations, extending the market, lowering production costs, and overcoming specific bottlenecks. Moreover, the pattern of secondary improvement is not arbitrary but, according to Kuznets and Burns, will evolve according to Wolff's Law.[24] The scope for improvement in any technology is bounded, and approach to this limit is slowed down by the increasing difficulty in generating successive incremental improvements. As corollaries to this, the more significant improvements tend to cluster close to the original time of innovation; many of these will not be foreseen at the time of innovation; and the cumulative impact of the secondary improvements will often exceed the impact of the initial innovation. Retardation in the rate of secondary innovation is thus a dominant factor in the overall tendency to retardation in the growth of a new industry.

Impulse, diffusion and profitability

We shall now relate this more general perspective on industrial growth to the standard model of the diffusion of innovation. Our discussion is meant to be suggestive, not exhaustive.

Imagine that the innovation in question is a new industrial material, eg rayon, which has certain advantages relative to existing competing materials. These advantages determine the equilibrium market for the innovation, and will depend on its technical properties and price relative to those of competing materials. Given an equilibrium market, we shall assume that the process of inter- and intra-firm adoption of the new material generates a logistic pattern of physical demand growth, thus

$$g_d(t) = b[m(p) - y(t)], \text{ and } y(0) = z \qquad (2)$$

where $g_d(t)$ is the proportionate rate of growth of demand at t, $y(t)$ is the rate of demand at t, b is the 'adoption' coefficient, z is the initial rate of demand, and $m(p)$ is the equilibrium market demand which, for pedagogic convenience, we assume depends upon p as follows

$$m(p) = c - a\,p(t) \qquad (3)$$

A lower price for the new material extends the equilibrium market, making it profitable to adopt in an increasing range of production processes. The parameters of Equation (3) will depend on several factors, including the price of competing materials and the comparative performance characteristics of the new and competing materials. We shall call Equation (3) the equilibrium demand curve for the innovation with adjustment to equilibrium determined by Equation (2). It is clear that the adjustment path depends upon the behaviour of the innovation's price, for as $p(t)$ changes, so the profitability of adopting the new material also changes.[25]

But how is $p(t)$ to be determined? Is the price path arbitrary, an exogenous component of the diffusion story, or is it (as a Schumpeterian perspective would suggest) determined by the process of diffusion? To answer this question we must turn to the supply side, the growth of capacity to produce the new material.

The crux of our argument is that the rate of increase in capacity depends on the profitability of producing the new material, in that the supply of capital to finance capacity expansion, from ploughed back funds and external funds, increases as the rate of return on investment increases. Under capitalism, profit is both the inducement and the means to effect the transition between different equilibria. The profitability of this new activity depends upon three elements: the price of the innovation (as yet undetermined), the technology of its production, and the price of the necessary material and labour inputs. Let the technology be described by a capital:output ratio v and a unit input requirements coefficient ℓ—inputs of labour and raw materials here being treated as a composite, for expositional convenience alone. Let w be the price paid for this composite input, and let d denote the capital replacement or depreciation ratio. The rate of profits r, the ratio of profits to capital invested, is then defined by:

$$r = \frac{p(t) - w(t)\ell - dv}{v} \tag{4}$$

Following Burns and Kuznets,[26] we shall assume that as the output of the new activity expands, so the price which has to be paid for the composite input increases, and we assume, again for simplicity, that this relation is linear:

$$w(t) = w_0 + w_1\, x(t) \tag{5}$$

Combining Equations (4) and (5), we see that the rate of return is smaller the larger the scale of production, $x(t)$. Note that we have assumed constant returns to scale, though the extensions to decreasing and increasing returns are obvious.

One final element is needed, the relation between profit and capacity growth. Let π be the fraction of the internally generated profits ploughed back in capacity expansion, and let μ be the ratio of external to internal funds invested at any time. It follows that the growth rate of capacity g_S is linked to the rate of profits by:

$$g_S(t) = \pi(1+\mu)r(t) \tag{6}$$

The values of π and μ will reflect alternative uses of corporate resources and barriers to entry into the new industry. Combining Equations (4), (5) and (6), we have:[27]

$$g_S(t) = \frac{p(t) - h_0 - h_1 x(t)}{k} \tag{7}$$

One implication of Equation (7) is immediate. Given technology and the price of the new material, the growth of capacity will follow a logistic curve, reaching a saturation level when prime cost is driven into equality with the given price.[28] From both the capacity growth and the demand growth perspectives, we therefore find a logistic process to govern diffusion, in which the price of the innovation plays a key role. Price not only determines the equilibrium market, it also determines the rate of growth of capacity to supply that market.

In our closed economy, the growth of demand and the growth of capacity cannot stray out of step for long. Profit-conscious producers will not want to forgo profitable investments or to build up permanent excess capacity. This immediately suggests a function for price, to balance the growth of demand with the growth of productive capacity, not necessarily at each point in time but certainly on a secular basis.[29] We shall call such a transition path a balanced diffusion path.

If we hold constant the fundamental parameters of the diffusion process, relating the supply of finance, the innovation's technology, the equilibrium demand curve, and the supply conditions of the composite non-capital input, we can trace out the balanced diffusion paths for output, prices, costs, and profits as follows. Along the balanced path, we have $g_d(t) = g_S(t) = g(t)$, and $y(t) = x(t)$.

Substituting Equation (7) into (3), and the outcome into (2), we have:

$$g(t) = b[c - a(kg(t) + h_0 + h_1x(t)) - x(t)]$$

or

$$g(t) = B[C - x(t)] \qquad (8)$$
$$x(0) = z$$

with

$$B = \frac{b(1+ah_1)}{1 + abk} \text{ and } C = \frac{c - ah_0}{1 + ah_1}$$

The balanced diffusion process is a logistic diffusion process with parameters which reflect the joint dynamics of demand growth and capacity growth. The saturation output level C depends on technology and the parameters of the equilibrium demand and input supply schedules in Equations (3) and (5).

The balanced adoption coefficient B depends on technology and the coefficients b and k, which govern the dynamics of demand and capacity growth. Figure 8.1 illustrates the main implications to be drawn from this simple framework. The innovation of the new material acts as an impulse in the economy creating a potential for growth equal to C with a post-innovation adjustment gap of C–z. This adjustment gap is eliminated by a diffusion process in which the demand for the innovation and the capacity to produce it grow in step. Figure 8.1a shows the logistic path followed by the output of the new material. The balanced diffusion coefficient is given by the product BC, and this is independent of h_1, and thus of the elasticity of supply of the composite input. By contrast, the standard diffusion model is one in which the price of the new material is constant over time, with a perfectly elastic supply of materials, labour, and external capital finance to the industry so that capacity growth adjusts passively to the growth of demand. In our more general framework, we find that as diffusion proceeds, the price changes and, with it, the profitability of adopting and producing the new material.

Figure 8.1b shows the path for the growth rate of output, that familiar curve of retardation with growth dropping asymptotically to zero as diffusion proceeds. Retardation occurs even with a fixed technology. Because the growth rate declines continuously, it follows from Equation (6) that the rate of profits on capital invested in the industry also declines continuously, and here we can see the relevance of Schumpeter's viewpoint on profits and innovation. Profits are the reward for entrepreneurial activity, accruing to those who innovate and disturb the pre-existing economic structure. But the innovator's reward is temporary, destined to be eliminated during the diffusion process. Profit[30]

attaches to the creation of new things, to the realization of the new value system. It is at the same time the child and victim of development.

The harmony between Schumpeter's view of profits and retardation is closely connected with the role of competition in the diffusion process. Figure 8.1c shows the effect of the balanced diffusion process upon unit prime cost, price and profit margins—the excess of price over unit prime cost. As the industry expands, the increasing price of the composite input pulls up unit prime cost along the path $h(0) - h(t)$, tending toward the stationary level $h_0 + h_1C$. The price of the new

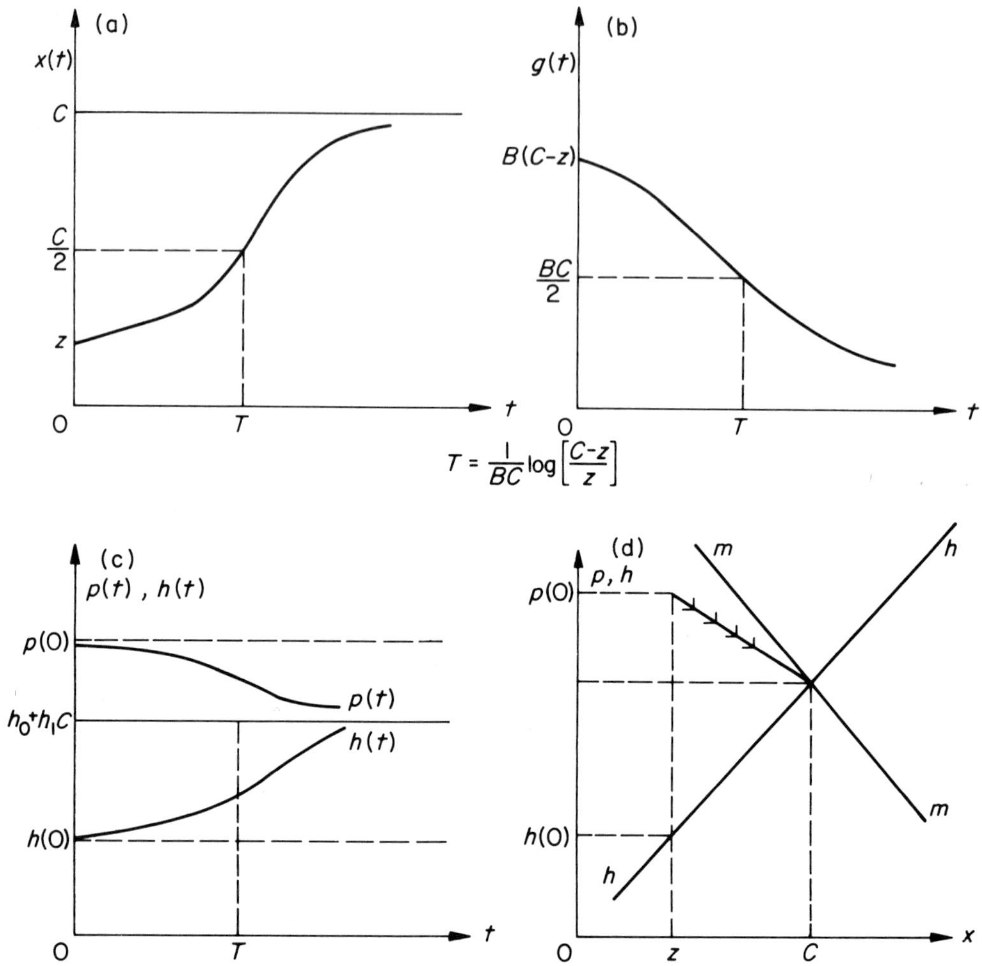

Figure 8.1. Patterns of diffusion.

material, however, is shown falling along the path $p(0) - p(t)$ toward the same stationary level so that, over time, profit margins are progressively squeezed.[31] For adopters, the reductions in price imply an increase in the profit to be gained by switching to the new material, and provide scope for competitive pressure to be put on the remaining firms to hasten their adoption decisions. From this we see how ambiguous is the relation of diffusion speed to measures of profitability. Profitability is perceived differently by the producers and the adopters of the innovation, and for both it varies systematically, and in opposite directions, over time. To relate a hybrid measure of diffusion speed to a single measure of adopting profitability hides more than it illuminates.[32] The pace of diffusion depends on supply-side constraints just as much as it does on constraints to adoption.

Finally, Figure 8.1d plots the implied relations between price and output. The curves *m–m* and *h–h* are the adopters' equilibrium demand curve and the producers' equilibrium supply curve, respectively, and they intersect at output level *C*. From the initial price $p(0)$ and output level z, price and output follow the arrowed path until *C* is reached and the new material is part of the stationary economic routine.

To summarise. Given the fundamental data, the balanced diffusion path is governed by two factors, the adjustment gap *C–z*, and the dynamic elements in demand and capacity growth, as summarised by *B*. This potential for growth can be eliminated only because it is profitable to make the transition to *C*—profitable, that is, to adopt the new material *and* to build up the capacity to produce it.[33]

The purpose of this framework is to serve as a starting point from which further questions can be asked. A great many of these will already have occurred to the reader, but I confine myself to the following observations. For example, this framework allows us to investigate the effect of differences in fundamental data on the paths of balanced diffusion. In practice, changes in fundamental data are likely to occur throughout the diffusion process, continually shifting *m–m* and *h–h* in Figure 8.1d, redefining *B* and *C*, and producing new balanced growth curves. Only by fluke is the outcome likely to be capable of summary by a logistic trend. Economic growth, for example, will normally shift the curve *m–m* upwards so that the saturation level *C* rises steadily over time. If general growth is fast enough, we might even experience acceleration in the diffusion rate rather than retardation.[34]

In a modern industrial economy, *C* may also change because of other complementary and competing innovations which stimulate or contract the market for the new material. In the latter case, growth may be cut short, and turned into decadence, long before the full adjustment potential of the innovation is exploited. Diffusion is always relative; it reflects the comparative merits of competing technologies so that one must always look beyond the specific innovation to gain a full understanding of its diffusion. Notice also that the process of diffusion extends the market as price falls. The often made observation that perceptions of the market at the time of innovation understate the final market, then falls into place. The diffusion environment is not static; it changes as a very consequence of diffusion.

Turning now to the important dimension of post-innovation improvements in technology, we can see how our framework helps sift relevant research questions. One may first distinguish improvements in the utilisation technology from improvements in the production technology. Improvements in the former shift the equilibrium demand curve, and thus *C*, while improvements in the latter shift the unit prime cost schedule, changing *C* and, to the extent that improvements influence the capital:output ratio, *B*. A question which is immediately raised is how the relative balance of these different types of improvements changes during the diffusion process. May it not be that the nature of improvements and the rate at which they occur are a function of the diffusion environment and thus of the pace of diffusion?[35]

There is a further important aspect to this. The pace of diffusion will influence the rate at which plant designs can exploit economies of scale, and the rate at which producers and adopters can mutually improve technology by way of learning by doing. The pace of diffusion will also determine the incidence of supply bottlenecks, and thus the direction of secondary innovative activity; different diffusion speeds may generate different sequences and types of improvement. In other words, the innovation which is being diffused is itself changing as a consequence of the diffusion process. Further considerations will be the position of the innovation in the input–output structure, and the maturity of the technologies to which it is related. These will influence the types of bottlenecks faced, and the possibility of induced technical changes in related sectors.

We have seen how profit margins will be highest in the early stages of a new activity, and this will provide the incentive to seek improvements and reinforce the likelihood that they will be clustered early in the innovation's life. The slackening of technical progress may then reflect not only the operation of Wolff's Law, but also the exhaustion of the profit potential of the new industry as it approaches its equilibrium size.[36] As technology changes during the diffusion process a further factor may be expectations of future technical change. As Rosenberg has argued,[37] such expectations may depress the growth of demand or reduce propensities to invest in the new productive technology and on either count the effect will be to reduce the balanced adoption coefficient, *B*, and slow diffusion. Finally, the 'steamship effect' cannot be ignored. The displacement of old technologies by an innovation may stimulate competing improvements in the former, with impacts on the diffusion of the innovation which should now be clear.

On each of the points raised above, there is scope for linking together analysis of the diffusion of innovation with analysis of the inducement to innovation. But this is simply another way of pointing out the limitations of the static diffusion model.

Conclusion

We have attempted to provide a more complete perspective on the diffusion process by integrating the growth of productive capacity with the more traditional focus upon the diffusion of demand. We have shown how this wider perspective links diffusion research with important contributions to the study of industrial growth, and fits in neatly with Schumpeter's emphasis on the transient nature of the profits to be gained from innovation. The impulses generated by technical change give rise to adjustment potentials. How these are determined and eliminated is central to an understanding of diffusion. We have also suggested that the pace and direction of technical change is a function of the diffusion process.

It cannot be stressed too strongly how tentative this framework of analysis is. Restrictions on space make it impossible to incorporate some of the important insights of Nelson and Winter,[38] to allow for technology differences among producers of the innovation, to incorporate barriers to entry such as patent restrictions, or to allow for foreign trade and investment. Such developments are properly subjects of separate papers.

Notes and references

1. For the most recent studies of diffusion see, for example, E. Mansfield *et al*, *The Production and Application of New Industrial Technology* (New York, W.W. Norton, 1977), chapters 6 and 7; S.

Davies, *The Diffusion of Process Innovations* (Cambridge University Press, 1979); L. Nabseth, "The diffusion of innovations in Swedish industry", in B.R. Williams, ed, *Science and Technology in Economic Growth* (Macmillan, 1973); A.A. Romeo, "The rate of limitation of a capital-embodied process innovation", *Economica*, February 1977, *44*, 1, pages 63–70; A.A. Romeo, "Inter-industry and inter-firm differences in the rate of diffusion of an innovation", *Review of Economics and Statistics*, August 1975, *57*, 3, pages 311–319; G. Rosseger, "Diffusion and technological specificity", *Journal of Industrial Economics*, September 1979, *28*, 1, pages 39–53; S. Globerman, "Technological diffusion in the Canadian tool and die industry", *Review of Economics and Statistics*, November 1975, *42*, 4, pages 428–434; and papers by Baker, Gold, and Hayward in M.J. Baker, *Industrial Innovation* (Macmillan, 1979).

2. See L. Nabseth and G. Ray, *The Diffusion of New Industrial Processes* (Cambridge University Press, 1974); B. Gold, "Technological diffusion in industry: research needs and shortcomings", *Journal of Industrial Economics*, March 1981, *24*, 3, pages 247–269; and Gold in Baker (ed), reference 1.
3. I shall not be directly concerned with the links between impulse, transition, and long waves, though their interconnection will be obvious to the reader of J.J. van Duijn, "The long wave in economic life", *De Economist*, 1977, *125*, 4, pages 544–576.
4. An important exception is N. Rosenberg, *Perspectives on Technology* (Cambridge University Press, 1976), Chapter 11.
5. It is only part of the answer because the outcome also depends on the rapidity of intra-firm diffusion of the innovation.
6. See Davies, reference 1, chapter 2, for a full treatment.
7. If $z > n/2$, we would only observe the upper branch of this curve.
8. Let $y(t) = (x(t)-n)/x(t)$; then in a regression of the form $\log y(t) = a + bt$, we find that $a = \log[(n-z)/z]$ and $-b = n\beta$, the diffusion coefficient.
9. Papers by Mansfield and Griliches pioneered the role of profitability and capital cost variables in explaining diffusion speed; see E. Mansfield, "Technical change and the rate of imitation", *Econometrica*, October 1961, *29*, 4, pages 741–766, and Z. Griliches, "Hybrid corn: an exploration in the economics of technical change", *Econometrica*, October 1957, *25*, 4, pages 501–522.
10. See Mansfield, reference 1, chapter 7; Romeo, reference 1 (both articles); and Davies, reference 1, chapter 8. The latter provides the most exhaustive treatment to date of the effect of industrial concentration on the pace of diffusion. The central hypothesis appears to be that information spreads more quickly in a concentrated industry.
11. Absence of a logistic curve should not therefore be equated with irrelevance of the logistic information/epidemic process.
12. See Gold, "Technological diffusion . . .", reference 2.
13. See Davies, reference 1.
14. See Gold, reference 2, page 250.
15. See Rosenberg, reference 4.
16. See Rosenberg, reference 4.
17. See S. Kuznets, *Secular Movements in Production and Prices* (Boston, Houghton Miflin, 1930); the summary in chapter 9 of S. Kuznets, *Economic Change* (London, Heinemann, 1954); A.F. Burns, *Production Trends in the United States since 1870* (New York, National Bureau of Economic Research, 1934); J.R. Hicks, *Capital and Time* (Oxford, Clarendon Press, 1973); and J.R. Hicks, "Industrialism" in *Economic Perspectives* (Oxford, Clarendon Press, 1977).
18. J. Schumpeter, *Business Cycles*, Volume 1 (McGraw Hill, 1939), pages 101–102.
19. See Kuznets, *Secular Movements*, reference 17, pages 4–5.
20. Burns went as far as to call retardation a "law of industrial growth"; reference 17, pages 169–173.
21. Burns, reference 17, pages 127–128; see also J. Schumpeter, *The Theory of Economic Development* (Oxford University Press, 1961; first published 1934), page 87; and Kuznets, *Secular Movements*, reference 17, pages 258 and 266.
22. Kuznets presents a detailed discussion of how the influence of supply bottlenecks depends on the position of the innovating industry in the input–output structure of the economy. Kuznets, *Secular Movements*, reference 17, pages 41–19.
23. See, in particular, Hicks, *Capital and Time*, reference 17, and the earlier version in Chapter 10 of *Economic Perspectives*.

24. Kuznets, *Secular Movements*, reference 17, pages 11–41; and Burns, reference, 17, pages 141–143.
25. See, for example, G. Chow, "Technical change and the demand for computers", *American Economic Review*, December 1967, *57*, pages 1110–1130.
26. See also Schumpeter, reference 18, page 100.
27. The coefficients in Equation (7) are to be interpreted as follows: $h_0 = w_0 \ell + dv$ is prime cost at a zero level of industry output, including in w_0 the appropriate element for a normal rate of return on invested capital. $h_1 = w_1 \ell$ is marginal prime cost. $k^{-1} = \pi(1+\mu)/v$ is the amount of profit required to expand capacity by one unit, given that capital funds come from internal and external sources.
28. This saturation level equals $(\bar{p}-h_0)/h_1$, for price $\bar{p}$.
29. The importance of the closed economy assumption will here be obvious.
30. Schumpeter, reference 21, pages 153–154, and reference 18, pages 104–105.
31. Note that $p(t)$ need not follow a falling path, the outcome depending on the relative dynamics of demand and capacity growth. However, the excess of $p(t)$ over $h(t)$ always declines with diffusion.
32. See Gold, reference 2, on the general inadequacy of profit measures.
33. The total profit earned by the producers of the innovation, the area between the curves $p(t)$ and $h(t)$ in Figure 8.1c, is proportional to the adjustment gap. Note that total profit is independent of the path of adjustment.
34. Considerations of this nature go some way to explaining Gold's negative comments on the Kuznets–Burns retardation thesis; B. Gold, "Industry growth patterns: theory and empirical results", *Journal of Industrial Economics*, November 1964, *8*, 1, pages 53–73.
35. See Utterback for related discussion of the pattern of innovation over a product's life cycle; J.M. Utterback, "The dynamics of product and process innovation in industry", in C.T. Hill and J.M. Utterback, eds, *Technological Innovation for a Dynamic Economy* (Pergamon, 1979).
36. This point is raised in J. Schmookler, *Invention and Economic Growth* (Harvard University Press, 1966).
37. N. Rosenberg, "On technological expectations", *Economic Journal*, September 1976, *86*, 3, pages 523–535.
38. See R.R. Nelson and S. Winter, "In search of useful theory of innovation", *Research Policy*, January 1977, *6*, 1, pages 36–76; and R.R. Nelson and S. Winter, *An Evolutionary Theory of Economic Change* (Harvard University Press, 1982).

CHAPTER 9

INNOVATION, AUTOMATION AND THE LONG-WAVE THEORY

R.W. Coombs

The author considers the role played by technological innovation in the theoretical frameworks found in the literature on long waves. Some writers place emphasis on consumer goods innovations, some on capital goods innovations, some on new technologies which affect consumer and capital goods, while others take no view on the relative roles of consumer and capital goods innovations. It is suggested that capital goods innovations embodying automation may have been important in the mechanism of the current long wave. Data on employment, rates of return, and value-added for the engineering industries lend support to the view that expectations and propensity to innovate may have been high in the capital goods sector before the upswing of the current long wave.

THE ECONOMICS of technological change has experienced several gusts of fresh air in recent years which promise to take the topic well beyond the elegant but narrow formalism of the neoclassical production function. The work of Nelson and Winter[1] on evolutionary theories of growth has offered a new view of the relationship between the complexity of technical change at the micro level and its aggregated effects at the macro level. Their work[2] on the search process which characterises innovation at firm level has opened up a new approach to the puzzle of different rates of technical change and productivity in different industries. Neil Kay[3] has added to our understanding of the way firms determine R and D budgets and the integration of that task into the strategic planning activities of modern firms. In the area of the diffusion of innovations, Davies[4] has offered a new way of analysing the process based on subjectively rational actions in response to changes in profitability and competitive environment rather than on simple epidemic models.

All these contributions fall broadly under the heading of behavioural approaches to economic theory, which, many feel, promise a significant step forward from the more restrictive assumptions intrinsic to orthodox marginalist economic theory. It is encouraging to students of technical change to see this latest phase in the evolution of economic theory adopting technical change as a central, endogenous variable, in the tradition of classical political economy.

R.W. Coombs is a lecturer in innovation at the Dept of Management Sciences, University of Manchester Institute of Science and Technology, Sackville Street, PO Box 88, Manchester M60 1QD, UK.

The influence of Schumpeter is also visible in much of this work and has contributed to a renewal of interest in broader historical analysis of the linkages between technical and economic change. Walsh *et al*[5] and Freeman,[6] developing the ideas of both Schumpeter and Schmookler, have cast new light on the old technology-push/demand-pull debate. Their data on the chemical industry give considerable force to the notion of the product/industry life cycle as the mechanism through which new technologies first enter and then become incorporated by the economy.

Perhaps the most fascinating body of ideas to re-emerge in recent years, however, is the theory of Kondratiev cycles. Interest in such ideas derives not only from their actual substance but also from their broad compatibility with the behavioural thrust and Schumpeterian flavour of many other new ideas mentioned above. There are many suggestive empirical and theoretical connections between, on the one hand, evolutionary growth models, 'natural trajectories' of technical advance, industry life cycles, and diffusion models, and, on the other hand, long waves of economic change imbricated with major changes in technology.

Such a fluid body of ideas creates many possible areas for investigation. This paper is devoted to a particular area of uncertainty in the current literature on long waves—the relative roles of technical change in the capital goods, consumer goods and intermediate goods industries in the mechanism of the long wave. General familiarity with long-wave theories is assumed. The paper reviews the attitudes of different writers to the specific question of the role of capital and consumer goods innovations, presents data on this topic for the current long wave, and draws conclusions.

Consumer and capital goods innovations in long-wave theories

Kondratiev[7] observed in his original paper that during the downswings of long waves there were increases in the rate of discovery and invention of techniques of production and communication which were then applied on a large scale in the subsequent upswing. He was therefore observing that the diffusion of innovations in the capital goods sector of the economy was associated with the upswing. At that point in the development of his ideas he did not formulate an explanatory theory of the relationship between these phenomena. Later however, following his sharp exchange with Trotsky[8] on whether the waves were genuine cycles, or contingent trends, he developed his position towards a fuller model of the process. In this later work he suggested that certain durable capital goods (eg canals, railways and buildings) generated a long re-investment cycle similar in form to the short-term cycle analysed by Marx. These early formulations of the long-wave theory are characterised, then, by a developing emphasis on technical change in capital goods as a prominent feature of the mechanism.

Following the purging of Kondratiev the long-wave theories were not prominent again in the literature until Schumpeter's *Business Cycles*.[9] In these writings however, the role attributed to technical change is somewhat different. The 'clustering' of innovations brought about by flurries of entrepreneurial activity in Schumpeter's theory does not imply or require that the innovations have any special characteristics other than being profitable relative to existing lines of business. The theory appears to have no special view on the relative roles or significance of consumer or capital goods innovations. However in outlining his 'first approximation',[10] Schumpeter does imply that the first effects of innovations are to stimulate *production* of capital goods. This might be expected to influence innovation in that sector.

Recent literature on long waves has continued to display this diversity of opinion on whether technical change in long waves is homogeneous or heterogeneous with respect to the main sectors of the economy. However, the increasing sophistication of the theories is now making the issue of more central importance. Four authors can be cited who hold varying views on this point.

Mensch[11] appears to follow Schumpeter in making no differentiation. His 'basic innovations' are both capital and consumer goods. It is the fact of their clustering and their aggregate stimulus to the growth of the economy which drives the upswing in his view. In complete contrast, van Duijn,[12] in common with Forrester,[13] proposes a quite specific relationship between the two major sectors of the economy. A summary of their mechanism is as follows. A 'cluster' of innovations has as its initial effect the stimulation of *consumer* demand. This creates demand for capital goods which cannot be immediately satisfied because the capital goods sector has to use some of its own output as input in order to expand capacity. This builds a 'lag' into the behaviour of the economy as it responds to the 'shock' of the innovations. Computer simulation based on these boundary conditions suggests a cyclical trend to the growth of aggregate output in keeping with the duration of the long waves. This view tends to suggest that the rate of innovation might be greater in the consumer goods sector in the earlier part of the upswing and greater in the capital goods sector subsequently. However, this implication is not mentioned by these authors.

Mandel's[14] interpretation—one of the earliest in the current phase of the literature—is rather different. In this interpretation of long waves, 'technological revolutions' in the capital goods industries are associated with the upswings of the long waves. The combined political, economic and social effects of the downswings create circumstances which eventually give rise to expectations of sustained periods of higher profits. Previously underinvested capital is then drawn into the creation of new branches of production in the capital goods industries which are centred on innovations more radical than those which had been marketable during the preceding depressed period.

Mandel's mechanism is in fact considerably more articulated than this, but the important point for this discussion is that he sees the focus of technical change in the upswing of the current long wave as the use of electronics, mechanical handling equipment and continuous flow processes in a major impetus to the progress of automation. The capital formation implied by the creation of factories and capacity to make these new technologies, and the monopoly profits (or 'technological rents') which they confer, 'drive' the initial phase of the upswing.

Lastly, the work of Clark, Freeman and Soete[15] contains yet another emphasis on the nature of technical change in the long wave. They emphasise the effects of a range of applications—in a variety of sectors—of a number of a new *forms* of technology. Principal candidates are electronics and synthetic materials. These technologies create a 'web' of interlocking products and firms which generates secondary waves of growth in associated supplying and using sectors and favourable market conditions for consumer goods with high income elasticities. In an earlier paper, Freeman[16] notes that these new technologies will 'most likely' be found first in machinery, transport or communications sectors. However, there is no necessity in his model for the nature of technical change to differ across sectors in the upswing of the long wave. The vital thing—as with the van Duijn model—is for a powerful new industry life cycle to be generated which can have major 'ripple' effects on other sectors. Only in the downswing does technical change become potentially more asymmetric as rationalisation investment overtakes the creation of new capacity.

Significance of clusters of innovations

Let us explore further the implications of these differing foci on the nature of the technological change involved in the long-waves. If we take a simple view of the clusters of innovations as new branches of profitable production, we restrict our view of their economic impact to the profits created, and the resources reallocated, at the *input side* of the innovations. At the other extreme, we can take the view that it is not just the products themselves, but the *economic consequences of their use*, which are important. In other words we have to examine the output side of the innovations. This leads immediately to the observation that capital goods and consumer goods innovations would have broadly similar effects on the economy at the input side, but would have quite different effects at the output side.

We know that mechanisation and automation have been broadly continuous trends throughout the past 150 years. Indeed, most 'long-wave believers' have pointed to the increasing rationalising aspect of investment in the downswing of the current long wave. If we are to believe Mandel, however, then it seems that the long waves may be linked to 'stages' in the progress of mechanisation and automation. The restructuring of the capital goods industries around the production, and the profit expectations, connected with new machinery systems is one aspect of the mechanism. The long-term effects of the diffusion of this equipment are another.

Clark *et al*, on the other hand, place less emphasis on the output-side effects of using the new technologies, and more emphasis on the input-side effects. In order to attempt a clarification of the different roles of technical change in these mechanisms, it would be valuable to distinguish between the input and output characters of the new technologies. There is broad agreement among all writers that electronics, synthetic materials, and a variety of consumer durables, are involved in some way in the current long wave. Diffusion curves for transistors, TV sets, plastics, man-made fibres, motor cars, and other broad indicators of changes in the structure of output all correlate well with the upswing of the long wave. But the uses of transistors are enormous, and the electronic capacity which they have been involved in augmenting has found its way into an extremely wide range of industrial and commercial equipment. Likewise, the impact of plastics had been not only to create growth for the chemical industry, but to provide a raw material which breaks down many technical barriers to automation and mechanisation in those processes where it is used to create final products.

Data on automation

In Figure 9.1 we present a preliminary attempt at indicating one aspect of the extent of these output-side effects of technical change in capital goods. The data presented in Figure 9.1 were constructed as follows. For both the UK and USA, *Census of Production* returns for machinery industries were examined and disaggregated where possible according to the following criteria. An attempt was made to identify those products or product groups which embodied, or contributed to, or were direct examples of: increased mechanical handling, continuous flow industrial processing, or electronic control. These were then designated as 'more automated'. These three types of technological property are those cited by Mandel as being centrally involved in the 'third technological revolution'.[17] For each year the value of the machinery possessing these qualities was calculated and expressed as a fraction of the total value of all categories of machinery delivered in that year.

This indicator is therefore a measure of the degree of 'automaticity' of additions to the capital stock in each year. It is, of course, open to the criticism that it is based on a fairly broad definition of 'automation'. Furthermore the *Census of Production* categories were not designed for this kind of operation. For instance, 'travelling tray baking ovens' (which appear in the numerator) and

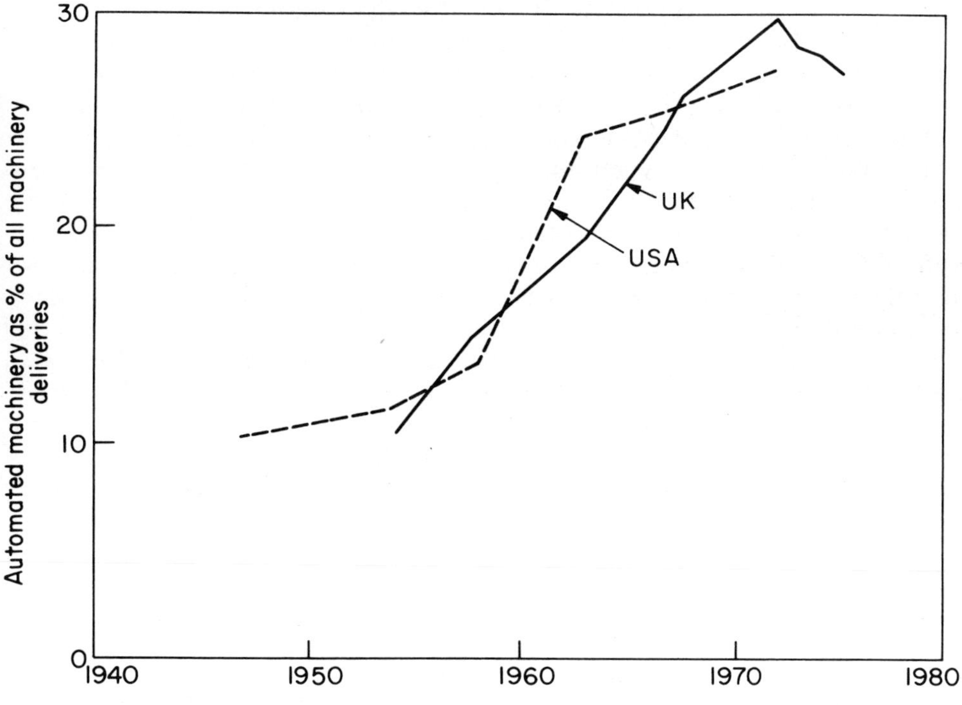

Figure 9.1. The diffusion of automated equipment.
Source: Calculated from UK and US *Census of Production* data.

'ordinary baking ovens' (which appear in the denominator) may not be an exhaustive set of categories for describing baking technology! It does, however, capture something of importance concerning the application of certain technologies of control to this process.

This process of disaggregation is therefore fraught with difficulty since the census categories are often not ideally suited to the purpose. There will undoubtedly be instances where relevant equipment is 'lost' in some broader category, and other cases where categories which are broadly 'automated' contain some equipment which is in fact of an older 'technical vintage'. However, it was possible to identify nearly 100 product categories in the UK data which could be allocated to the numerator of the fraction, and a substantial majority of these could be identified back to the beginning of the time period covered. The corresponding figure for the US data was 60. The resulting curves show a rise from around 10% in the 1940s to around 30% in the 1970s.

Now these data do not in themselves 'prove' anything about the role of automation in the long wave. However, they do seem to suggest that the method of identifying rapidly diffusing clusters of products can produce a variety of inferences depending on the clusters chosen. These products have shared characteristics which relate to their *purpose in use* rather than simply to the technologies from which they are constructed. Everything we know about the influence of demand on the nature of innovation should tell us not to be surprised by the fact that new technologies (electronics etc) find their applications channelled partly in this direction. The size and shape of the diffusion curve of these products (or process technology characteristics) can be seen as just as important a feature of the 1950s and 1960s as the more familiar diffusion curves of transistors, TV sets, and plastics. Furthermore, a dramatic increase in the public perception of the rate of development of automation characterised this period. The literature has been well documented by Cohen.[18]

Although the fears of labour displacement were not realised then, the scope and breadth of the discussion must surely have reflected real changes in the development of automation. To pursue this point further it would be necessary to examine the actual histories of production technology in the industries which were the principal purchasers of the products identified above, and to analyse their trends in productivity and share of labour costs in total costs. These issues must be deferred to a separate paper.

We have established that there was a substantial increase in the degree of automaticity of additions to the capital stock during the upswing of the current long wave, and that this was in part a consequence of the impetus delivered by the presence of new technological principles which we know to be more generally associated with the wave. But others have emphasised the role of these technologies in the consumer goods sector. Does this mean then that the impact of these technologies was distributed in a broadly equal manner between the consumer and capital goods sectors of the economy? To explore this further it is interesting to consider the extremities rather than the middle portion of the curve. Although the data are limited, there is a strong suggestion that a significant amount of this equipment for automation was in production considerably earlier than the main upswing of the long wave. This may be consistent with the suggestions of Freeman,[19] that in the depression of the previous wave there can be significant steps towards basic innovation, prototype production, and even a small amount of labour-intensive investment and employment generation in those product areas which may be important in the following upswing. Also important in this connection is the role of the rate of profit in different sectors and the *expectations* concerning future profit rates and the industries where they will be most favourable. Aspects of this part of the problem can be illuminated by more data.

Capital goods industries before and during the upswing

Tables 9.1 and 9.2 show that engineering employment, which we are using here as a proxy for employment in capital goods production, rose both absolutely and as a proportion of total manufacturing employment in the period *before* the upswing of the current long wave. The contrast with total manufacturing employment is particularly stark in the period 1924–35, and is not complicated by the period of wartime control of the economy as in the 1935–48 period. The contrast persists through the 1950s and 1960s and would be consistent with the interpretation that labour-intensive production of automated equipment was creating employment in the capital goods sector and beginning to displace labour in other industries. After 1970 employment declines in both sectors and engineering's share of total employment appears to stabilise.

The data therefore suggest that expectations and general buoyancy were picking up earlier in the engineering sector than elsewhere in the economy as the long downswing continued. We are again reminded of Schumpeter's 'first

TABLE 9.1. GROWTH IN EMPLOYMENT IN UK ENGINEERING AND MANUFACTURING (%)

	1924–35	**1935–48**	**1948–70**	**1970–78**
All engineering	19.0	83.0	47.0	–10.5
All manufacturing	0·5	25·6	13·5	–10·3

Source: UK *Census of Production.*

TABLE 9.2. UK ENGINEERING EMPLOYMENT AS A SHARE OF TOTAL MANUFACTURING EMPLOYMENT (%)

Year	$\frac{\textbf{Eng empl}}{\textbf{Tot man empl}}$ **%**
1907	10.0
1924	11·1
1930	12·1
1935	12·1
1948	19·2
1951	19·8
1954	20·1
1958	21·5
1963	23·0
1968	24·3
1970	24·8
1976	24·3

Source: UK *Census of Production.*

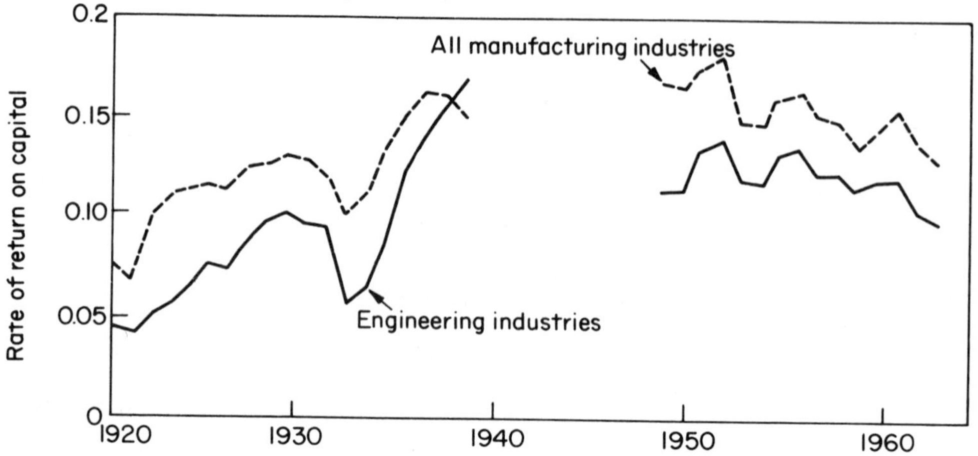

Figure 9.2. Rates of return in UK engineering and manufacturing industries.
Source: Adapted from P. Hart, *Studies in Profit, Business Saving and Investment in the UK, 1920–1962*, 2 vols (Cambridge University Press, 1965 and 1968).

approximation'. The acceleration in the 1950s and 1960s is undoubtedly helped by the virtuous circle created by expanded consumer demand, as Forrester's model suggests, but it seems inconsistent to identify the movement in the capital goods sector as being entirely contingent on the *general* upswing. Rather it seems to have been *leading* the general upswing and to have then become accelerated by it.

The principal components of engineering which experienced this growth of employment before World War 2 were electrical and instrument engineering. This is consistent with the interpretation of Landes,[20] Hobsbawm,[21] Pollard[22] and other historians who identify electrical technologies, vehicles, and some parts of the chemical industry as being the healthy and expanding branches of the inter-war UK economy. What we are concerned to emphasise, however, is not simply these established views that industry was being restructured toward new growth points during the Depression, but that this restructuring involved a substantial shift in the technological and commercial prospects of the machinery industries. This then suggests a degree of linkage in the trends in quality improvement of capital goods output between the long waves, despite the shifts in the technical principles by which those improvements were achieved.

This view can be further illustrated by examining data on profit rates and value-added (Hart[23]). Figure 9.2 shows rates of return on capital in current prices for total manufacturing and for engineering industries. Several points are worthy of mention. First, the general rate of profit is higher in the upswing period than in the inter-war period, and the general trend is for it then to fall. This is consistent with the general theory of the long wave, suggesting that inflationary pressures, and the competing away of monopoly profits, eroded the initial high rate of return.

Second, the increase in profit rates starts from the early low point of 1920 and continues throughout the inter-war period, though with the interruption of the severe depression in the early 1930s. Thus, although production was growing slowly in this period, profit rates appear to have been rising, contrary to the expectations of Mandel (who, it should be said, would wish to measure profit differently). It seems that the entire long-wave mechanism may be characterised by a distinct phase lag between the movements of production growth and rates of return.

Third, and most relevant to the present discussion, the rate of return in engineering is tending to catch up with that in manufacturing as a whole during the 1920s, and even overtakes it in the late 1930s. Subsequently, it falls to a lower level in the post-war period. Interpretation of this phenomenon is difficult, but it does raise the possibility that in the late 1930s the engineering sector was already experiencing the early effects of the stimulus from producing a new cluster of products which later become more visible in the 'diffusion curve' of Figure 9.1.

This increase in the rate of return in engineering relative to other industries is further reflected in Figure 9.3. This shows value-added per unit of output in engineering divided by value-added per unit of output in all manufacturing. This indicates that engineering had already improved its relative ability to create added-value in the economy *before* the long boom of the 1950s and 1960s. To summarise: the data on employment, value-added and rates of return in UK engineering are here used to suggest that the restructuring of UK industry between the wars had a particular component relevant to the question of the role of innovation. The gradual increase in concentration and scale, the emergence of some 'mass production' industries, and the extra impetus of the war-time influences on the economy, all contributed to an environment in which technological advances were likely to be sought for, and applied in, the area of the improvement of machinery. This was surely an important component of the gradual emergence after World War 2 of expectations of, and investments in, a virtuous circle. Increasing production of labour-saving machinery fuelled cost reductions in industries producing intermediate and

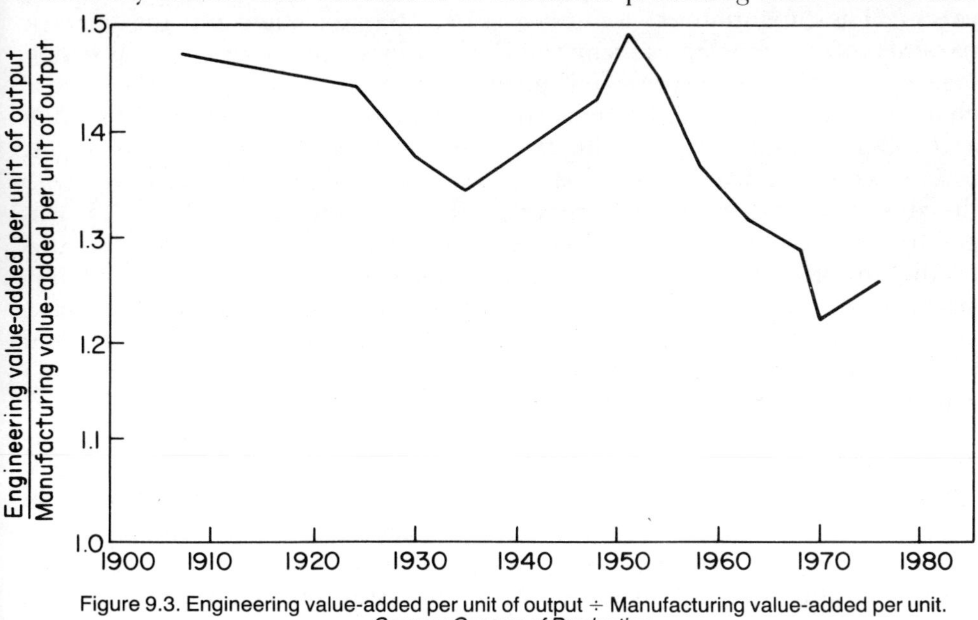

Figure 9.3. Engineering value-added per unit of output ÷ Manufacturing value-added per unit. *Source: Census of Production.*

consumer goods; this in turn expanded demand, but simultaneously set in train the process of heightened competition which eventually created the inflationary pressures which stifled the boom. Hence, as Wragg and Robertson[24] note, from 1924 up to 1950, half of productivity growth in UK industry was due to reallocation of resources towards more productive industries; but after 1950 the bulk of productivity advance is created within the individual branches of industry themselves.

Conclusions

This paper has explored the relationship between innovation in capital and consumer goods in order to clarify the apparently conflicting views held by long-wave protagonists. The data on the diffusion of automated machinery and on the circumstances which lead up to that diffusion can now be used to evaluate the views of the authors cited at the beginning of this paper.

The data seem inconsistent in some respects with the implications of the model of Forrester and van Duijn. The upswing in production and in innovation in the capital goods sector does not come after the consumer boom. Its main growth period seems contemporaneous with the consumer boom, and there is even evidence of an earlier 'preparatory' phase.

The data do seem consistent with several aspects of Mandel's views, although the anomaly concerning the movement of profit before the upswing, and the general complexity of Mandel's model, mean that this cannot be seen as support for the model as a whole but rather for one of its components.

The data are not inconsistent with the views of Clark *et al.* Rather, they suggest a refinement of the notion of a technological web of interconnected innovations. Nelson and Winter[25] have argued that certain 'natural trajectories' can be identified which figure prominently in the rate and direction of innovation. They mention economies of scale, mechanisation and the intrinsic parameters of technologies themselves (thrust to weight ratios etc). It is tempting to use these broad categories to help interpret the complexity of technical change in the long wave. It seems possible that in the downswing of the 1920s and 1930s the relatively enduring 'trajectories' of mechanisation and economy of scale were significant biases in the nature of technical change. However, with the stimulus of conditions of higher profitability after World War 2, it was possible for those trajectories to be 'reinforced' by the new trajectories created by the powerful and often transferable technologies of control, manipulation and new materials. Thus in the upswing of the wave the stimulus of a new phase of automation facilitated by new technologies was perhaps at least as important to growth as the final consumer products to which those technologies contributed.

Clearly further work is necessary to establish whether the suggestions made in this paper are valid. Empirical analysis of production technology innovations, patent statistics, and industry productivity data could all play a part in this. More understanding of the relative roles of consumer and producer innovations in the long wave would contribute to the debate over the possibility of a new counter-cyclical policy, and to the assessment of new growth products today, such as robotics, computer-aided manufacture, and biotechnologies.

Notes and references

1. R. Nelson and S.Winter, "Evolutionary versus neoclassical theories of growth", *Economic Journal*, March 1974, *89*, pages 866–905.
2. R. Nelson and S. Winter, "In search of useful theory of innovation", *Research Policy*, 1977, *6*, 1, pages 36–76.
3. N. Kay, *The Innovating Firm* (London, Macmillan, 1979).
4. S. Davies, *The Diffusion of Process Innovations* (Cambridge University Press, 1979).
5. V. Walsh *et al*, "Trends in Invention and Innovation in the Chemical Industry" (mimeo), University of Sussex, Science Policy Research Unit (SPRU), 1979.
6. C. Freeman, "The determinants of innovation", *Futures*, June 1979, *11*, 3, pages 206–215.
7. N. Kondratiev, "The long waves in economic life", *Review of Economic Statistics*, 1935, *17*, pages 105–115.
8. For an account of this exchange, see R. Day, "Mandel, Trotsky and Kondratiev", *New Left Review*, 1976, *99*.
9. J. Schumpeter, *Business Cycles* (New York, McGraw-Hill, 1964) (abridged version).
10. *Ibid*, pages 105–114.
11. G. Mensch, *Stalemate in Technology* (Cambridge, MA, Ballinger, 1979).
12. J. van Duijn, "The long wave in economic life", *De Economist*, 1977, *125*, 4, pages 544–576.
13. J. Forrester, "Business structure, economic cycles and national policy", *Futures*, June 1976, *8*, 3, pages 195–224.
14. E. Mandel, *Late Capitalism* (London, New Left Books, 1976).
15. J. Clark, C. Freeman and L. Soete, "Long waves and technological developments in the 20th century" in D. Petzina and G. van Roon, eds, *Konjunktur, Krise, Gesellschaft: Wirtschaftliche Wechsellagen und sozialer Entwicklung im 19 und 20 Jahrhundert* (Stuttgart, Klett-Cotta, 1981) pages 132–169.
16. C. Freeman, "The Kondratiev long waves, technical change and unemployment", in *Structural Determinants of Employment and Unemployment*, Vol 2, reports prepared for Experts Meeting in Paris, 7–11 March 1977 (Paris, OECD, 1979), pages 181–196.
17. Mandel, reference 14, chapter 6.
18. J. Cohen, "The Automation Debate in Perspective", MSc Thesis, Manchester University, 1975.
19. Clark *et al*, reference 15, page 28.
20. D. Landes, *The Unbound Prometheus* (Cambridge University Press, 1969).
21. E. Hobsbawm, *Industry and Empire* (London, Pelican, 1969).
22. S. Pollard, *The Development of the British Economy 1914–1967* (London, Edward Arnold, 1969).
23. P. Hart, *Studies in Profit, Business Saving and Investment in the UK, 1920–1962*, 2 vols (Cambridge University Press, 1965 and 1968).
24. R. Wragg and S. Robertson, *Post-war Trends in Employment* (London, Department of Employment, Research Paper No 3, June 1978).
25. Nelson and Winter, reference 2, page 56.

CHAPTER 10

INNOVATION AND ECONOMIC CHANGE

Jay W. Forrester

The MIT System Dynamics National Model represents the physical and human processes involved in the long-term growth of a national economy, and can provide a theoretical explanation for the generation of long waves. The phases of the long-wave cycle strongly influence the climate for innovation: towards the end of a boom minor improvement innovations in established industries tend to predominate, but managerial and political innovations in a depression can ultimately lead to an upsurge of radically new technologies. A balance between human population and environment may now be the most fundamental requirement for political and social innovation, taking precedence over the proliferation of technical innovation for its own sake.

TECHNICAL innovation shaped today's industrial world. Products and processes reflect new developments of the past four decades. Technical innovation has become recognised as the driving force behind a rising standard of living. Innovation is accepted as a weapon in corporate and national competition. However, many people perceive that the innovative process is faltering. Conferences are held to analyse innovation. Governments appoint task forces to encourage innovation. But is the concern properly directed? Innovation is not a constant process. Innovation shifts in both character and intensity in response to the long sweep of economic change.

Economic conditions do change, not only with short-term fluctuations of the business cycle, but also in far longer patterns. For example, three economic eras divide the past 50 years: first, the Great Depression of the 1930s; second, the capital-investment boom of the 1950s and 1960s; and third, the present period of rising social pressures and growing economic uncertainty.

As economic conditions change, so do opportunities for innovation. Corporate strategy, to be effective, must reflect the fluid environment within which

Professor Jay W. Forrester is Germeshausen Professor at the System Dynamics Group, Alfred P. Sloan School of Management, Massachusetts Institute of Technology, 50 Memorial Drive, Cambridge, MA 02139, USA. The paper is based on the keynote speech at a Massachusetts Institute of Technology Symposium on Technology, Innovation and Corporate Strategy for executives of European corporations, London, 16 November 1978.

innovation takes place. As the needs of society vary, and the challenges confronting corporations shift, the nature of innovation must also change to fit the circumstances.

There are times during which technical innovation should focus on improving products already developed, and times to prepare daring new products for the future. There are times for technical innovation, and times for managerial innovation. There are times offering opportunities within a corporation, and times imposing threats from outside. There are times for innovation in science and engineering, and times for innovation in society and government.

There seems to be an alternating tide in economic affairs, spanning some 45 to 60 years, that determines the climate for innovation. A typical 50-year pattern of long-term economic change includes a decade of depression, 30 years of technical innovation and active capital investment, and finally, ten years of economic uncertainty while the growth forces of the past subside. The 50-year repeating rise and fall of economic activity is called the 'long wave', or Kondratiev cycle.

Long-term economic change has received little attention. People in management and politics focus on business-cycle behaviour and the accompanying recessions at intervals of three to seven years. But a business cycle is not an adequate time horizon. There exist larger and slower changes of greater importance to corporate strategy and to understanding the role of innovation.

In Western industrial economies, capital investment has been concentrated in periods of economic excitement lasting about three decades. Such periods of active new construction have been interrupted by major depressions. Three such major capital-investment cycles have occurred since 1800. Vigorous economic activity has been terminated by severe depressions in the 1830s, 1890s, and 1930s. Only a small body of economic literature treats the long wave. The long wave has been observed in the form of historical episodes supported by some statistical analysis of economic behaviour.

The MIT System Dynamics National Model

In our research at MIT on economic behaviour, we have been drawn into examining the economic long wave through our work on the System Dynamics National Model.[1] The National Model represents the physical and human processes found in any national economy. The model replicates the policies and structure that cause the unfolding progression of economic changes. Such a model brings social and economic structure into the laboratory where the relationship of policies to behaviour can be identified.

The National Model is being built up in stages with different configurations of industrial sectors, such as consumer durables, capital equipment, energy, agriculture, and building construction. Each industrial sector of the Model is constructed to represent a typical business firm in that sector of the economy. The Model represents production processes in comprehensive detail, and acquires the many inputs to production on the basis of inventories, prices, costs, order backlogs, growth rate, marginal productivity, liquidity, profitability, return on investment, and regulatory restraints. Each production sector of the Model contains a full accounting system that handles accounts payable and receivable, generates a balance sheet and profit-and-loss statement, pays taxes, and computes indices of financial performance. The market clearing function, which balances supply and demand, responds not only to price but also to availability of output product. This availability, or delivery delay, corresponds to market behaviour in the real economy, where many prices change slowly and supply and demand are partially balanced by allocation and delays in filling order backlogs.

In similar detail, the Model contains a labour mobility network for the movement of people between sectors, a banking system, the national monetary authority, household-consumption sectors, a government sector, and a demographic sector. Such a model is a translation into computer language of the knowledge people have about organisational structure and operating policies surrounding their daily activities.[2,3]

Few people suspect the degree to which the puzzling complexities of business cycles, unemployment, depressions, and inflation arise from interactions between well-known and well-understood parts of the economic system. When a simulation model is constructed from policies, organisational structure, and physical processes that would be familiar to any businessman, the model produces the same troubling modes of behaviour experienced in real life. Actual economic behaviour is puzzling, not because of insufficient information about the parts of an economic system, but because, until recently, it has not been possible to show how well-understood parts interact to produce the baffling behaviour of the whole system.

The System Dynamics National Model differs substantially from the more familiar econometric models.[4] The National Model is built up from the operating structure within corporations, rather than from macro-economic theory. It is derived from management policies as observed in the detailed, practical, working world, rather than from statistical correlations derived from broad time series representing aggregate economic behaviour. The National Model is intended for evaluating alternative corporate and national policies, rather than for short-term forecasting. The National Model is a computer simulation model that plays the roles of economic actors to produce the same kinds of behaviour that are generated by real economic systems. By bringing economic structure into the laboratory where controlled experiments can be performed, the System Dynamics National Model allows one to untangle complex socioeconomic interactions. The Model deals with the fundamental processes of management, production, and consumption, so it is applicable not only to the USA but to all industrial economies.

The National Model and long waves

The National Model was not undertaken for the purpose of studying long-wave behaviour. In fact, I was unaware of the scant literature on the long wave while we were initially formulating the Model. But when we assembled a consumer durables sector along with a sector that produces capital equipment, we found that the Model exhibited strong fluctuating growth and collapse in the capital sector with about 50 years between peaks of capital output.

When such unexpected behaviour is encountered in a simulation model, the prudent investigator first questions the structure and plausibility of the model itself. In the early stages, anomalous behaviour often points to dubious assumptions in model structure. However, as the quality of a model improves, unanticipated model behaviour is more likely to suggest new insights about actual economic behaviour. So it was with the unstable rise and fall of the capital sector. After we analysed the reasons for the 50-year mode of behaviour in the National Model, we concluded that the underlying assumptions still seemed reasonable. It was only then that the literature on the Kondratiev cycle became a part of the investigation.

Literature on the Kondratiev cycle is filled with debate and conflicting assertions. Economic evidence has been interpreted differently by different observers. There has been no cohesive theory to explain how an economic pattern spanning a half century could be systematically and internally generated. Because no theory of the long wave existed to show how the many aspects of reality could fit into a unified pattern, controversy was unavoidable.

We believe the National Model now provides a theory of how the economic long wave is generated.[5] The process involves an overbuilding of the capital sectors in which they grow beyond the capital output rate needed for long-term equilibrium. In the process, capital plant throughout the economy is overbuilt beyond the level justified by the marginal productivity of capital. Finally, the overexpansion is ended by the hiatus of a great depression during which excess capital plant is physically worn out and financially depreciated on the account books until the stage has been cleared for a new era of rebuilding.

The process of decline and revival produced by the long wave can be traced by thinking back to the 1920s: 50 years ago, prices had been rising; interest rates were high; investment opportunities had diminished; cities were afflicted by excess office space; land prices had been rising sharply; heavy debts had been incurred, and the financial system was overextended.

Under economic stresses that reached a climax in the 1920s, previous trends faltered and reversed in the 1930s. The earlier pace of capital investment collapsed. Overcapacity appeared in much of the industrial world. Prices fell under the pressure of oversupply, and wages were driven down by falling profits and high unemployment. Expansion ceased because agriculture and consumer sectors had excess capacity, so demand fell for construction and capital equipment. For 15 years the industrial economies coasted on the physical capital plant that had been accumulated before 1930. Defaults and bankruptcies cleared out the excess debt load.

By the end of World War 2, physical capital at all levels had been depleted, in some places by age alone, in other places also by war damage—consumers needed housing and durables; industry needed factories and machines; society needed school systems and highways; and so a great rebuilding began. Capital sectors drew labour from consumer sectors, thereby producing a tight labour supply and still more incentive for consumer sectors to become more capital-intensive. Demand for capital was self-reinforcing: in order to expand, capital sectors themselves required new capital plant, thus creating further demand on capital sectors. A long, powerful regenerative process drove the expansion of the capital sectors.

Capital plant was rebuilt in a rather brief 20 years. To meet the high demand during these two decades, construction capacity in the capital sectors became far greater than would later be required merely to replace continuing physical depreciation.

Once capital plant had been rebuilt, the capital-construction process would not immediately slow to a sustaining rate. Assume that an appropriate level of capital plant was reached in 1965, as indicated by declining return on investment in new capital plant, and by progressively deeper business-cycle recessions since then. Nevertheless, the capital accumulation process did not stop in 1965 because tremendous momentum had been established in the previous years—capital investment had become stylish; labour unions wanted to build more highways and public buildings; banks had loaned money successfully on earlier construction and sought new opportunities to make loans; national monetary authorities had provided credit expansion in the 1950s and 1960s without producing serious inflation, and assumed the process could continue. For more than a decade, from 1965 to 1978, physical capital has been forced into many of the world's economies beyond the point where capital earns an adequate return. But such growing investment in the face of falling incentive cannot continue indefinitely. Eventually, an excess rate of capital construction saturates the need for more capital plant. Such excess capacity on a worldwide basis is now appearing in steel production, shipping, diesel-engine manufacture, synthetic textile fibres, chemicals, electronics, and automobiles.

From the 1920s to the 1970s is a typical 50-year long wave, beginning and ending with fully developed technologies and saturated capital markets. At both the start and finish of the 50-year period, opportunities had declined for attractive investment in the maturing technology of the times. For example, in the 1920s the world had substantially completed its railroad system, and now in the 1970s it has correspondingly completed its air-transport systems.

People often question the idea that a long-wave economic mode could persist for nearly 200 years, despite the major changes that have occurred in society and technology. But the policies and structures that generate the long wave have changed very little. The long wave depends on production methods that use capital equipment, on the life of capital equipment and buildings, and on the sluggish pace with which people move between sectors of an economy. The long wave is accentuated by how far ahead people plan and the length of their memories of past economic disasters—both of which are substantially determined by the length of a human lifetime. None of these factors that give rise to the long wave depends significantly on faster communications or details of technological change. The policies and industrial structure that generate the long-wave capital-construction cycle have changed very little since 1800.

Long-wave behaviour, as revealed in the National Model, seems to explain many things now happening around the world. Current economic conditions are much like those that the National Model exhibits at a peak of the long wave. At such a peak one should expect a decline in new capital investment, rising unemployment, a levelling out in labour productivity, high interest rates, rising prices, falling return on investment, increasing amplitude of business cycles, and reduced innovation from maturing of the current wave of technological advance. Such conditions fit today's situation. Similar conditions last occurred in the 1920s at the previous long-wave peak.

Long waves and innovation

I believe that the long wave strongly influences the climate for innovation. That statement reverses the relationship between innovation and the long wave that is often argued in the economic literature. There has been a theory that innovation, in the form of great new inventions, causes the long wave. The source of such a theory is easy to see. Successive periods of major economic expansion have each been characterised by a unique mix of technologies. During successive cycles transportation has been provided by canals, later railroads, and currently aircraft. In energy, industrial society has been through the wood-burning wave, the coal-burning wave, and now the oil-burning wave.

But I do not see innovation as causing the long wave. In fact, the National Model generates long-wave behaviour even without technological change. Instead, I see the long wave as compressing technological change into certain time intervals and as altering the opportunities for innovation.[6]

Each major expansion of the long wave grows around a highly integrated and mutually supporting combination of technologies—for example, energy sources determine the design of factories; transportation is highly interrelated with the pattern of living; communication and the pace of business move to the same tempo; education is geared to the existing technology; and financial institutions come to understand opportunities that lie within the pattern of the times.

After such an integrated pattern of economic development becomes established, it rejects incompatible innovations. A major innovation that breaks sharply from the existing status quo is perceived as an impractical idea. Those trained in the old technology do not comprehend a major innovation: they are more comfortable making marginal improvements on the current technology. Bankers who have succeeded by backing familiar ideas want modest innovations within conventional patterns; they no longer think in terms of innovative ideas that must be supported by new research, new education, and a new perception of market demand.

So, for the last half of a 30-year upswing of a long wave, radical innovation remains outside the circle of acceptance. For example, I believe we are in the late stage of a long-wave expansion and one sees few technical innovations today in the mainstream of business that are not step-by-step routine improvements on ideas that have been around for many years. In fact, I do not know of any innovation that was first created in the past 20 years that has become a major, commercially successful product comparable to the earlier inventions in computers, solid-state electronics, nylon, television, or aircraft. That does not mean innovation is dead; it means the climate is not receptive for adopting radical innovations into commercial channels.

Furthermore, the environment for commercial exploitation of technical innovation does not improve during the decade following a peak in the long wave. In the ensuing major depression lasting a decade or more, the climate for innovation remains unfavourable: the social structure is in disarray; profits are low; those who managed the last stages of the growth period are still in control but are discouraged; fundamental innovation continues but is left dormant; and the process of using up and wearing out the old technology runs its course.

Then, the door begins to open on a bright new future. Hope flourishes. As rebuilding begins, the old technology is not reestablished. The reservoirs contain 30 years of stored innovations; and accumulated new ideas are tried and developed. During the early stage of a new expansion wave, new products and new businesses appear.

Returning to the contrasting times in the innovative environment, there are times during which technical innovation should focus on improving products already developed, and times to prepare daring new products for the future. I believe the industrial countries are in transition from the former to the latter. For the past 15 years R and D has been commercially exploiting the major innovations that were pioneered between 1925 and 1960. Those developments are reaching maturity. The manufacturing processes have become common knowledge on a worldwide scale. Capital investment to produce the products has been overexpanded. An executive of a major retailing corporation told me recently that, in every product line they carry, the largest supplier could disappear from each industry without significantly diminishing the availability of products. Prices are under competitive pressure. Customers are demanding quality and economy more than further innovation.

It is time to liquidate and recover the investment in many current product lines and rebuild investment along those new patterns that will become the technological wave of the year 2000. To do so requires looking across a 20-year gap and imagining a world as different from today as today is from 1910.

The road for technological innovation is now branching. The old technology with its hordes of technical experts will continue for a time through a succession of minor improvements, while the competitive situation becomes more acute. At the same time and in parallel, a small but gradually increasing number of people will develop the radical new mix of technology to replace that which the first group is exploiting to obsolescence. That is the picture I want to draw for technological innovation. But there are other, more timely kinds of innovation.

Technical innovation, managerial innovation and political innovation

There are times for technical innovation, and times for managerial innovation. Technical innovation comes to a discontinuity when refining old products will no longer suffice, but before the full force of society can be harnessed to major innovations of the next technological wave. During the break between technological waves, which can last up to 20 years, is a critical time requiring managerial innovation—old industries are mature; top management attention has shifted from technical to legal and financial aspects of business; research is geared to a past technology; organisations have grown too large for the decentralised competition needed to establish a new technological wave. Traditionally, the downturn after the end of a major wave of capital investment leads to failure of many corporations. Bankruptcy is a form of managerial innovation that shakes the economic system into a new shape. Companies that want to remain in control of their own destinies must anticipate major changes in organisation, style of innovation, markets, and international commerce. Although there is less reliable precedent for managerial innovation than for technical innovation, the challenge must be met. We are entering a time when creative management, more than creative technology, can make the difference between corporate death and survival.

There are times offering opportunities within a corporation, and times imposing threats from outside. At the beginning of a new technical wave, as from 1945 to 1960, opportunities are internal and lie in developing technology, human skills, and manufacturing methods. In the late stages of a growth wave, opportunities shift to coupling internal capability of a corporation to growing markets. After economic growth has stopped—during that interval between waves that has historically been a time of economic depression—the needs of society can no longer be met by innovation within corporations. Unemployment spills out of the old technology faster than jobs are created in the embryonic new technology of the next cycle. Those trained in the old technology are cast adrift. Not only have corporations offered little of relevance at such a time, but corporations are blamed by the public and politicians for growing economic difficulties. Corporate management retreats into a defensive position. Initiative shifts to radical innovators in the political realm. The needs of society appear not in technical terms, but in social and political terms.

There are times for innovation in science and engineering, and times for innovation in society and government. I believe we will soon enter another time of political innovation corresponding to the 1930s. Not all technical innovations are successful. Even fewer political innovations survive the test of time. Central Europe tried a major political innovation in the 1930s; but that new social order survived not for the proclaimed 1 000 years but for less than 20. Other Western industrial countries innovated in the 1930s by creating social welfare programmes; most have survived until now, but with threatening strains from inflexible bureaucracy and rising governmental costs. Many countries are now trying the political innovation of monetary inflation in search of short-term solutions to political pressure, but the history of that road is not promising. In the next few years, political innovation will be needed; it will be tried, and we should see that it is guided constructively.

Looking beyond the current political unrest and economic uncertainty, what might be the nature of future innovation? For those who try to address the technological future, the challenge is one of selection, not one of inventing. In general, each new technical wave has developed around ideas that had long been known, even for as long as a hundred years. What currently existing ideas will be drawn together into a new internally consistent and reinforcing infrastructure? How can one separate the ideas that will succeed in the next wave of economic development from crackpot ideas that will fall by the wayside? The answer probably does not lie in the merits of any idea by itself. Instead one must visualise the fabric of the future society—how it will live, travel, eat, communicate, and govern. Only then can a picture of compatible innovations be perceived. Oil and the automobile produced the suburbs; an energy shortage may force a return to more concentrated urban living. Costly energy may defeat economies of scale in centralised manufacturing and favour more self-sufficient and decentralised socio-industrial patterns.

In fact, one should ask: will the focus of innovation in the future continue to be technological? For 200 years humanity has clung to the hope that technical innovation could deliver the good life. But, after several waves of technical advancement, more people are hungry than ever before, pursuit of energy and resources is leading to greater political stress, deserts are expanding, forests are disappearing, and pollution is becoming more intractible. Should we continue to hope for a solution from technology? Or, does the next frontier for innovation lie in new approaches to managing the relationships of man to man and man to Nature? Most problems that technology has been asked to solve are a consequence of rising population. Perhaps the most fundamental innovation would be to find the social and political incentives necessary to stabilise population without starvation and war.

We should consider even longer dynamic changes than the Kondratiev cycle. The life cycle of growth describes the processes of youth, growth, and maturity. There can be a life cycle of population growing to fill geographical capacity.[7] Might there not be a life cycle of technological innovation? Past growth in technology may be a transient that could reach a reasonable maturity. Some recent technical innovations seem to be little more than for the sake of innovation, rather than for a lasting contribution to human well-being. Supersonic air transport seems to have overreached human need and economic justification. Atomic power is beset by growing difficulties. Space flight can scarcely solve the problems on earth. Perhaps the next innovation is not to rely on technology but to look at ourselves. It would be a true innovation to discover enough about the behaviour of social and economic systems to allow humanity to live in a happy and sustained balance with the capacity of the planet.

Notes and references

1. For background on the system dynamics approach to modelling see: Jay W. Forrester, *Industrial Dynamics* (Cambridge, MA, MIT Press, 1961); Jay W. Forrester, *Principles of Systems* (Cambridge, MA, MIT Press, 1968); Michael R. Goodman, *Study Notes in System Dynamics* (Cambridge, MA, MIT Press, 1974); Jay W. Forrester, "Information sources for modelling the national economy", *Journal of the American Statistical Association*, September 1980, *75*, 371, pages 555–574.
2. For a more complete description of the System Dynamics National Model, see Jay W. Forrester, Nathaniel J. Mass and Charles J. Ryan, "The System Dynamics National Model: understanding socio-economic behavior and policy alternatives", *Technological Forecasting and Social Change*, July 1976, *9*, 1 and 2, pages 51–68.
3. For a description of behaviour of one of the early assemblies of the Model, see Jay W. Forrester, "Business structure, economic cycles, and national policy", *Futures*, June 1976, *8*, 3, pages 195–214.
4. Jay W. Forrester, "An alternative approach to economic policy: macrobehavior from microstructure" in Nake M. Kamrany and Richard H. Day, *Economic Issues of the Eighties* (Baltimore and London, The Johns Hopkins University Press, 1979).
5. For a more complete discussion of the theory of the long wave as manifested in the System Dynamics National Model see Jay W. Forrester, "Growth cycles", *De Economist* (The Netherlands), 1977, *125*, 4, pages 525–543.
6. Alan K. Graham and Peter M. Senge, "A long-wave hypothesis of innovation", *Technological Forecasting and Social Change*, 1980, *17*, 4, pages 283–311.
7. Jay W. Forrester, *World Dynamics* (Cambridge, MA, MIT Press, 1973).

CHAPTER 11

LONG WAVES IN ECONOMIC DEVELOPMENT

Causes and empirical evidence

Hans H. Glismann, Horst Rodemer and Frank Wolter[1]

This paper is part of the newly arising "Kondratiev" literature. Before embarking upon the empirical analysis the paper delineates ideas about the moving agents of the Kondratiev from prototypical theories advanced in the literature. It argues that all economies, developed or not, socialist or capitalist, exhibit long waves of economic activity. This proposition is tested empirically for a limited sample of industrial countries, and it is argued that the observed fluctuations are caused by corresponding waves of investment, which, in turn, are induced by long-run changes in incentives. The hypotheses are tested that the development of incentives is endogenous to long-term changes in the distribution of factoral incomes, in the level of government interventions and in the intensity of international competition.

Explanatory hypotheses for long waves

Long waves in economic thought

IN HIS seminal *Business Cycles* Schumpeter strongly argued for the simultaneous presence in economic development of numerous waves of different frequencies and amplitudes which partly dampen, partly reinforce each other.[2] Drawing on the empirical studies of Kondratiev,[3] Schumpeter recognised, among others, long waves of approximately 50 years' length and called them "Kondratiev cycles", although other researchers—among them Pareto[4] and van Gelderen[5]—had posited the existence of long waves before Kondratiev.

Prototypical hypotheses for the phenomenon that economic development is not only subject to short- and medium-run cycles, but in addition follows a long-run wave-like pattern, can be classified under the headings: structural capital shortage, structural overproduction, and innovation.

The authors are economists working at the Kiel Institute for World Economics.

Kondratiev viewed periodically appearing capital shortage as a significant cause of long waves in economic development.[6] He maintained that during a long-run development cycle, investment activity eventually wanes because borrowed capital becomes scarce and expensive. In the further course of the cycle, falling prices would lead to an increase in savings and in the supply of capital via a rise in the real value of the money stock (Pigou effect). Long-run profit expectations thereby rise, investment activity becomes more buoyant, and a new upswing ensues.

Spiethoff advocated the hypothesis that gradually increasing overproduction causes the economic ups and downs.[7] In his view, overinvestment emerges in certain branches during the upswing because of overestimation of future increases in domestic demand. This leads to accumulation of inventories, underutilisation of physical capacity, and export of the "surplus". As a result, business profits decline, and an enduring abatement of investment activity sets in. Finally, the long-run decline comes to an end; reversal towards a long-term phase of prosperity is indicated by a rise in the demand for investment goods.[8]

The innovation hypothesis goes back to Schumpeter.[9] He saw the preconditions for a lasting rise in investment activity and for an acceleration in economic growth in the cumulative appearance of entrepreneurial pioneers who carry out innovation across a broad range of economic activity. Of course, innovative activity occurs during all phases of a long-term cycle. But a pronounced burst of basic innovations, brought about by daring entrepreneurs, precedes the prosperity phase, and constitutes the ignition for the subsequent upswing. During the upswing, the entrepreneurial pioneers are followed by a growing swarm of imitators. This finally leads to boom-like excesses, followed by an enduring decline in economic activity. Only the renewed appearance of entrepreneurial pioneers, who often exhibit irrational courage in taking on the risks of innovative activity, succeeds in achieving the breakthrough to a new phase of prosperity. With the appearance of new entrepreneurs on the scene, the elites of industrial society are exchanged as well.[10]

Schumpeter's hypothesis constitutes the central element in van Duijn's recent attempt to integrate and further develop various explanations for long-term economic fluctuations.[11] Besides Schumpeter's innovation hypothesis, further building blocks are Forrester's overinvestment hypothesis[12]—which in its essentials is similar to Spiethoff's explanation—and Kuznets' industrial cycles hypothesis:[13]

> The point of departure is the cumulative appearance of Schumpeterian entrepreneurial pioneers, who carry out a host of fundamental innovations and call forth imitators.[14] These innovations induce a number of industry cycles, which lead to a spurt in demand for investment goods. With the growth of the new industries, unfilled orders in the capital goods sector mount. Efforts to expand capacity in this sector serve to further tighten bottlenecks. From the steadily growing pile of unfilled orders, producers of investment goods derive too optimistic expectations for the future. They do not recognize the signals of saturation which emerge in the late phase of the industry cycles. As saturation is approached, the demand for investment goods unexpectedly declines, and working off the stock of unfilled orders in the capital goods sector begins. That sector's demand for investment goods subsequently declines also.

In the face of a rapidly dwindling stock of orders, entrepreneurs in the capital goods sector now evaluate future prospects too pessimistically. Later, the process leads to disinvestment. The more deeply the economy gets caught in the downswing, and the longer the downswing takes, the less profitable would be investing in the received technology. Finally, in the advanced stages of the decline, a new generation of entrepreneurial pioneers emerges.

All these theories are based on a common mechanism insofar as they connect growth cycles primarily with changes in investment activity. They differ, however, in the factors behind changes in investment activity. In our view, none of these mechanisms is completely convincing.

According to the capital shortage hypothesis, investment cycles are primarily attributed to long cycles in interest rates on the capital market. The existence of a relationship between interest rates and investment activity can hardly be denied—in the short run, the rate of interest is surely the decisive macro-economic investment demand regulator. In the long run, however, when expected profits themselves are variable, interest rates react to rather than determine economic development.

The overproduction hypothesis imputes persistently inaccurate expectations to producers of investment or other industrial goods. However, systematic wrong perceptions and decisions on the part of entrepreneurs in the capital goods sector (or in any other sector) as a result of a long-term rise or decline in unfilled orders are theoretically implausible (on account of the neglect of the price mechanism) and, in addition, empirically undetected[15]—even by Forrester who bases his whole argument on this behavioural assumption.

The innovation hypothesis claims that innovations regularly appear in clusters. The degree to which growth and investment cycles are attributable to innovation cycles is difficult to test empirically. A major problem is identifying the innovations that sustain an upswing. Even if one adopts Schumpeter's (intuitively appealing) classification of innovations, the resulting fit with respect to the hypothesised regularities is not all that tight.[16] More recent attempts to detect systematic clustering of innovations remain unconvincing too.[17] This lends weight to a hypothesis of Kuznets that innovations are randomly distributed over time.[18] Nevertheless, the possible existence of innovation cycles cannot be said to have been disproved.

A model based on distribution

In contrast to the explanatory approaches of Kondratiev, Spiethoff, and Schumpeter, long-term fluctuations in investment activity can also be interpreted as resulting from systematic price distortions and rectifications on product and factor markets which directly impinge upon profit expectations.[19] Growth-constraining distortions can ensue, for example, from an increase in the degree of monopoly on product and factor markets, or from carrying the protection of economically weak industries, regions, or groups of people too far. There is no lack of indications that growth-retarding price distortions have developed on the product and factor markets of numerous countries for quite some time. In this regard, particular attention has been focused on the substantial increase in labour costs, on rising protection in international trade, and on the surge of unproductive transfers.

The labour costs hypothesis has been primarily propagated by the German Council of Economic Experts.[20] The Council has shown that during the first half of the 1970s, the development of labour costs no longer had a neutral effect on the level of production costs, hence in the medium term a need has developed for correction of the real wage position—real wages adjusted for labour productivity, capital costs, and the terms of trade.

In public discussion, the growth-retarding effects of recently intensified protectionism are frequently pointed out.[21] It is not only within the European Community that there are attempts to extend the model of agricultural protection to other industries, such as raw materials, textiles, steel, or shipbuilding.

Exponents of the government consumption hypothesis attribute sluggish economic growth at least in part to increasing bureaucratic paralysis of entrepreneurial activity, which is thought to have proliferated in the 1970s.[22] In particular, bureaucratic forms of organisation have increasingly spread into fields of economic activity in which the market constitutes the more efficient form of organisation.

The redistribution hypothesis has two variants: the first relates to the burden of direct taxation on private enterprises and households; the second points to public payments to households that are unrelated to output (transfers). Both lower incentives in the private sector.[23]

Undoubtedly, these factors are not the only determinants of investment and growth. In addition, the hypothesised direction of causation is subject to controversy.

With respect to wage increases, it is argued that an aggressive wage policy stimulates investment and promotes growth through an induced increase in labour productivity. Without doubt, wage increases that lower the return on capital will in general induce an increase in labour productivity. This effect primarily occurs because more labour is replaced by capital than would have been the case without the wage increase. While the induced rationalisation measures ensure that the level of production costs is lower than immediately after the wage increase, the original cost level is not reattained:[24] profits are reduced, marginal supply does not appear on the market, additional unemployment results.[25] Therefore, it would be erroneous to expect more rapid economic growth from an aggressive wage policy.

The government consumption hypothesis will fail to find adherents among those who propound an expansion of public production of goods and services, and who see in public planning an escape from the "chaos of the market". Admittedly, the long-run relationship between government activity and economic development is difficult to evaluate, because no operational measurement concept like the real wage position is available. Economic development can certainly be promoted by improvements in the legal framework and a complementary supply of public goods.[26] The issue, however, is whether the government does not have a tendency to overdo it in legislation and in production. Indeed, governments may create and increase differences between private and social returns, rather than restricting themselves to removing these. Thus, government can crowd out private initiative through administrative rules, fiscal measures and (other) discriminatory interventions. Examples abound.[27]

As far as the protection argument is concerned, one might claim that protection raises the expected profits of weak industries and, hence, their investment activities. The other side of the coin is, however, that the industries that are actually growing have to pay the costs of the protective measures, directly or indirectly. The negative effect on economic growth occurs through the shifting of production factors from relatively productive to relatively unproductive activities.[28]

It is hardly controversial that incentives are reduced when private enterprises and households are burdened with wage and profit taxes. On the other hand, unrequited personal transfers may improve the conditions for economic growth, eg if these transfers serve to secure internal peace. However, transfers may well go beyond the point where the marginal contribution to improving macro-economic performance is outweighed by the disincentive effects on those who have to pay.

In spite of such objections, it therefore seems plausible to posit that long-run economic development is primarily determined by the development of incentives as we have described. However, in order to arrive at an explanation for long waves, one must go one step further.

The basic problem of every theory of long swings consists of explaining why the agents in the model react to changes in market signals only with a long lag.[29] Starting from the observation that common interest organisations (organised special interest groups) are a shared feature of economies, the basic idea behind our approach is that institutions—in contrast to individuals—react slowly and sluggishly. It seems to us that this is the actual cause of long-run fluctuations in growth.

The hypothesis of institutional sluggishness in reacting to market signals can be justified on various grounds. One cause lies in the life cycle of organisations. If the need for a collective representation of individual interests arises, it takes time—not least because of the free rider problem—until the common interests are organised. Further time passes until new organised interest groups have penetrated into the political arena against the resistance of already established interests. Once they have gained their positions, they will defend them as long as possible, even if the original interests have ceased to exist.

A further cause for the sluggishness of public and private institutions can be seen in the nature of bureaucracy. Organisations have a vested interest in expanding their field of activity as far as possible. The older and larger organisations become, the longer and more tenuous the communication channels between the organised and the organisations, whether because the individual no longer counts for much, or because of "bureausclerosis".

Enduring, counterproductive behaviour of organisations, particularly of public administration, may also be based on a self-escalating process of societal misinterpretation of market signals, ensuing (counterproductive) intervention and renewed misinterpretation of ever more ominous market signals.[30]

A diagrammatical representation of the basic structure of this mechanism might clarify our approach (Figure 11.1). The right-hand side of the figure describes the part of the mechanism coming from organisations with an

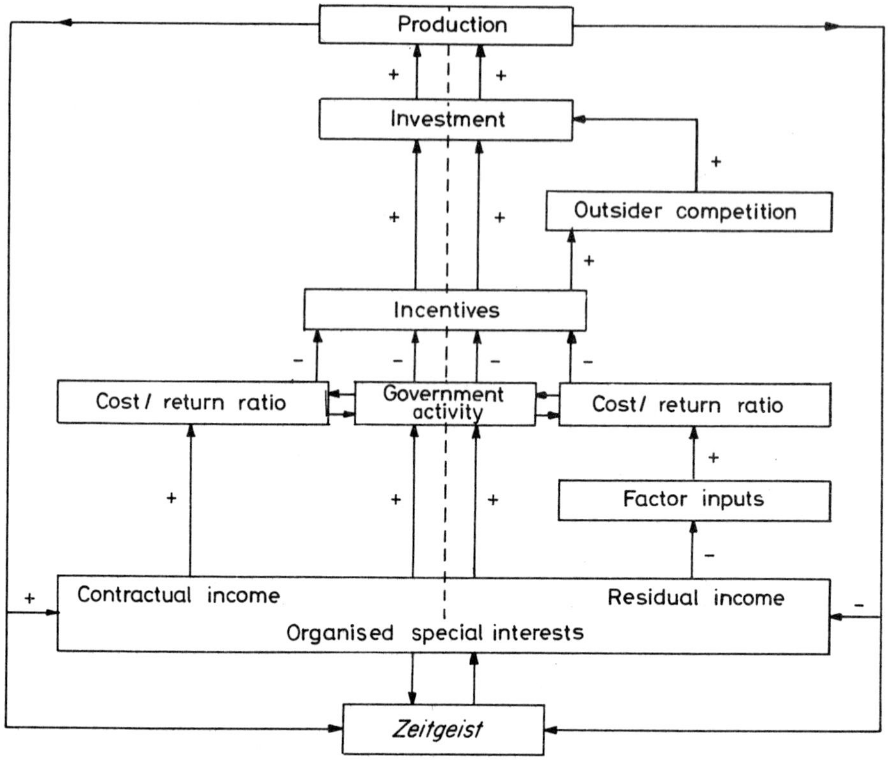

Figure 11.1. On the theory of long waves in economic development

interest in raising product prices or reducing factor prices; the left-hand side refers to organisations interested in low product prices, high factor prices, or transfer payments. The first set will usually be composed of business interests; the second set will contain organisations of employees or pensioners and the like. Adjustment processes occur slowly insofar as they depend on collective decisions, quickly insofar as they depend on individual behaviour. The diagram is restricted to the basic principles of the long-term interrelationships. Thus, foreign economic relations, for example, have not been explicitly shown, although they play a role at all the levels mentioned. Also, the existence of feed-back effects between all the elements of the scheme can be surmised. The interrelationships between government activity and factor and product markets seem to be of particular importance, because significant retarding elements may be found above all in government influence. This holds eg if firms call for—and receive—public assistance when competitive pressures rise. On the other hand, labour unions might attempt to pass on increased taxes in collective bargaining agreements. The scheme can be described as follows.

Let a disturbance impinge on the initially existing equilibrium growth path (all public and private organisations that are required in such a situation, or are compatible with it, exist). The disturbance may appear on the level of the "organised special interests", perhaps because certain groups have gradually acquired monopoly power. This will have consequences on factor markets (in the form of persistent upward cost pressure, for example), on product markets (in the form of supply restrictions, for example), or in the field of governmental activity (in the form of subsidies, for example). In whatever way the tendencies toward monopolisation change, from an overall economic point of view incentives will change.

Insofar as incentives deteriorate (as in the case of persisting upward cost pressure), investment activity abates and production declines (relatively). If, on top of that, government gives in to the demands of organised interests and voters for compensating action—such as partial or sectoral employment guarantees, subsidisation of private activities, or expansion of social services—it will often aggravate the overall economic decline and intensify the downswing. The longer the contraction continues, the more those interest groups that were the original source of the disturbance lose power. This may result from the progressive circumscription of the interest groups' financial resources in the course of the economic decline, or because members more and more lose their trust in and their loyalty to their organisations. Thus, it would appear to be in the organisations' self-interest to learn from the experience of the contraction and restrict the struggle over distribution. Changes in societal beliefs—the *Zeitgeist*—push in the same direction. These can even end the decline on their own if in the course of revolutionary developments the received social framework is burst asunder.[31] Removal of the original disturbance tends to reduce costs relative to returns for the producers.

To the degree that individual economic agents' profit expectations are raised (in the short run) by cartelisation of important product markets, a long-run contraction can be initiated as well. The cause for this may be seen in the tendency of a cartel to opt for the beaten track internally, and to discriminate against outside producers (and individual cartel members) by securing its position domestically and abroad. User industries in particular are weakened by cartel prices. These, in turn, will combine in order to receive protection against outside competition. The same complaints can be expected from the cartels' supplying industries as their sales decline, as well as from all others who have to finance the protection. Supply reductions spread throughout the economy; the production structure becomes rigid. In the long run, the cumulative group egoism on the entrepreneurial side that drives the economy into the downswing is broken by the always existing, steadily growing danger of outsiders. Since no one wants to be thrown out of business, outsider competition leads to a renewed increase in investment and production.

In both cases a new upswing begins. With the increasing duration of the upswing, more and more scarcities appear on product and factor markets, which encourage monopolistic tendencies. The same social and institutional inertia that was effective during the downswing now prevents interest groups from reacting to their newly acquired power by rapidly changing their behaviour, which is still marked by the last crisis.[32]

Our explanatory approach is basically endogenous, as was Schumpeter's, even if exogenous disturbances are not ruled out. Nevertheless, two differentiating characteristics bear emphasis. First, while we maintain that labour cost developments and government activities constitute the central agents in long waves, according to Schumpeter[33] they contributed to economic decline after World War 1 but were more the result than the cause of long-run fluctuations. The central element in Schumpeter's explanatory approach is the systematic cyclical clustering of innovations and imitations, which leads to a rise and fall of pioneers' profits. Second, while we certainly are far from denying the growth- and employment-stimulating effects of innovations and imitations, innovations randomly distributed over time are equally compatible with our approach as are clusters. Innovations may be randomly distributed if the—according to Schumpeter—"often irrational" entrepreneurial pioneers act out of a creative drive (ie independently of economic incentives). That would differentiate them from imitators, who try new ways only when investment conditions have fundamentally improved. In our model, on the other hand, clusters of innovations could be explained by the accumulation of a stock of unused knowledge during the declining phase of the long-run cycle. This gets economically used only when the distributional struggle has eased off, and the government once more concentrates on its complementary role. But, as mentioned earlier, clustering of innovations is difficult to detect.

International comparison of growth and investment fluctuations

The development of national product

The long-run development of national product can be described by a growth path beset by wave-like movements of various frequencies and amplitudes. Calculations of the deviations of smoothed national product series from the long-run trend reveal that in Germany and in other countries there have been long-run fluctuations in the development of national product. To be sure, these fluctuations varied in intensity from country to country (Figure 11.2). The wave pattern of economic development can also be seen if one looks at rates of change in national product (Figure 11.3).[34] In these calculations the war years and early post-war years were omitted in the case of the two world wars. The hypothesis was that wars—independently of the general development pattern—lead to economic setbacks, and that in the aftermath it takes a few years to regain pre-war development.[35]

If one includes the war years and post-war years as far as data permit, no serious changes in the cyclical pattern emerge (Figure 11.4). Nonetheless, it is noteworthy that

- in all countries, cyclical peaks are shifted;
- in the case of France, the United Kingdom, Italy, and the United States, the amplitude of the waves increases;
- in the case of Germany, an additional cyclical trough appears in 1950.

Even the last observation does not seriously cast doubt on the findings thus far, since the path of national product in the other countries shows that this trough is more a disturbance in the growth pattern that reflects the war-associated production losses since the middle of 1944.

Kondratiev's dating of the long waves[36]—which Schumpeter termed "industrial Kondratiev", "bourgeois Kondratiev", and "neomercantile Kondratiev"—is approximately supported by the present results where the sample periods coincide.[37] The analysis also shows that during the coinciding sample periods, international developments in long-run economic ups and downs were in part synchronised.[38] However, international synchronisation

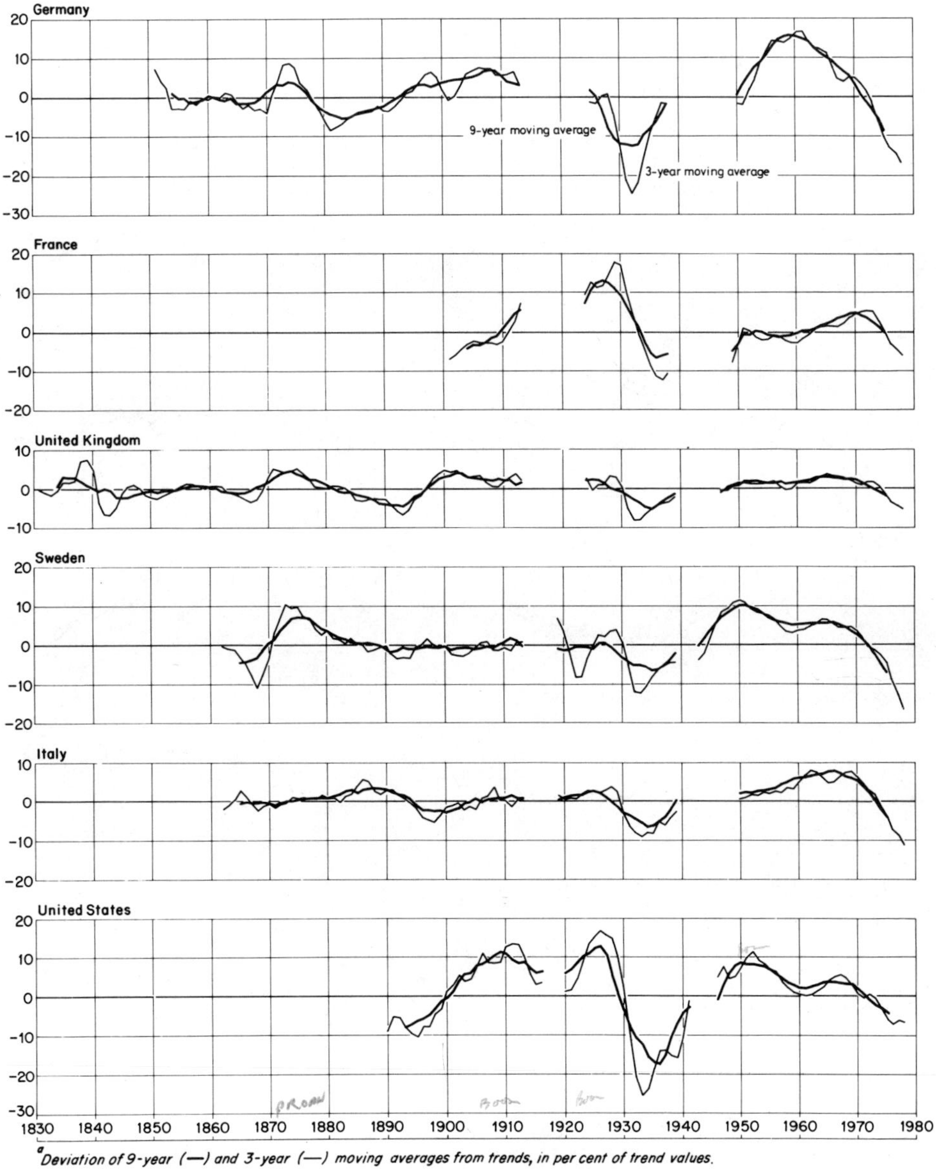

[a]*Deviation of 9-year (—) and 3-year (—) moving averages from trends, in per cent of trend values.*

Figure 11.2. An international comparison of long waves in aggregate output (excluding war years),[a] 1830–1979 (%)

Source: Kiel Institute for World Economics

was apparently not as close as Kondratiev and Schumpeter supposed. This is less true of Kondratiev's sample period (up to 1920) than for the period thereafter. Thus in France, the upswing of the neomercantile Kondratiev reaches far into the 1920s.[39] In the most recent Kondratiev, which one may perhaps call the neoclassical Kondratiev, significant international differences both in the beginnings and in the peaks are apparent. In all the countries,

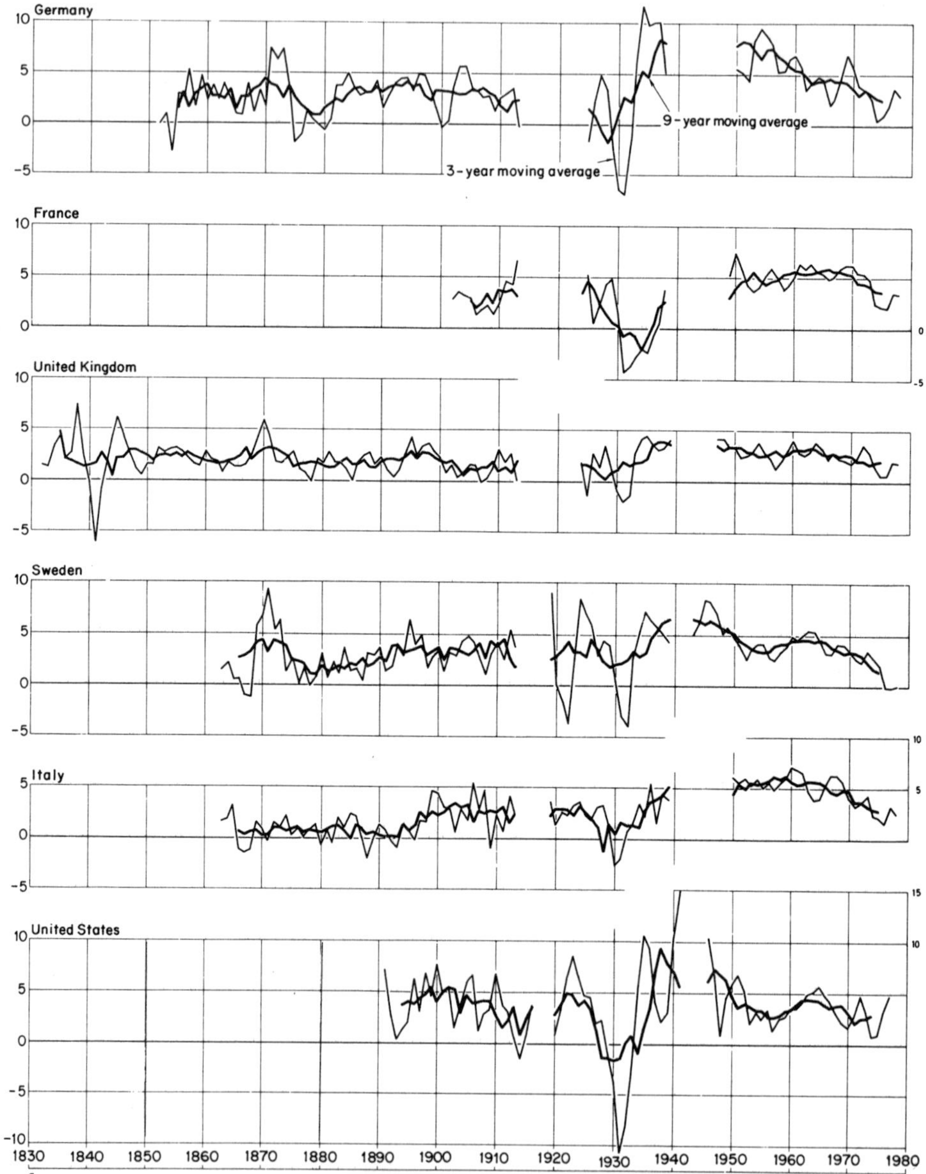

[a]*Annual rates of change of moving averages of national product.*

Figure 11.3. An international comparison of long waves in aggregate output: rates of change (excluding war years),[a] 1830–1979 (%)

Source: Kiel Institute for World Economics

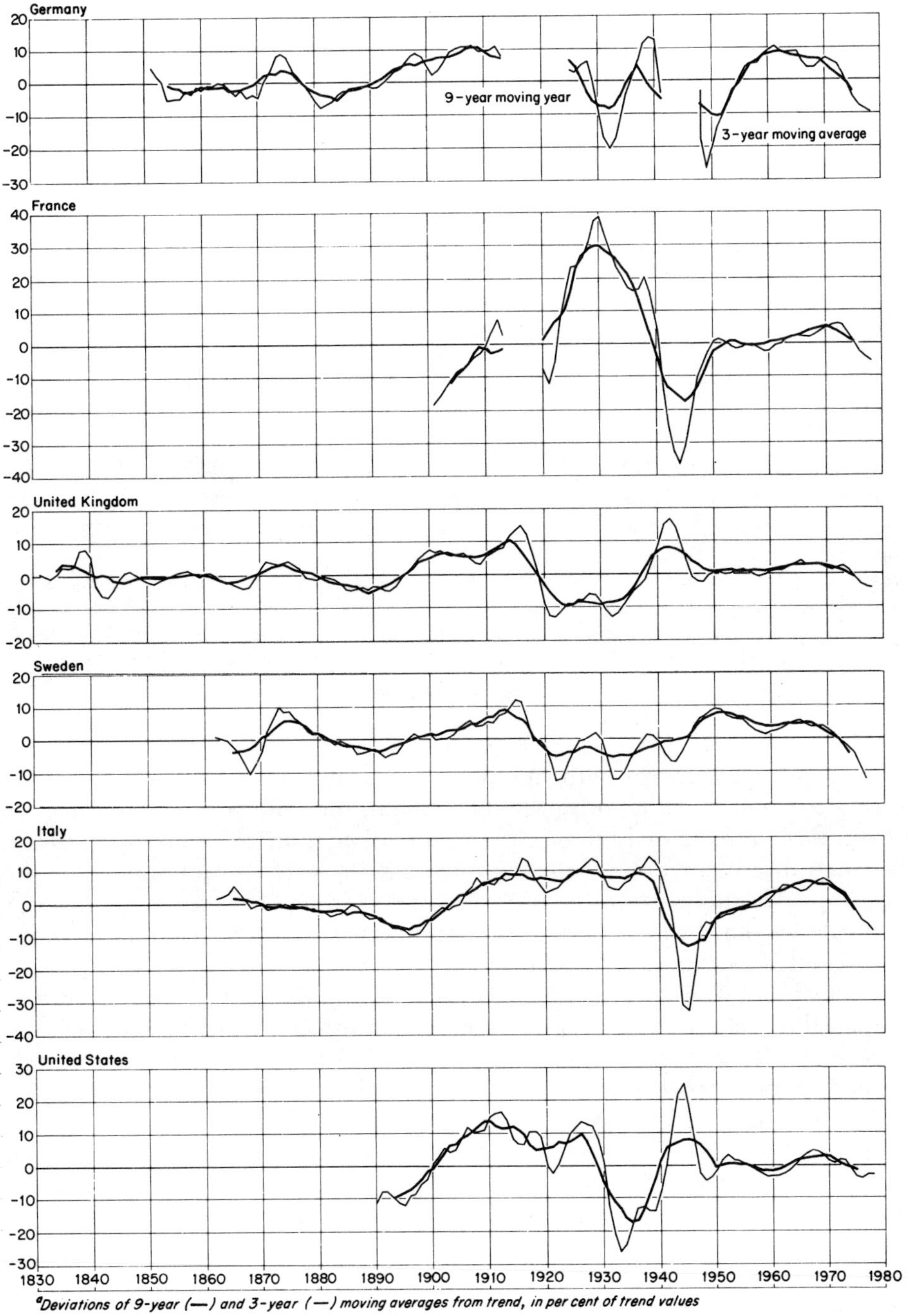

[a]*Deviations of 9-year (—) and 3-year (—) moving averages from trend, in per cent of trend values*

Figure 11.4. An international comparison of long waves in aggregate output (including war years),[a] 1830–1979 (%)

Source: Kiel Institute for World Economics

however, the present development cycle seems to have already passed its peak.

Within the individual cycles, economic expansion, both in international and in inter-temporal comparison, is rather diverse (see Table 11.1). Developments in Italy and the United Kingdom are particularly noteworthy.

TABLE 11.1. CYCLE-SPECIFIC GROWTH RATES OF AGGREGATE OUTPUT IN SELECTED COUNTRIES[a] (% p.a.)[b]

Country	Bourgeois Kondratiev			Neomercantile Kondratiev			Neoclassical Kondratiev		
	Beginning	End	%	Beginning	End	%	Beginning	until	%
Germany	(1854)	1882	2.6	1882	1932	2.0	1932	(1975)	4.0
France	–	–	–	(1905)	1938	1.1	1938	(1975)	3.7
United Kingdom	(1834)	1893	2.1	1893	1933	1.2	1933	(1975)	2.4
Sweden	(1864)	1892	2.6	1892	1934	2.8	1934	(1975)	4.0
Italy	(1864)	1899	0.8	1899	1934	1.9	1934	(1975)	3.6
United States	–	–	–	1893	1934	2.7	1934	(1975)	3.9

[a] The cyclical troughs were identified according to Figure 11.2. Dates in parentheses indicate uncertainty. [b] On the basis of 9-year moving averages of aggregate output series.

Source: Own calculations.

In Italy, economic growth apparently has increased from cycle to cycle. During the bourgeois Kondratiev, the Italian economy seems to have grown substantially more slowly than those of the other countries examined. Only in the two subsequent development cycles did the rate of economic growth in Italy approximately attain the international average. The cause for this might primarily be that broadly based industrialisation in Italy began later than in the United Kingdom or in Germany.

In the United Kingdom, the tempo of economic expansion never again reached the international average after the bourgeois Kondratiev. During the neomercantile and neoclassical Kondratievs, the United Kingdom relinquished its previous leadership more and more, and finally fell to one of the lower ranks among Western industrial countries.

These observations might be taken as an indication that economically backward countries can avail themselves of relatively substantial scope for economic growth, and that economically advanced countries face great difficulties in maintaining their lead. However, catching up processes on the part of less developed countries probably cannot be exclusively explained by the scope for imitation on the part of laggards and with symptoms of ageing on the part of the advanced countries. Otherwise, the United Kingdom, for example, would again have experienced growth rates more closely corresponding to the international average once it had substantially fallen behind other countries in the level of economic development.

The hypothesis of synchronous economic development among market economies is not completely rejected by the present observations, since there in fact has been some long-term synchronisation. On the other hand, Kondratiev's claim that long-term fluctuations in the development of national product are a phenomenon confined to capitalist systems is dubious. Although no comparable analyses can be carried out for the socialist countries

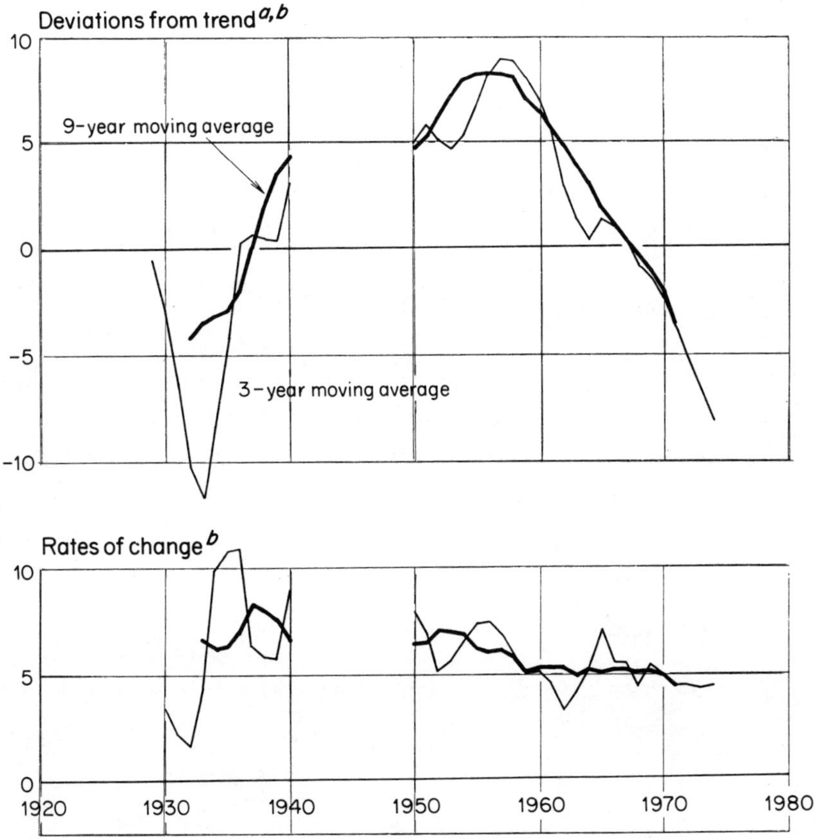

[a] *In per cent of trend values.* [b] *1937 prices*

Figure 11.5. Development of national product in the Soviet Union (excluding war years), 1928–1975 (%)

Source: Kiel Institute for World Economics

because they have not yet existed sufficiently long, the example of the Soviet Union also shows persistent acceleration and deceleration of economic growth (Figure 11.5).

Investment and employment

During the sample period, investment activity has been subject to fluctuations similar to those of national product, whether one measures investment activity as the share of investment in national product (Figure 11.6)[40] or as

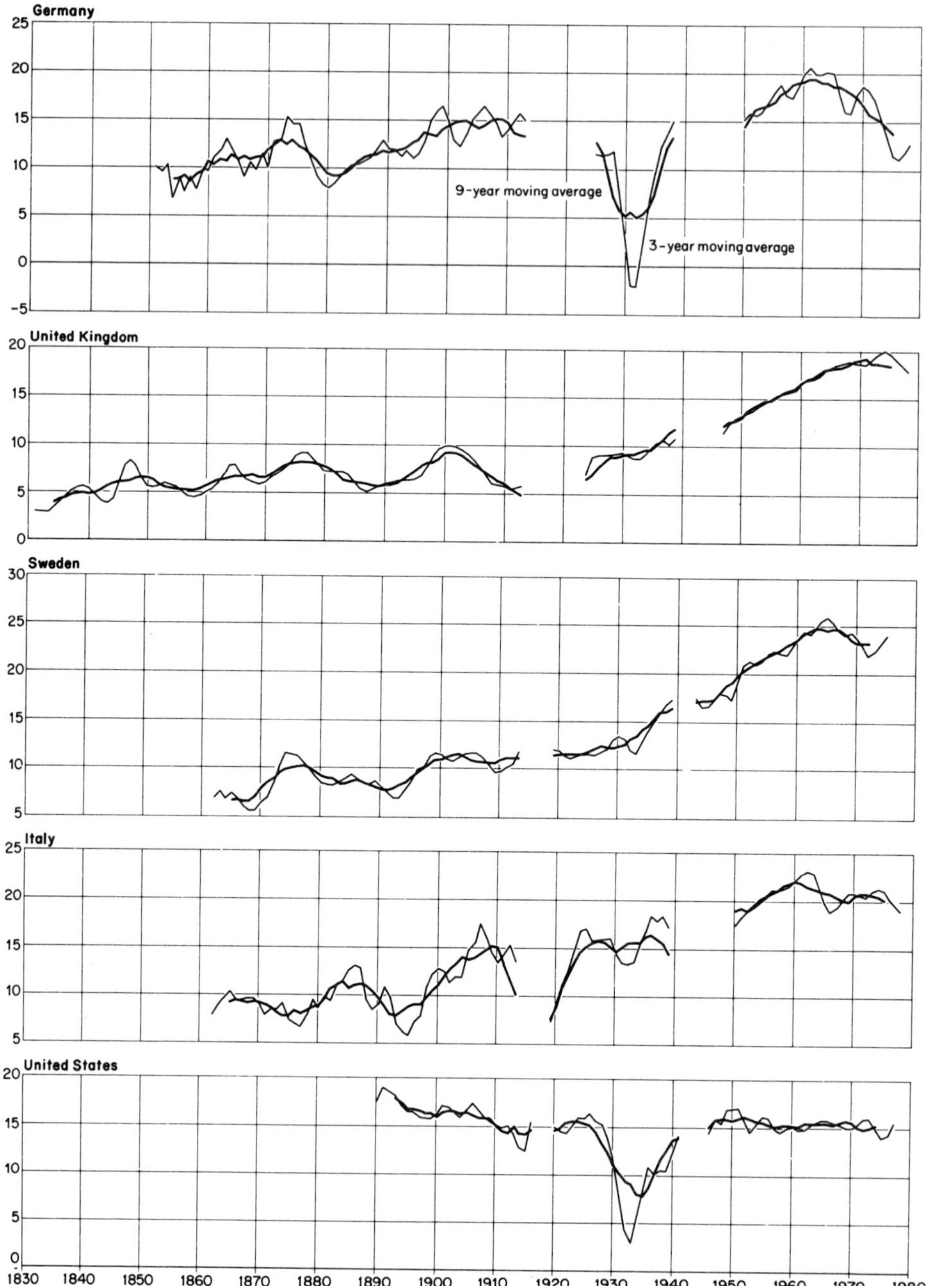

[a] *3 and 9-year moving averages of nominal investment shares*

Figure 11.6. An international comparison of long waves in investment activity: investment shares (excluding war years),[a] 1830–1979 (%)

Source: Kiel Institute for World Economics

deviations of real investment from its long-term trend.[41] It is true, however, that the relationship between economic growth and investment activity has not always been equally pronounced. In a few cases, peaks in investment lag peaks in national product. This result is not implausible, as capital productivities are unstable over time.

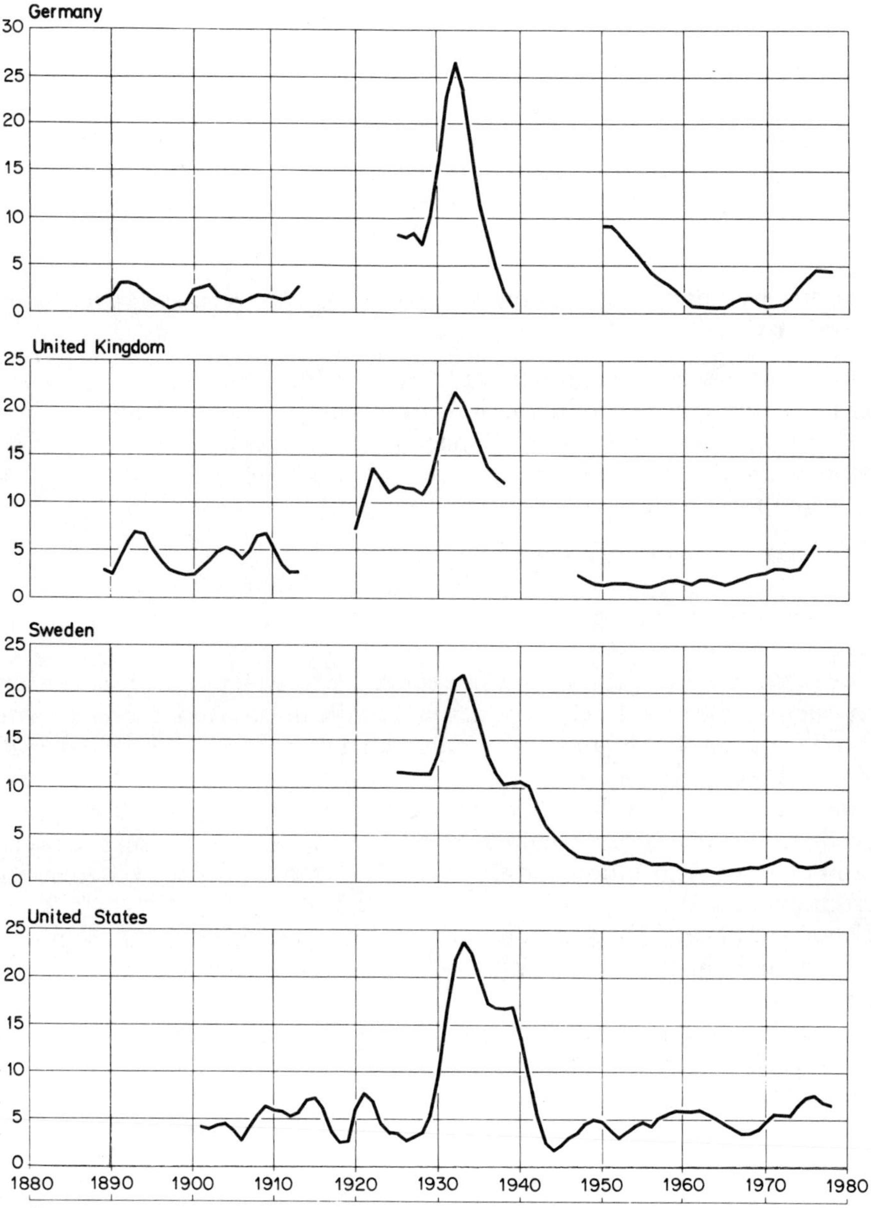

[a] *3-year moving averages of the unemployment rate.*

Figure 11.7. Development of unemployment rates in selected industrial countries,[a] 1887–1979 (%)

Source: Kiel Institute for World Economics

Except for the United States, the long-term trend of the share of investment in national product is upward in all countries. In the case of the United States, however, it should be noted that for statistical reasons only private investment could be drawn upon.

Finally, the present observations suggest that, in the long run, the rate of employment and growth cycles correspond rather well to each other (Figure 11.7).[42] It is noticeable, however, that during the crisis of the early 1930s the connection between national product and unemployment is apparently tighter than at the following peak. This asymmetry could be based on a tendency for labour temporarily to receive less than its marginal product during a secular upswing, even after reaching full employment. Higher wage increases than is consistent with stable costs threaten full employment only when real wages begin to exceed the marginal product of labour.

Economic performance and the struggle over distribution: the example of Germany

Germany can serve as an example with which empirically to test the hypothesis that long-term fluctuations in incentives cause the observed cycles in investment and growth. In this analysis, the following proxies will serve as indicators for the development of expected profit, of government consumption, of the tax burden and redistribution, and of wage developments:

- The share of undistributed profits in net non-wage income of firms is to indicate expected profits (ie the investment climate).[43] Insofar as data on undistributed profits are unavailable, but data on total profits are available, the latter are used.
- The development of the share of direct taxes, as well as the share of transfer payments, in net national product is to indicate to what degree incentives faced by households and firms have been strengthened or weakened by government activity.
- The development of the share of government consumption in net national product is to indicate whether private initiative has been restricted or promoted by legislation, administration, or other types of government production.
- The share of imports plus exports in net national product is to show how protectionist measures have affected the degree of Germany's international integration.
- Finally, the development of the real wage position of employees is to show whether positive or negative incentives for investment activity have emanated from wage developments.[44]

Expected profits and investment activity

A comparison between total profits and undistributed profits (Figure 11.8) shows that the two indicators are nearly synchronous[54]. The pronounced cyclic pattern is remarkable, and it generally corresponds to that of investment activity: upswings and downswings coincide; troughs and peaks occur in

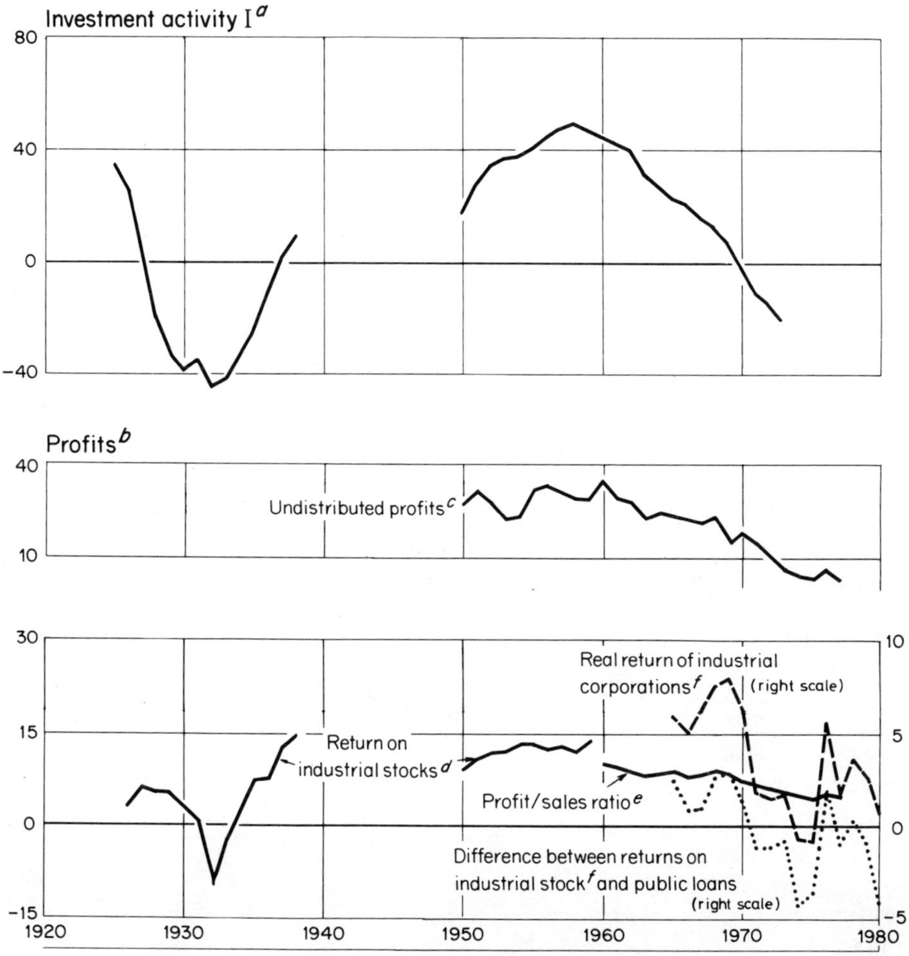

[a]Deviations of a 9-year moving average of real net investment from trend, in per cent of trend values. [b]Changing territory. [c]Undistributed profits in per cent of net non-wage income. [d]According to Hoffman. [e]In per cent of production. Estimated from Sachverständigenrat. [f]According to Institut der deutschen Wirtschaft, Cologne.

Figure 11.8. Expected profits and investment in Germany, 1850–1980 (%)

Source: Kiel Institute for World Economics

the same year. The long-run relationship between growth, investment, and expected profits is clearly visible. The development of the difference between the rate of return of industrial corporations and the interest rate on government bonds[46] points to increasing business risks in the 1970s, and may be regarded as additional evidence of a retardation of profit expectations in the 1970s as compared to the 1960s.

The role of protection

If one looks at the degree of Germany's international economic integration, which can be traced back to 1880, the cyclic pattern appears only relatively

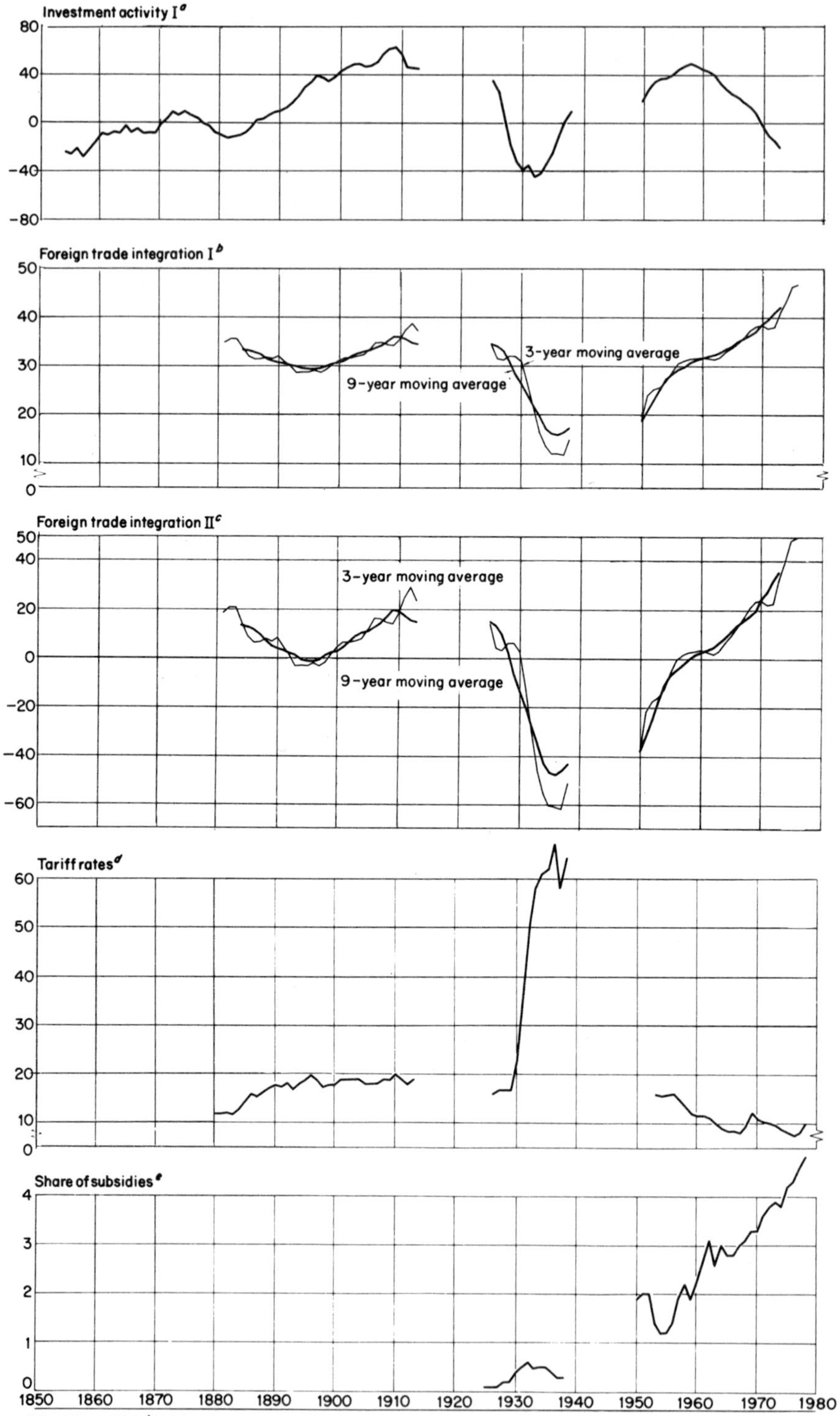

[a]See Figure 11.8. [b]Total trade in per cent of NNP, in current prices; changing territory. [c]Deviations of moving averages from trend, in per cent of trend values. [d]Tariff revenue in per cent of dutiable imports. [e]Subsidies in per cent of NNP, changing territory

Figure 11.9. Protection and investment in Germany, 1850–1979 (%)

Source: Kiel Institute for World Economics

indistinctly (Figure 11.9). Nonetheless, the year 1896 could well have been a trough in international economic integration.[47] The next trough shows up at the end of the 1930s.[48] Actually, the degree of international economic integration would seem to have fallen further in the years after 1938. In the post-war years, a new cycle began which apparently has not yet run its course. It is not even clear whether this cycle has passed its peak yet.[49] The pronounced increase in foreign trade intensity observable in 1974 is presumably due less to increased international integration and more to price increases for crude oil and other raw materials. Indeed, since 1974, foreign trade protectionism among the industrial countries has increased.[50]

Comparing the development of the degree of integration on the one hand, and the investment and national production cycles on the other, the following interpretation suggests itself:

- The expansion gets underway even though international economic integration is still declining further. Only after the upswing has lasted a number of years does liberalisation set in, and the degree of integration increases.
- The contraction sets in, although integration into the international divisions of labour still continues for a longer period. Only after the downswing has lasted a number of years and has penetrated into public awareness does the degree of integration decline in response to rising protectionism.

Since economic upheavals occur in spite of countercyclical foreign trade policy, one can infer that investment and growth cycles are not caused by cycles in protection and integration. Rather, it appears that integration cycles are a result of investment and growth cycles: once the boom has come to fruition (and unemployment has been eliminated), the forces in favour of a more liberal stance on foreign trade have increasing influence, and eventually gain the upper hand. As a result, the dynamics of the upswing are reinforced. The situation during the contraction is reversed: if, through rising unemployment, the public becomes generally aware of an enduring deterioration of the economic situation, the conviction that additional protective measures are an appropriate antidote to the crisis gains ground. During the 1920s and early 1930s this view led to international competition in protectionism, which substantially accelerated the economic decline.[51] At present, too, in numerous Western countries—particularly in the countries of the European Community—the belief that national economic difficulties can be overcome by additional protection is gaining ground.

The role of government consumption

In Figure 11.10 long-term development cycles in government consumption are evident. This applies both to the moving averages of the government share, as well as to the corresponding detrended values.

Comparing the cycles in government consumption with the cycles in investment activity and national products, it is noticeable that the cycles generally mirror each other. Investment activity and national product always approximately reach a cyclical peak (trough) when government consumption is at a cyclical trough (peak). It is remarkable that cycles in government consumption always begin before cycles in real investment and national product. The lag relationship between government consumption and the development of investment and national product is not quite as pronounced during the contraction.

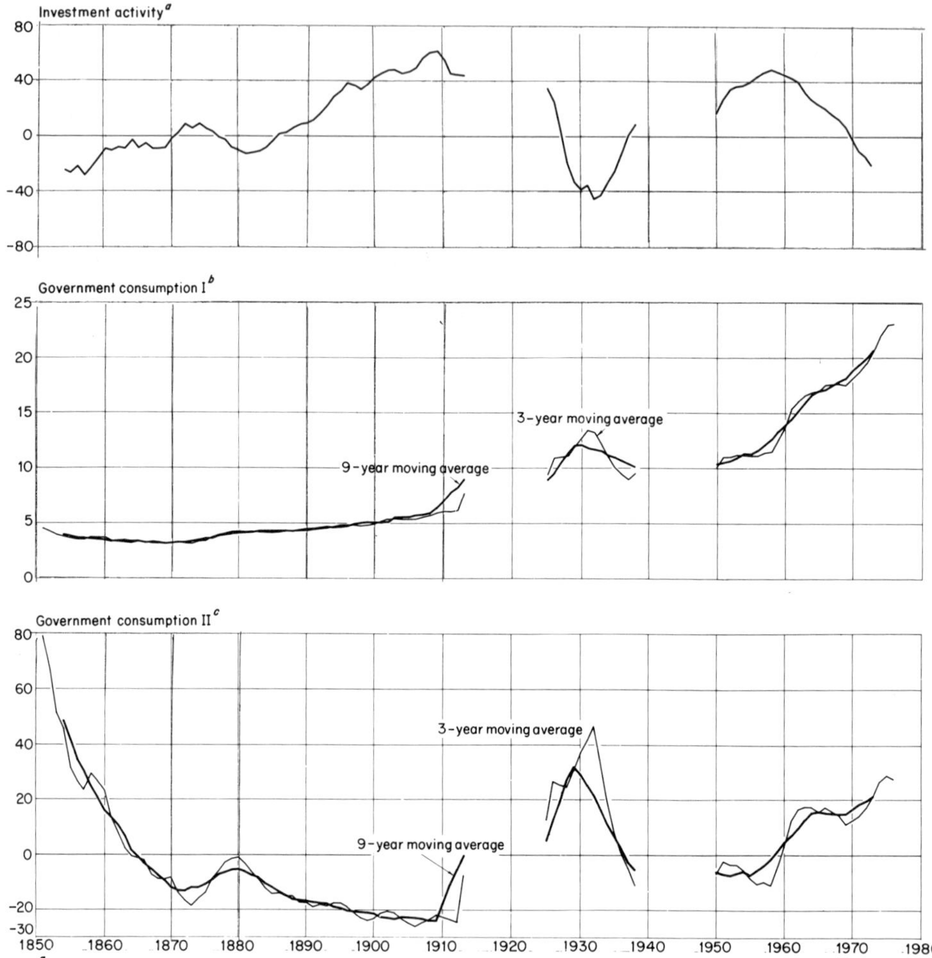

[a]Deviations of 9-year moving average, of real net investment from trend, in per cent of trend values; territory of the Federal Republic of Germany, 1961. [b]Civilian government expenditures in per cent of NNP, current prices. [c]Deviations of moving averages of investment share in NNP, in per cent of trend values, current prices; changing territory

Figure 11.10. Government consumption and investment activity in Germany, 1850–1977 (%)

These results indicate that there is indeed a relationship between the development of public consumption and the development of investment and national product. The observation that a long-term (relative) decline in government consumption is regularly accompanied by a longer-run prosperity phase (and vice versa) suggests that, on average, government has done more in the field of legislation and in the production of public goods than would have been advantageous from the point of view of economic growth.

Otherwise, at least occasionally a reversed sequence between the development of government consumption and national product would have to have shown up. Against this background, it becomes plausible that (relatively) increasing government consumption weakens investment by cutting expected profits, and reduces economic welfare, and that in the opposite case of reduced government consumption economic growth is promoted.[52] Of course, this does not exclude the possibility that it might be growth promoting if the production of complementary goods by government increases more rapidly than private sector output.

The role of redistribution and the tax burden

Figure 11.11 shows that for redistribution, which can be traced back to 1925, persistent fluctuations have also occurred that mirror the development of national product. In accord with expectations, redistribution increased until the trough of the neomercantile Kondratiev and then declined. Developments since 1950 do fit into the picture insofar as the share of redistribution, which shows a positive trend, initially remained stable, and later rose substantially. Of course, the overall extent of redistribution was far greater after World War 2 than before. The need to pay transfers to those who had sustained injuries, war widows, war orphans as well as other war victims led to substantial transfers that scarcely had disincentive effects. For natural causes the share of war-related transfers in the national product has continuously declined. Performance-reducing transfers could be increasingly substituted for neutral transfers, without requiring an increase in the share of redistribution. It may

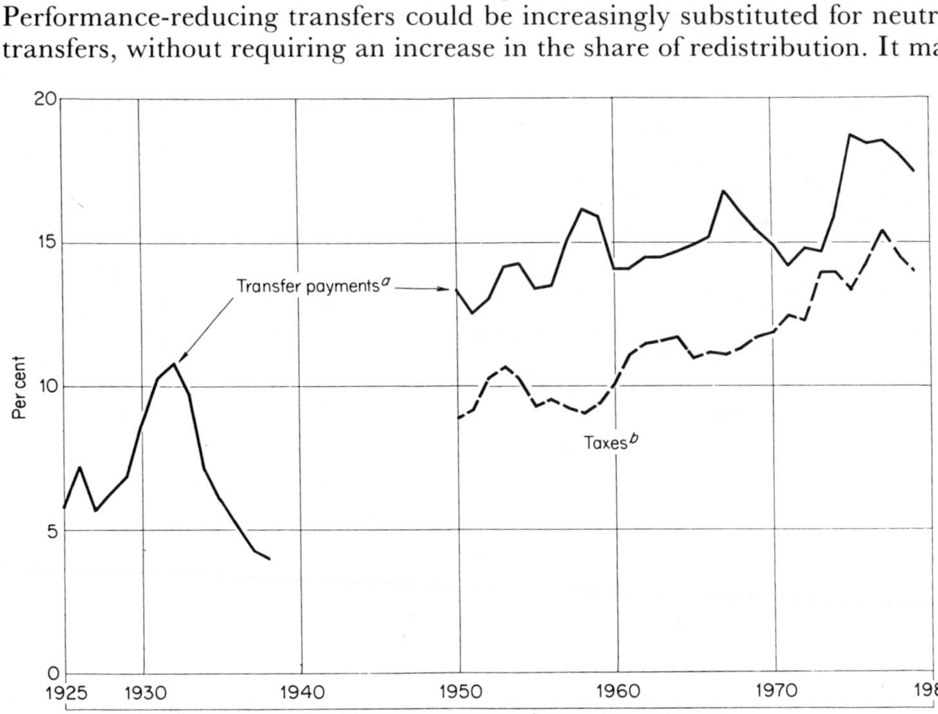

[a]*Government transfer payments to individuals in per cent of NNP.* [b]*Direct taxes in per cent of NNP.*

Figure 11.11. Redistribution and taxation in Germany, 1925–1979 (%)

redistribution. It may be due to this phenomenon that the share of redistribution has increased substantially only well after the peak in national product.

The relationship between direct taxes and performance in the private sector could be examined only as of 1950. Figure 11.11 shows that the burden of taxation has noticeably increased since the end of the 1950s. Given a constant progressive tax rate structure, the increase in the tax share has slowed in times of reduced economic growth. It follows from the reverse development in Figure 11.11 that the burden of direct taxation has increased.

The role of wages

The degree to which wage developments have contributed to the emergence of investment and growth cycles will be analysed with the help of the real wage position concept. However, data with which to implement this concept were only available as of 1925.

Figure 11.12 shows that the real wage position,[53] too, was subject to pronounced fluctuations during the period 1925 to 1977. While no complete cycle can be identified during the sample period, the year 1929 appears to mark an upper turning point of wage development. The wage cycle beginning in 1929 has apparently not yet run its course. It seems to have passed through its trough during the years 1955–58.

Comparing the cyclic pattern of wages with the cycles in investment activity and national product, a mirror image again emerges, except for a short phase displacement:[54]

- A long-run deterioration in the real wage position is accompanied by a long-run economic upswing and, with a certain time lag, a reduction in unemployment.
- A long-run improvement of the real wage position is accompanied by a long-run economic decline and, with a lag, an increase in unemployment.

These results suggest that, at least during the period from 1925 to the present, wages were a decisive determinant of profit developments and thereby contributed to the emergence and course of investment and growth cycles.

The degree to which the explanatory approach attempted for Germany is sound for other countries as well, and the nature of the mechanism transmitting fluctuations in economic growth internationally, remain open questions. As far as the second issue is concerned, it is conceivable that synchronised ups and downs in economic activity are the result of values changing from generation to generation, of the *Zeitgeist* spreading worldwide. The cyclical synchronisation in the countries examined would then be based on a cultural synchronisation, which makes itself felt economically in evolving investment and growth conditions. Our analyses are restricted to the description of an economic mechanism. A key to a more basic explanation of the phenomena described might be found in drawing on Pareto's theories of collective monopolies and of the circulation of elites.

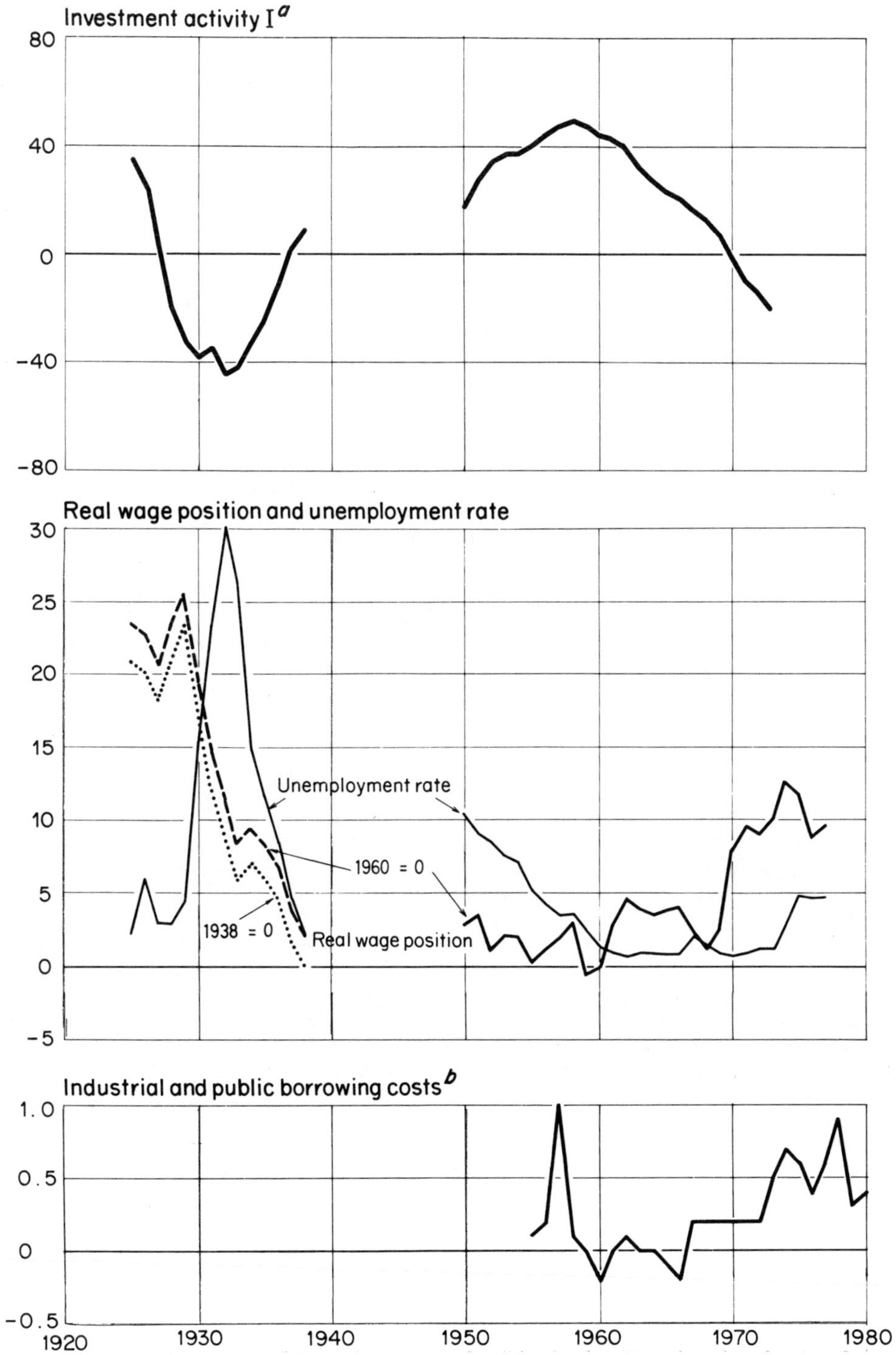

[a]see Figure 11.8. [b]Interest rate differential between private and public loans.

Figure 11.12. Real wage position and investment activity in Germany, 1925–1980 (%)

Source: Kiel Institute for World Economics

Notes and references

1. This paper is essentially an English version of two earlier publications in German. In the first one (Hans H. Glismann, Horst Rodemer, Frank Wolter, *Zur Natur der Wachstumsschwäche in der Bundesrepublik Deutschland—Eine empirische Analyse langer Zyklen wirtschaftlicher Entwicklung*, Institut für Weltwirtschaft, Kiel Discussion Paper No. 55, June 1978) we tried to throw some light on the causes of Germany's slow economic growth in the 1970s. The second (Hans H. Glismann, Horst Rodemer, Frank Wolter, *Lange Wellen wirtschaftlichen Wachstums—Replik und Weiterführung*, Institut für Weltwirtschaft, Kiel Discussion Paper No. 74, December 1980) was an answer to some critical comments received. This English version, however, has been extended to include among other things the influence of the public sector in Germany. Some details relating to German developments and to statistical procedures have been omitted. For statistical and methodological information, the interested reader is referred to Hans H. Glismann, Horst Rodemer, Frank Wolter, *Zur empirischen Analyse langer Zyklen wirtschaftlicher Entwicklung in Deutschland—Datenbasis und Berechnungsmethoden* (Institut für Weltwirtschaft, Kiel Working Paper No. 72, 1978).
2. Joseph A. Schumpeter, *Business Cycles*, 2 volumes (New York and London, McGraw-Hill, 1939), pages 161ff.
3. Nikolai D. Kondratiev, "Die langen Wellen der Konjunktur", *Archiv für Sozialwissenschaft und Sozialpolitik* (Tübingen), 1926, *56*, pages 573–609; Nikolai D. Kondratiev, "Die Preisdynamik der industriellen und landwirtschaftlichen Waren", *Archiv für Sozialwissenschaft und Sozialpolitik* (Tübingen), 1928, *60*, pages 1–85.
4. Vilfredo Pareto, "Alcuni relazioni fra la stato sociale e la variazoni della prosperita economica", *Rivista Italiana di Sociologia*, September–December 1931, pages 501–548.. Quoted in Ulrich Weinstock, *Das Problem der Kondratieff-Zyklen* (Berlin, Munich, 1964).
5. Jan van Gelderen, "Springvloed: beschouwingen over industrieele ontwikkeling en prijsbeweging", *De Nieuwe Tijd*, 1913, *18*.
6. Kondratiev, "Die Preisdynamik . . .", reference 3, pages 37ff. In a different sense, Mandel can be counted among the supporters of a capital shortage hypothesis. While according to Kondratiev the long-term boom gets choked off by rising interest rates, according to Mandel the marginal productivity of capital falls during a long-run upswing. But during the long-term downswing, exogenous events, such as revolutions or gold discoveries, have to occur in order to start another upswing. See Ernest Mandel, "The heyday of neo-capitalism and its aftermath", *Socialist Register*, 1964; Ernest Mandel, "Explaining long waves of capitalist development", *Futures*, August 1981, *13*, 4, pages 332–338. Nonetheless, Mandel's explanation of the Kondratiev cycle is a Marxist one, "essentially based on long-term movements in the rate of profit determining . . . quicker and slower long-term paces in capital accumulation . . ." [Ernest Mandel, *Long Waves of Capitalist Development. The Marxist Interpretation* (Cambridge University Press, 1980) page vii]. Mandel wonders whether "subsequent attempts to 'marginalize' the role of profits and of capital accumulation, in favour of monetary, psychological, or purely inventive factors, will continue to be the rule . . ." [ibid, page viii].
7. Arthur Spiethoff, "Krisen", *Handwörterbuch der Sozialwissenschaften* (Jena), 1925, *6*, 4, pages 8–91. During the 1930s, Spiethoff had his hypothesis on cyclically occurring industrial overproduction tested through a number of industry studies.
8. The relationship between demographic cycles and cycles of economic development analysed by Lösch may also be termed an overproduction hypothesis: because the entrepreneurs overestimate future population growth during periods of rapid population growth, they put more capital goods in place than will actually be needed in the future. Such overproduction of investment goods, together with the unexpected labour shortage, leads to reduced profits and to bankruptcies as the decisive cause of economic fluctuations. See August Lösch, "Bevölkerungswellen und Wechsellagen. Erwiderung an Professor Johan Akerman", *Schmollers Jahrbuch* (Munich, Leipzig), 1937, *61*, pages 455–460.
9. Schumpeter, reference 2, chapter III. For a current interpretation of the innovation hypothesis see Gerhard Mensch, *Das technologische Patt. Innovationen überwinden die Depression* (Frankfurt/M., Umschau Verlag, 1975).
10. According to Schumpeter, the entrepreneur (who carries out innovations) must not be confused with the manager. The entrepreneur has recourse to greater flexibility and is

more prone to incur risk, differentiating him from the pure manager (bureaucrat) and from the owner of capital as well. See Schumpeter, reference 2, pages 102ff.

11. Jacob J. van Duijn, "The long wave in economic life", *De Economist* (Leiden), 1977, *125*, pages 544–576.
12. Jay W.Forrester, "New perspectives for growth over the next thirty years", in *Middle- and Long-Term Energy Policies and Alternatives*, 94th Congress, 2nd Session, Part 1, House of Representatives, Subcommittee on Energy and Power (Washington, 1976), pages 163–191.
13. Simon Kuznets, "Retardation of industrial growth", *Journal of Economic and Business History* (Cambridge, Mass.), 1929, *1*, pages 534–560.
14. Van Duijn calls this a "Schumpeterian shock". Whether the first concentrated appearance of entrepreneurial pioneers is a chance occurrence or results from the development process, is not discussed in his study.
15. In fact, empirical evidence, eg for the Federal Republic of Germany, on the longer-term relationship between orders and sales contradicts this hypothesis. See Bundesministerium für Wirtschaft, *Auftragseingang in wichtigen Industriezweigen in der Bundesrepublik* (Bonn), current issue; Statistisches Bundesamt, Fachserie D: Industrie und Handwer, series 9: *Auftragseingang in wichtigen Industriezweigen* (Stuttgart), current issue.
16. Schumpeter, reference 2, pages 252ff.
17. Such attempts have been made particularly by Mensch and Kleinknecht. See Gerhard Mensch, "Zur Dynamik des technischen Fortschritts", *Zeitschrift für Betriebswirtschaft* (Wiesbaden), 1971, *41*, pages 295–314; Mensch, reference 9; Alfred Kleinknecht, "Observations on the Schumpeterian swarming of innovation", *Futures*, August 1981, *13*, 4, pages 293–307. We can endorse most of the criticism of these investigations: see Klaus Brockhoff, "Zur Dynamik des technischen Fortschritts. Bemerkungen zu dem gleichnamigen Aufsatz von G. Mensch", *Zeitschrift für Betriebswirtschaft* (Wiesbaden), 1972, *42*, pages 283–291; John Clark, Christopher Freeman and Luc Soete, "Long waves and technological development in the 20th century", in D. Petzina and G. van Roon, eds, *Konjunktur, Krise, Gesellschaft: Wirtschaftliche Wechsellagen und sozialer Entwicklung im 19 und 20 Jahrhundert* (Stuttgart, Klett-Cotta, 1981), pages 132–169; Christopher Freeman, "Innovation as an engine of growth: retrospect and prospect", paper prepared for Conference on "Emerging Technology: Consequences for Economic Growth, Structural Change and Employment in Advanced Open Economies", Kiel, June 1981.
18. See Simon Kuznets, "Schumpeter's business cycles", *The American Economic Review* (Evanston), 1940/41, *30*, pages 257–271.
19. See Christopher Freeman, "The Kondratiev long waves, technical change and unemployment", in *Structural Determinants of Employment and Unemployment* (Paris, OECD, 1979), page 181.
20. Sachverständigenrat zur Begutachtung der gesamtwirtschaftlichen Entwicklung, *Jahresgutachten 1976/77. Zeit zum Investieren* (Stuttgart and Mainz, Kohlhammer-Verlag, 1976), No. 290.
21. Richard Blackhurst, Nicolas Marian and Jan Tumlir, *Trade Liberalization, Protectionism, and Interdependence*, GATT Studies in International Trade, No. 5 (Geneva, 1977).
22. Wolfram Engels, "Bürokraten verspielen unsere Zukunft", *Wirtschaftswoche*, 1977, *31*, 1/2, pages 60ff.
23. See Roland Vaubel, "Redistribution, social security and growth", in Herbert Giersch, ed., *Towards Explaining Economic Growth*, Symposium 1980 (Tübingen, Mohr-Verlag, 1981).
24. If this were the case, the substitution of capital for labour would have paid prior to the wage increase, and thus would have been carried out.
25. Indeed, without the rationalisation process, wage-induced unemployment would have been even greater.
26. Classical fields of activity in this connection are, for example, internal and external security or the administration of justice.
27. Wissenschaftlicher Beirat beim Bundeswirtschaftsministerium, Gutachten vom 8.1.1976, *Kosten und Preise öffentlicher Unternehmen* (Bonn, 1976); Monopolkommission, Erstes Zweijahresgutachten, *Mehr Wettbewerb ist möglich* (Cologne, Nomen Verlag, 1976); Jürgen

Müller and Ingo Vogelsang, *Zwischen Wettbewerb und Planung*, Lehren aus der amerikanischen Regulierung für öffentlich gebundene Unternehmen und wettbewerbliche Ausnahmebereiche (Berlin, Internationales Institut für Management und Verwaltung, 1971).

28. Such negative allocations and growth effects also occur when, in the absence of externalities or other market imperfections, growth industries are promoted at the expense of declining industries.

29. Until now, there have been only few theoretical approaches which permit the formulation of a falsifiable "long-wave hypothesis". Among them may be counted the theories of Kondratiev, Schumpeter, and Lösch. In the case of Spiethoff's, and still more Forrester's, overinvestment hypothesis, the specifically long-run approach is already hardly discernible. This also applies to approaches of Kaldor and Hicks, which are presumably intended to explain more short-term business-cycle phenomena. But Kaldor's approach—especially because it does not produce cycles artificially (such as by means of incorrect predictions on the part of entrepreneurs, or through technical constraints in the sense of Hicks)—could prove fruitful for the explanation of long waves: it claims that every upswing, once it has got going, must come to an end by itself because the overall capital intensity of the economy declines with the progressing upswing. According to Kaldor, this is because in the upswing real wages decline relative to interest rates. Therefore, investment occurs in relatively labour-intensive techniques. In consequence, rising real savings, accompanying rising real income, lead to deficient aggregate demand, as an additional increase in capital intensity does not compensate shortfalls in demand. Unless additional investment becomes increasingly labour saving, a labour shortage would soon appear. The marginal product of capital would decline, and the downswing would ensue. See especially Nicholas Kaldor, "Capital intensity and the trade cycle", *Economica*, February 1939, *VI*, 21, pages 40ff; "The trade cycle and capital intensity: a reply", *Economica*, February 1940, *III*, pages 16ff. John R. Hicks, *The Trade Cycle* (Oxford, New York, 1950).

30. See Ludwig Mises, "Interventionismus", *Archiv für Sozialwissenschaft und Sozialpolitik* (Tübingen), 1926, *56*, page 610.

31. Borchardt—admittedly without wanting to support the hypothesis of long cycles in economic development—attributes the decline of the German economy in the years after World War 1 to the fact that the German state "became far more than ever before a subsidization and redistribution state". The economic policy makers of the Weimar Republic "should have replaced the then apparently urgent issue of redistribution with what would have counted in the long run: the growth issue". However, according to Borchardt, at that time there was no leeway for such a solution. See Knut Borchardt, *Zwangslagen und Handlungsspielräume in der großen Wirtschaftskrise der frühen dreißiger Jahre: Zur Revision des überlieferten Geschichtsbildes*, Vortrag auf der Jahressitzung der Bayerischen Akademie der Wissenschaften (Munich, 2 December 1978), page 15; and "Trends, Zyklen, Strukturbrüche, Zufälle: Was bestimmt die deutsche Wirtschaftsgeschichte des 20. Jahrhunderts", *Vierteljahresheft für Sozial- und Wirtschaftsgeschichte* (Wiesbaden), 1977, *64*, 2, page 17.

32. For comparison, an attempt to represent key elements of Schumpeter's approach to long waves in a similar scheme might be as follows:

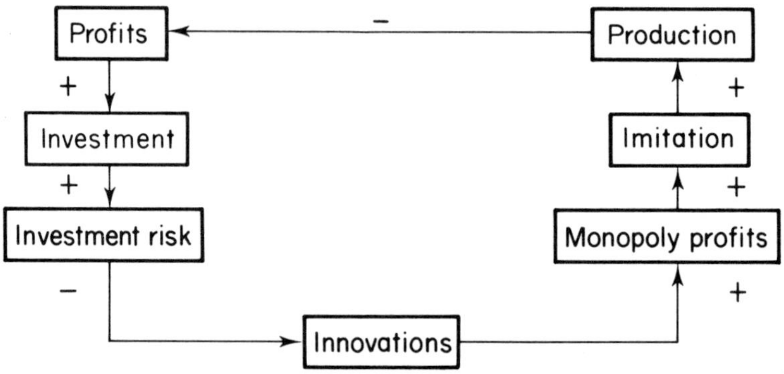

The signs at the transition from one level to the next should be understood multiplicatively. See Schumpeter, reference 2, pages 130ff.

33. Schumpeter, reference 2, page 708.
34. Since the rates of change are derivatives of the original series, upper and lower turning points are earlier than for national product.
35. In discussion on long swings in development, wars are often regarded as the cause of investment and growth fluctuations. In contrast to this position, the view is also taken that wars originate primarily in economic conditions. Whether they are a product of the upswing or the downswing is subject to disagreement, however. See Kondratiev, "Die langen Wellen . . .", reference 3. S. de Wolff, "Prosperitäts- und Depressionsperioden", in *Der lebendige Marxismus—Festgabe zum 70. Geburtstage von Karl Kautsky* (Jena, 1924); Karl Kautsky, "Krisentheorien", *Die Neue Zeit* (Stuttgart), 1902, *20*; Mensch, reference 9, pages 86ff. The empirical evidence for Germany does not suggest a systematic relationship between wars and growth cycles: for example, the wars of German unification in the 19th century fell (in Schumpeter's terminology) rather into the upswing of the bourgeois cycle; World War 1 fell into the declining phase of the neomercantile cycle and World War 2 fell rather into the upswing of the most recent development cycle. Accordingly, wars and their aftermath do not seem to constitute the cause or the consequence of long swings, but merely lead to disturbances of the general development pattern. This interpretation can also be found in Schumpeter, reference 2, pages 311ff. Of course, this does not mean that the origins of wars are unrelated to economic interests.

    It has been argued that the decline of economic growth in the Federal Republic of Germany is a return to a "normal growth path" after the post-war boom rather than part of a long-run cyclical movement. See Werner Abelshauser and Dietmar Petzina, "Krise und Rekonstruktion, Zur Interpretation der gesamtwirtschaftlichen Entwicklung Deutschlands im 20. Jahrhundert", report to the conference "Wachstumszyklen der deutschen Wirtschaft im 19. und 20. Jahrhundert. Ergebnisse, Methoden, Erklärungsansätze", Bielefeld, 25–27 May 1979.

    There can be no doubt, indeed, that World War 2 brought a massive break in economic development followed by a longer reconstruction phase both in Germany and in other countries. When phases of reconstruction are to be regarded as completed may, however, be debatable. In fact, only two criteria can be considered. In a narrow sense, reconstruction has been completed when national product has re-attained its pre-war level; in a wider sense, reconstruction would only have been completed when national product attains the level indicated by a continuation of the long-term trend of economic growth through the war and post-war years and beyond. Whichever criterion one prefers, reconstruction in the Federal Republic of Germany should have been completed toward the end of the 1950s, at the latest.

    The reduction in economic growth during the 1960s is more difficult to interpret. One might see in it "the return of the German economy to a growth rate familiar from a long term view of German economic history" (Abelshauser and Petzina, page 27). Nor does this turn of events contradict the hypothesis that the beginning of a contractionary phase has set in: every growth cycle still shows above average growth during the first years of the downswing, albeit at declining rates.

    Of course, economic development in the 1970s is hardly consistent with the implications of the reconstruction hypothesis. While it appears difficult for methodological reasons to test the hypothesis of a return to economic normality by empirically analysing the existence of "natural" growth rates, the enduring rise in unemployment in West Germany (as in other industrialised countries) can hardly be deemed as an expression of economic normality. Rather, these developments, too, support the diagnosis of a long-term contraction.
36. Timing of long cycles according to Kondratiev, "Die langen Wellen . . .", reference 3:

Cycle	*Expansion*	*Contraction*
Industrial Kondratiev	1787/93–1810/17	1810/17–1844/51
Bourgeois Kondratiev	1844/51–1870/75	1870/75–1890/96
Neomercantile Kondratiev	1890/96–1913/20	–

37. Differences in timing may also be attributed to Kondratiev's use of indicators other than aggregate output in identifying long waves.
38. An international and inter-temporal comparison of rates of economic growth is subject to severe limitations, particularly as the initial years of the bourgeois and the final years of the neoclassical Kondratiev cannot be determined with the available data.
39. During the pre-1920 period, the contractionary phase of the bourgeois Kondratiev terminated much earlier than in the other countries examined.
40. No information on investment activity is available for France prior to 1950.
41. Calculations of investment shares at constant prices revealed approximately the same pattern as the corresponding shares at current prices shown here. For evidence of deviations of real investment from its long-term trend see Glismann, Rodemer and Wolter, *Lange Wellen . . .*, reference 1, page 24.
42. A more precise analysis of the relationship between economic development and employment runs into serious difficulties on account of varying international and inter-temporal measurement concepts. Whereas international comparisons of degrees of employment are hardly possible with the present data, inter-temporal analyses of fluctuations in employment can be more readily undertaken. The latter, however, is not true for France and Italy, whose pre- and post-World War 2 data are incompatible. Hence, these countries were not included in Figure 11.7. If one examines those periods for which approximately consistent data are available in these two countries, a pattern similar to that of the other countries emerges.
43. This is the share of profits reinvested in the enterprises. The present analysis assumes that this measure reflects the long-run state of confidence of the investors.
44. See page 138 for a description of the real wage position. Annual changes in the real wage position of employees (in DM) were cumulated and expressed in per cent of the corresponding gross wages and salaries bill per employee. Positive (negative) changes of the indicator show that wage increases were above (below) a cost-neutral wage policy. On the method of calculating the real wage position of employees see Glismann, Rodemer and Wolter, *Zur empirischen Analyse . . .*, reference 1.
45. The relationship between expected profits and investment activity can only be examined for the period 1926–1979, as suitable prior data are not available. The calculations of Hoffmann on the development of corporations' return on capital in the business sector could be drawn upon. See Walter G. Hoffmann, *Das Wachstums der deutschen Wirtschaft seit der Mitte des 19. Jahrhunderts* (Berlin, Duncker und Humbolt, 1965), pages 500ff. This series was supplemented with calculations on the development of profit-type income and income from property between 1960 and 1977, as well as the real net return of industrial corporations between 1965 and 1979. The development of the share of undistributed profits in non-wage income, which is regarded as a better proxy for expected profits, could only be determined for the period 1950–1977. The indicators mentioned can be subjected to theoretical and statistical criticism. Each indicator taken for itself must therefore be interpreted very carefully.
46. This indicator could be drawn upon only for the period after the new German Stock Law took effect (1965).
47. The present analysis closely corresponds to the calculations of effective rates of protection for selected German industries undertaken by Sidney B. Webb, "Tariff protection for the iron industry, cotton textiles and agriculture in Germany, 1879–1914", *Jahrbücher für Nationalökonomie und Statistik* (Stuttgart), 1977, *192*, pages 336–357. Prior to 1896 an additional cycle may have occurred, but its beginning and peak cannot be identified with the available data.
48. The values for 1913–1925, as well as for 1938–1950, can hardly be compared on account of the pronounced changes in territory. This particularly affects the share of foreign trade.
49. See Blackhurst, Marian and Tumlir, reference 21.
50. Kindleberger notes that government measures which worsen the contraction accumulate during economic crises. See Charles P. Kindleberger, *Die Weltwirtschaftskrise 1929–1939* (Munich, 1973), pages 111ff.
51. Besides the development of the degree of integration, Figure 11.9 contains as additional indicators of protection the development of tariff rates (tariff revenue in per cent of dutiable

imports), and the share of subsidies (subsidies in per cent of net national product). Whereas the development of tariff rates approximately corresponds to the development of the degree of integration, fluctuations in subsidies correspond more to the investment cycles.

52. The government might also hinder economic growth by reducing the potential labour force through educational and social policy. In Germany, the participation rate rose substantially between 1950 and 1957, and thereafter declined continuously and, all in all, much more rapidly than was to be expected on demographic grounds.
53. On no account should the development of the real wage position be confused with the development of real wages. Real wages, as well as the wages and salaries bill, can rise, in spite of (or because of) a decline in the real wage position.
54. This phase displacement must not be interpreted as a reaction lag. Because the concept of a cost-neutral wage policy focuses on rates of change, a synchronous development of wages and growth rates of national product are to be expected. Such a relationship can indeed be observed.

CHAPTER 12

THE FIFTY YEAR CYCLE: A LOOK AT THE EMPIRICAL EVIDENCE

M. N. Cleary and G. D. Hobbs

This paper discusses the empirical evidence for the existence of long cycles in economic development. Using mainly US and UK data it argues that the evidence is strong in relation to price series but much weaker in relation to statistics of industrial production.

THE ECONOMIES of the Western industrialised world began to behave in an unexpected manner in the late 1960s and early 1970s. That it was happening did not really become apparent until the energy crisis of 1973 shook the world into a new state of awareness. Even then, and for some years after, many believed that the world's economic ills had their origin in the abrupt increase in the price of oil brought about by the OPEC cartel and that we would return to the trends of the 1950s and 1960s when that increase had been absorbed. However, the problems of declining growth, rising unemployment, high inflation and exchange rate instability persisted. As historical trends began to become clearer (Figure 12.1), it could be seen that many of the symptoms had been present even in the 1960s. Realisation dawned that perhaps the change had more fundamental origins—that the oil price rise had itself been an effect not a cause.

One facet of this realisation was a reawakening to the hypothesis, dormant for many years, that there existed a periodic variation in world economic activity which repeated itself approximately every 50 years. Parallels began to be drawn between the 1970s and the 1920s, and even with the Great Depression of the 1880s. The history of the 50-year wave hypothesis is long and dates back at least to 1847. One of the best-known studies of the phenomenon was by Kondratiev[1] and it is often referred to as "the Kondratiev cycle".

This paper describes some of the empirical evidence available to contemporary workers and assesses the extent to which it provides support for, or is in conflict with, the long-wave hypothesis. No claims are made that the survey is comprehensive. Nor is any attempt made here to report the results of a detailed statistical analysis. There are a number of reasons. Firstly, the hypothesis is at least 134 years old. It is believed to be true by many. It is therefore illuminating to sample the direct evidence that is capable of convincing these observers that the hypothesis is at least plausible—even if

The authors are with ICI, Corporate Management Services, Fulshaw Hall, Wilmslow, Cheshire, England. The views expressed are those of the authors and do not necessarily reflect those of ICI.

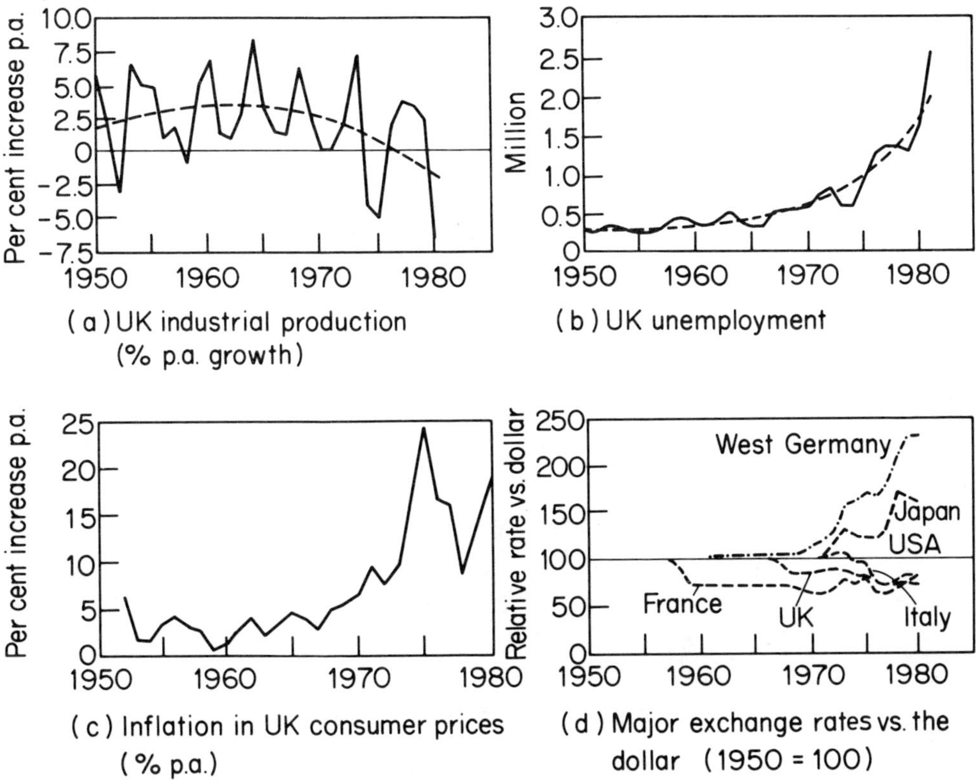

Figure 12.1. Post-war economic trends

the case has subsequently to be strengthened (or destroyed) by statistical methods. Secondly, because data exist for only three and at most four cycles, many of the techniques available to the researcher (eg spectral analysis) are of limited value. Thirdly, some of the simpler, more obvious methods of processing the data—smoothing by moving averages for example—can themselves introduce spurious cycles that are no part of reality.[2] While not disputing the value of statistical analysis techniques, in this paper we are asking readers to use their eyes and trust their own judgement.

The paper also concentrates almost exclusively on quantitative economic evidence. Additional evidence of a qualitative nature is available from the published writings of the times (some word frequency analyses have also been attempted[3]). However, both this and an analysis of the social and political changes that have occurred over the last 250 years are beyond the scope of the present exercise.

The number of continuous time series dating back more than 150 years is small and the data are of variable quality. To obtain long series it is frequently necessary to splice together series for different periods, which, although measuring broadly the same economic variable, nevertheless are different in detail, both in terms of definition and in the methods used for collection. It is therefore inadvisable to put too much weight on any one piece of evidence. The case for long waves rests much more on an ensemble of evidence.

The various theories of long waves are not explored in any detail. However, where the empirical evidence has relevance to a particular theory a brief mention is included.

Prices

It was in price series that the 50-year wave was first observed and in which some of the strongest evidence is found. Figure 12.2 shows the wholesale price

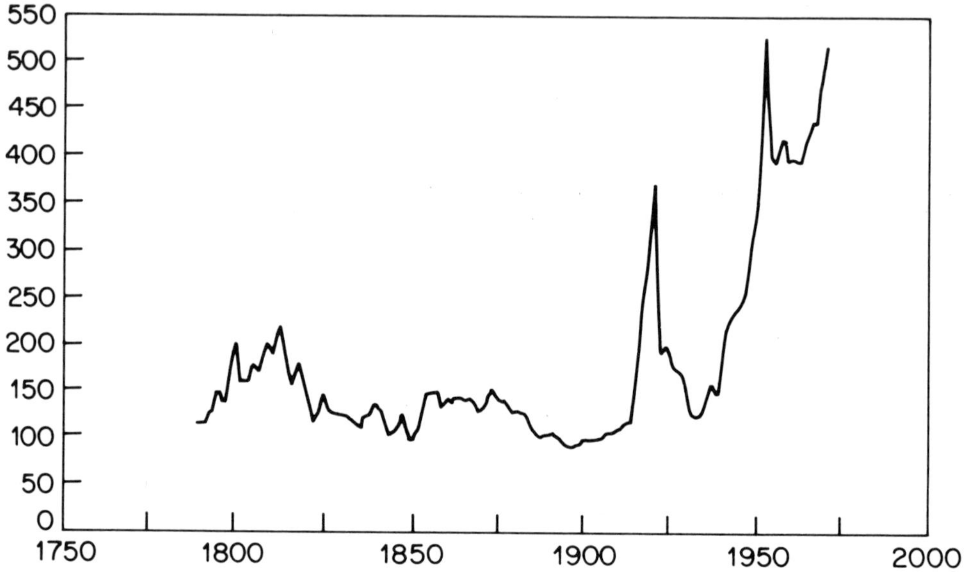

Figure 12.2. UK wholesale price index, 1790–1970 (1900=100)

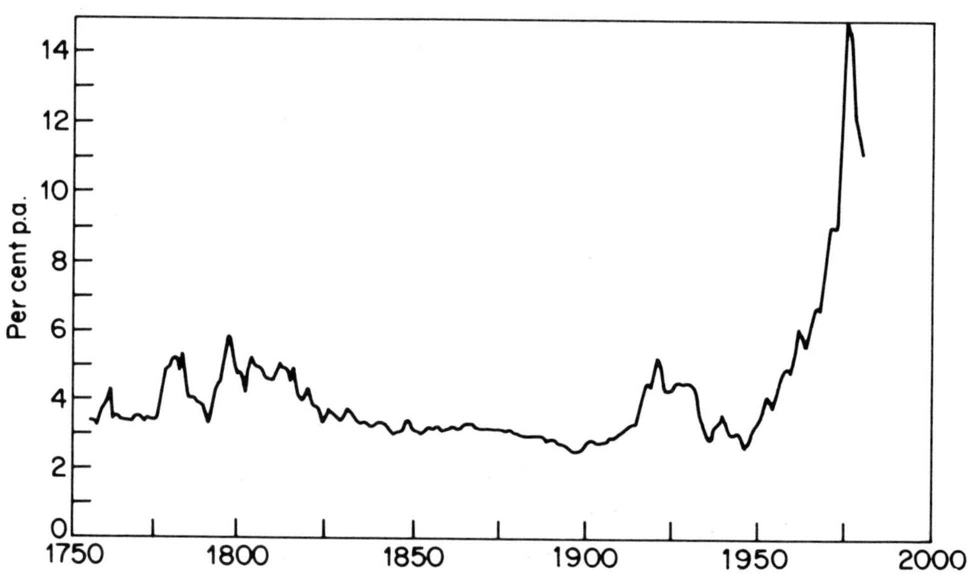

Figure 12.3. Yield on UK consuls, 1756–1979

index for the UK starting in 1790.[4] The series has been truncated at 1970 to prevent the inflation of the last decade overshadowing the price movements in earlier years (the index had passed a value of 2000 by 1979). The eye has no great difficulty in picking out a 50–60-year periodic movement in prices with peaks in the 1810s, the 1860s and the 1920s. Although of a totally different order of magnitude, the price rises of the 1970s are consistent in timing with a fourth peak in the 1980s. Corroborating evidence can be found in the UK wage index, the index of UK building costs, and the yield on UK consuls (Figure 12.3).

Displayed together in Figure 12.4 are the wholesale price indices (plus the yield on UK consuls) for five countries. The peaks and troughs are all broadly

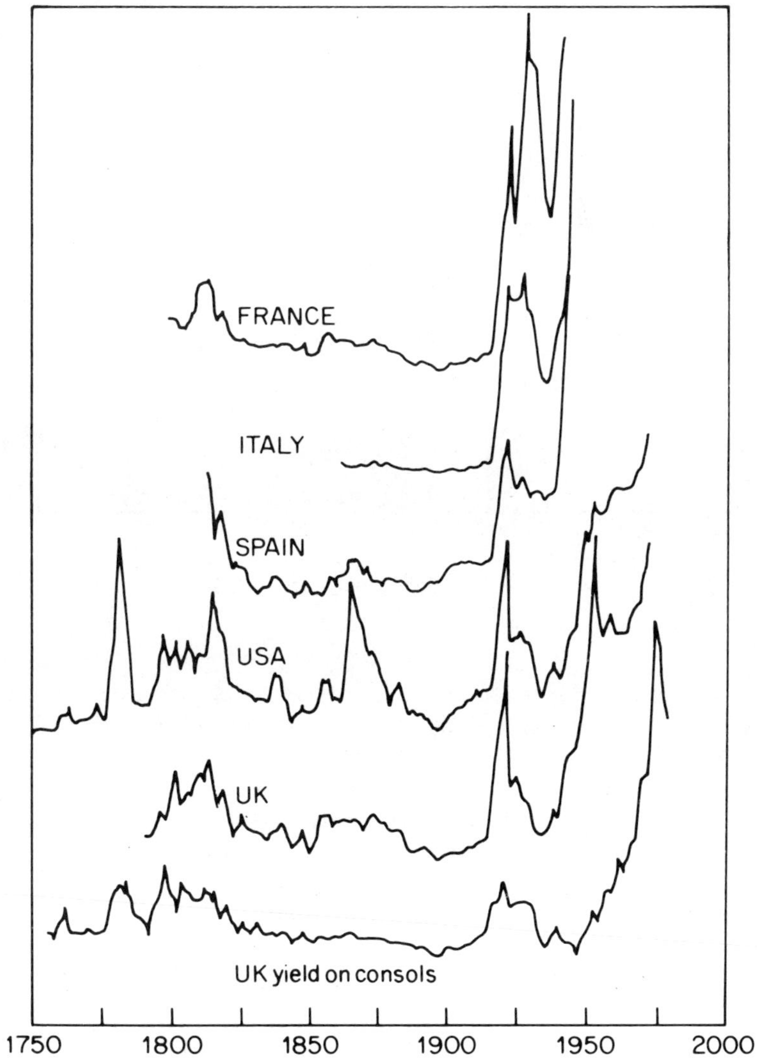

Figure 12.4. Wholesale prices in five major countries

coincident, the only 'peaks' missing from the set being those for Italy in the 1810s and 1860s. It is relevant to recall the massive changes that these countries will have seen over the 200-year period, particularly with regard to industrialisation, and the differences that would have existed between, for example, Spain and the UK at any particular time.

Additional data are available for individual commodities. Some can be found in Kondratiev's original paper[5] and others in the more recent studies by Rostow.[6] The price of energy is a particularly important example. Figure 12.5 shows the prices for coal and oil in the USA over the period 1800–1970. There is clear evidence of price peaks in the 1810s, 1860s and 1920s when, over a period of a few years, prices increased by as much as 100% or 200%.

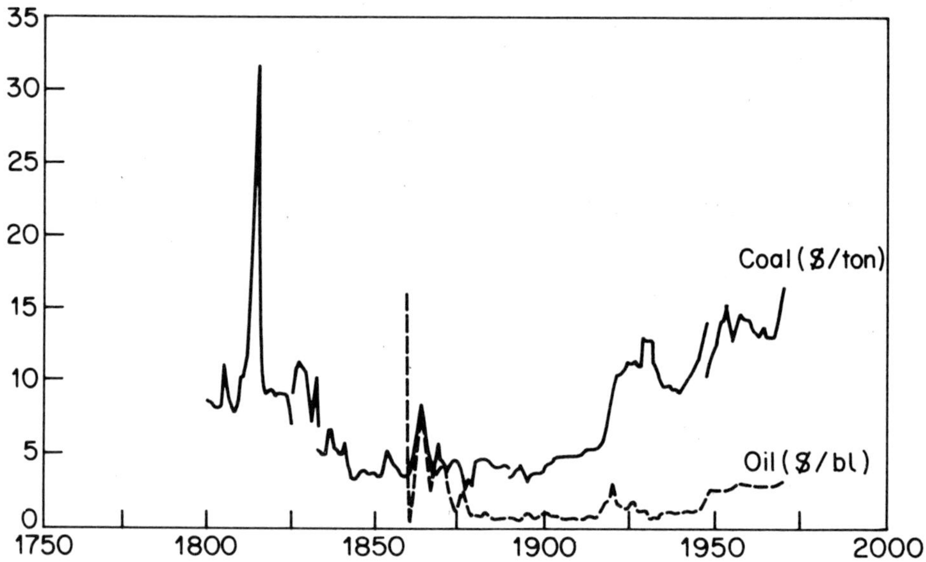

Figure 12.5. Energy prices in the USA

Rostow has analysed price series for foodstuffs and raw materials into "trend periods", ie periods during which, on trend, prices either rise or fall in relation to industrial goods. The results of his analysis are shown schematically in Figure 12.6. The timing bears a close resemblance to the broad pattern of changes shown in Figures 12.2–12.4, perhaps not surprisingly as wholesale prices are heavily weighted by food and raw materials. The third and fourth cycles differ from the first two in that (a) they are shorter (38 years compared with an average of 53 years) and (b) the fourth, which according to Rostow "peaks" in the early 1950s, appears to be an oscillation about an even longer-term inflationary trend and has a downswing (1951–1972) which is a period of stagnant rather than falling prices. An alternative interpretation is that the fourth and fifth cycles should be taken together as a single period of rising prices starting in the mid-1930s and with a peak still to come. Although neither interpretation fits very comfortably into a 50-year wave or with the

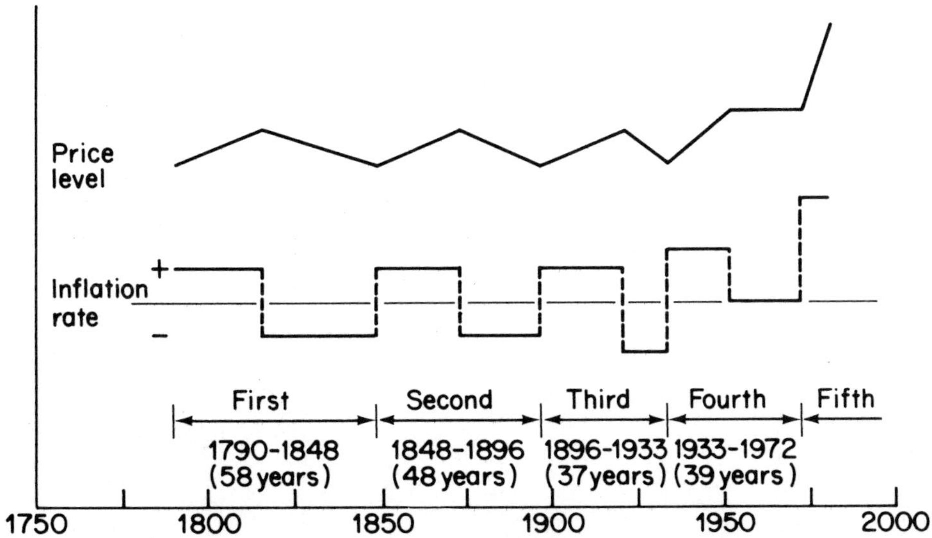

Figure 12.6. Phases of the long wave in prices (Rostow)

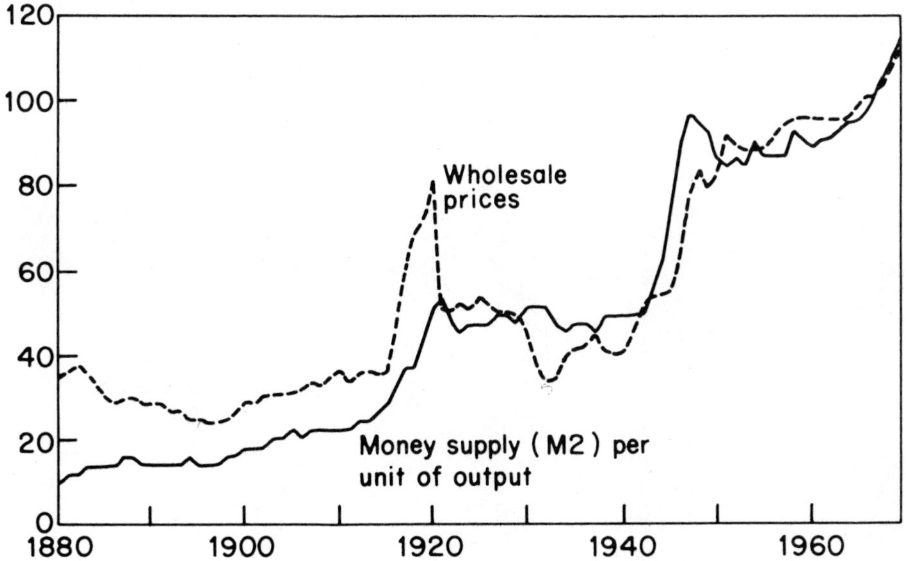

Figure 12.7. Indices of real money supply and prices in the USA (1967 = 100)

evidence for such waves found in other measures of economic activity, the latter seems to fit the overall picture best.

Some authors studying the long wave in the 1930s suggested that the major gold discoveries of the 1850s in California and Australia and those of the 1880s and 1890s in Australia, Alaska and the Transvaal caused the upswing in prices of the 1850s and 1900s. As most industrialised countries were on the gold standard at that time these discoveries increased the amount of money in circulation. This fact, coupled with the theory that prices will rise when in the long term the supply of money increases faster than the supply of goods, provides a plausible explanation of at least one influence on 19th century prices. However, gold discoveries by themselves do not help to explain the earlier price peaks around 1780 (US) and 1815 or the inflation of recent years. The gold standard effectively collapsed with the outbreak of World War 1 and

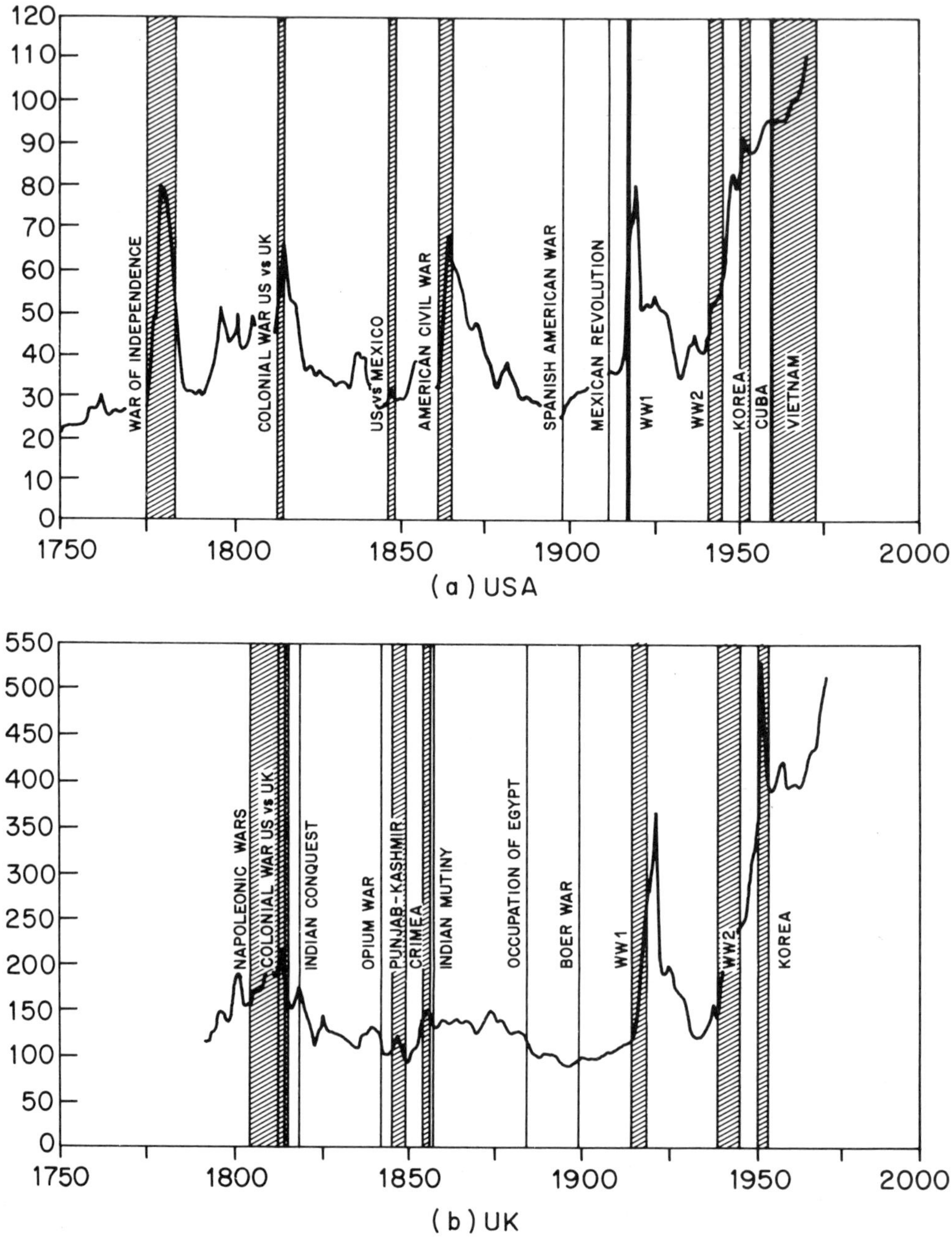

Figure 12.8. The effect of war on indices of wholesale prices

since then the proponents of this theory have looked increasingly to the general availability of money as the cause of price waves. The correlation between money supply/unit of output and wholesale prices is shown for the USA in Figure 12.7.

Another cause of rapid monetary expansion is war. There is a striking correlation between major wars and the sharpest peaks in wholesale prices, as illustrated in Figure 12.8. The War of Independence, the Napoleonic Wars, the American Civil War and World War 1 were funded, according to Galbraith,[7] by an expansion of the money supply, either directly by the printing of money (the pound first came into circulation in the Napoleonic Wars, and the "greenback" in the American Civil War), or indirectly by the suspension of specie (cash) payments and the subsequent issuing of notes.

It is therefore not implausible to theorise that all the observed variations in prices since 1750 have been caused by variations in the supply of money—war-time monetary expansion being responsible for the sharpest peaks, and gold discoveries or the printing of money being responsible for the slower movements. Even if this is accepted however, we are still left with a residual question: has this all been a random process which just by chance appears to be periodic, or have monetary conditions themselves, even war, been dictated by a cyclical behaviour having its origin in the management of the "real" economy?

Output

Industrial production

It is much more difficult to find evidence for long waves in measures of physical output. One of the problems is that all output series have strong underlying "exponential" growth (Figure 12.9a), a trend which makes any cyclical movement that may also exist very difficult to study. If attempts are made to remove the trend directly from the index, the cyclical residual is found to depend critically on the exact nature of the trend line chosen. A more satisfactory approach is to express the index in growth rate terms. The resulting series is unfortunately so 'noisy' (Figure 12.9b) that it defeats the eye in any attempt to find a long wave by examination. Simple smoothing (Figure 12.9c) can help to uncover the underlying movement, although the hazards associated with smoothing must constantly be borne in mind.

Several authors have attempted detailed analyses of output data to test the long-wave hypothesis, notably van Duijn[8] and Kuczynski.[9] Van Duijn's chronology for the long wave, based on an analysis of "crisis, recession and depression years" and supported by the published historical perceptions of the times, is shown schematically in Figure 12.10. Also shown are the phases of the long wave derived from Rostow's "trend period" analysis of commodity prices. Thus a period of prosperity is characterised by high output growth (and rising prices); recession by declining growth (stagnant or falling prices); depression by low growth (falling prices); recovery by rising growth (stagnant or rising prices). The prosperity phase is typically 15 to 20 years in length,

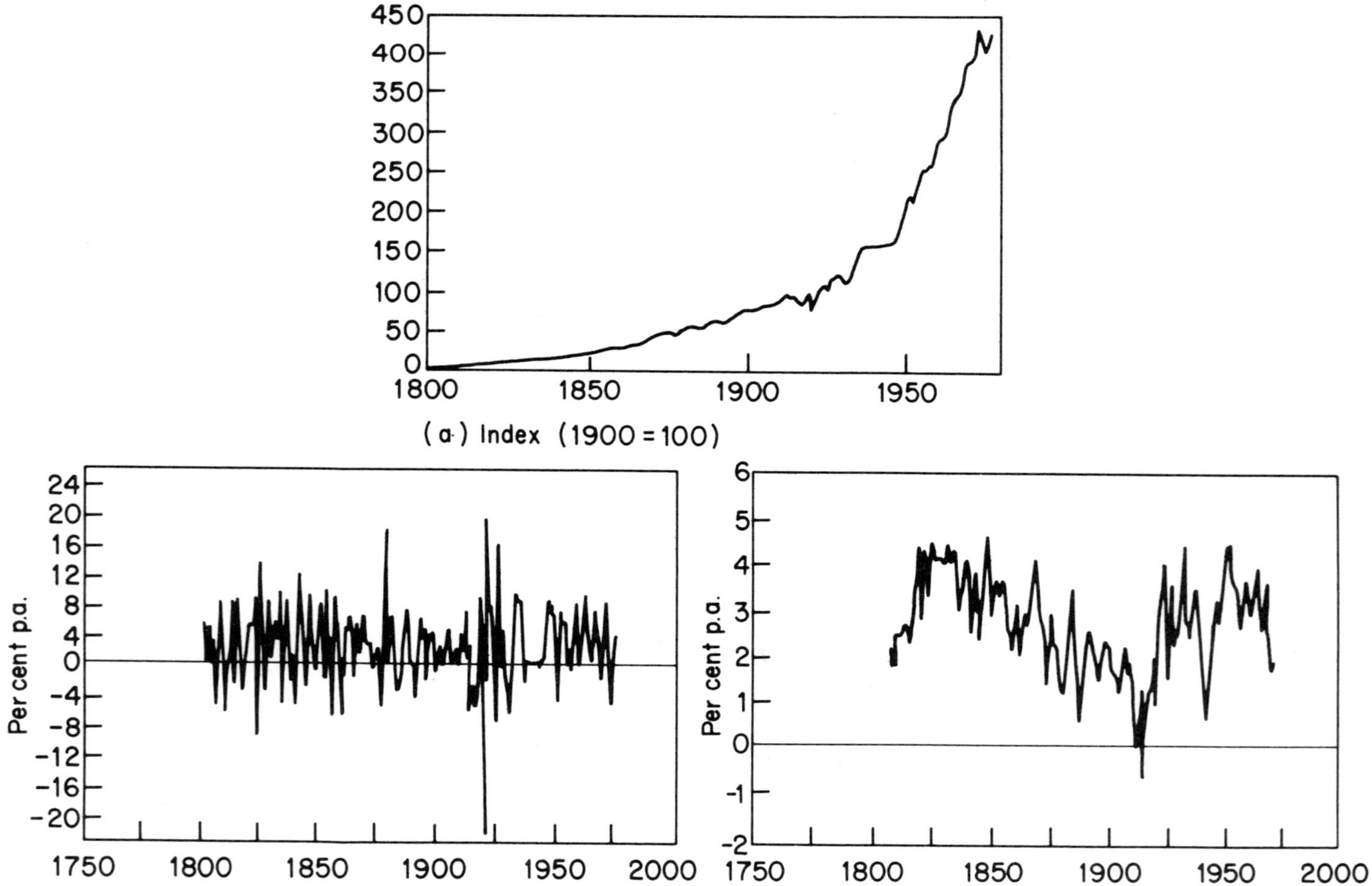

(a) Index (1900 = 100)

(b) Growth rate (% p.a.)

(c) 11-year moving average of growth rate (% p.a.)

Figure 12.9. UK industrial production (excluding dwellings)

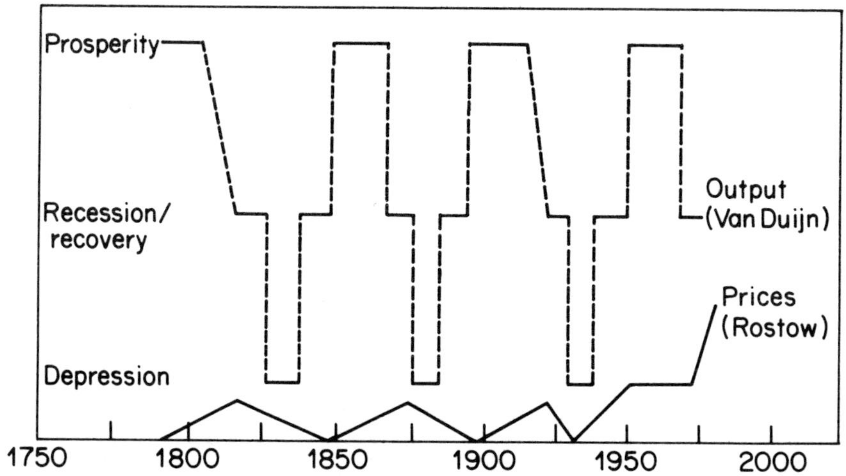

Figure 12.10. Phases of the long wave

recession 8 to 10 years, depression and recovery each 10 years. Two of the recession phases were delayed by wars (Napoleonic, World War 1). Van Duijn suggests that the current recession phase commenced in the late 1960s. On the basis of previous history a price peak would have been expected at about the same time. Rostow claims that it occurred in 1951, which is far too early to be consistent with van Duijn's chronology. If, on the other hand, Rostow's timing is accepted, it must be concluded that the output and price waves have moved totally out of phase. If the price peak is still to come, it is either late or the recession is unusually prolonged.

Van Duijn has analysed data on industrial production for a number of countries into Juglars (cycles of approximately 10-year periodicity) and has examined the variation in average growth rate from cycle to cycle. In aggregate, the data are very broadly consistent with the chronology of Figure 12.10. There are however several serious inconsistencies. One of these, for the UK, has already been illustrated in Figure 12.9c. Growth rose sharply until 1825 and then declined on trend for the next 100 years. Growth was again high during the depression phase of the 1930s. The picture is only slightly less inconsistent for the USA (Figure 12.11).

Kuczynski has analysed aggregate industrial production data for the capitalist world. The smoothed data are shown in Figure 12.12 and demonstrate how the variations in the growth of aggregate industrial production can be seen as broadly consistent with the long-wave hypothesis.

One of the oldest of the long-wave theories claims that the prime cause of long waves, particularly in industrial production growth, is major technological change and the innovation to which it gives rise. The theory was first formulated by Schumpeter,[10] was amplified by Kuznets[11] and has been given new momentum by the work of Mensch,[12] Freeman[13] and Clark, Freeman and Soete.[14] The theory's proponents argue that innovation, ie the creation of new products and business out of previous inventions, proceeds at an uneven pace. Mensch claims, on the basis of his analysis of UK, US and German

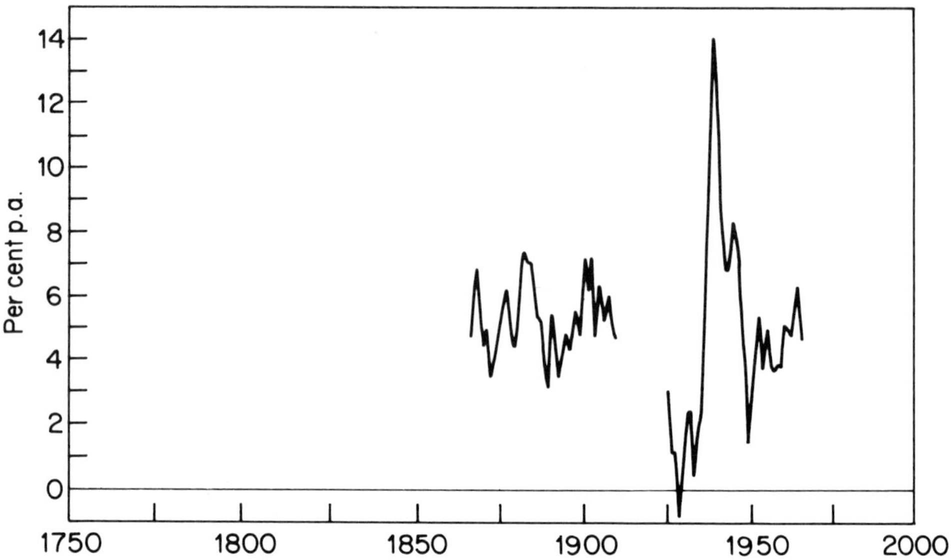

[a]Growth rate in per cent p.a., 11-year moving average.

Figure 12.11. US index of manufacturing production[a]

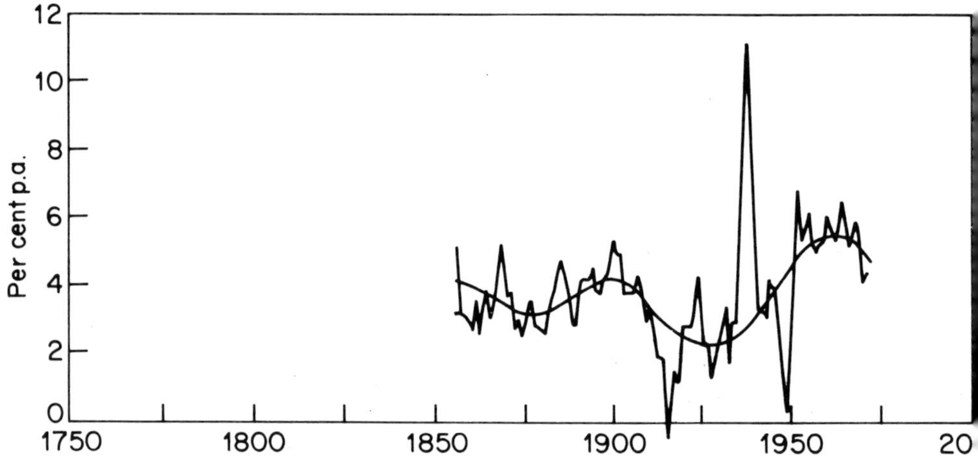

[a]Growth rate in per cent p.a., 11-year moving average.

Figure 12.12. Capitalist world industrial production[a]

data, that "basic" innovations have clustered at 50-year intervals, the last cluster falling in the 1930s (Figure 12.13). The innovation clusters are broadly coincident with the depression phase of van Duijn's long-wave chronology. It is the subsequent rapid development of these innovations into major new industries that gives rise to the buoyant economic climate of the prosperity phase. The innovation surge ultimately exhausts itself (recession) and creative effort is increasingly diverted into process innovation, ie to improving existing products and making them more efficiently.[15] The new industries have now entered the maturity phase of their life cycles. It is during this phase that labour is displaced by capital, unemployment rises (see later paragraphs), markets saturate, excess capacity develops, profitability and investment decline and the depression phase ensues.

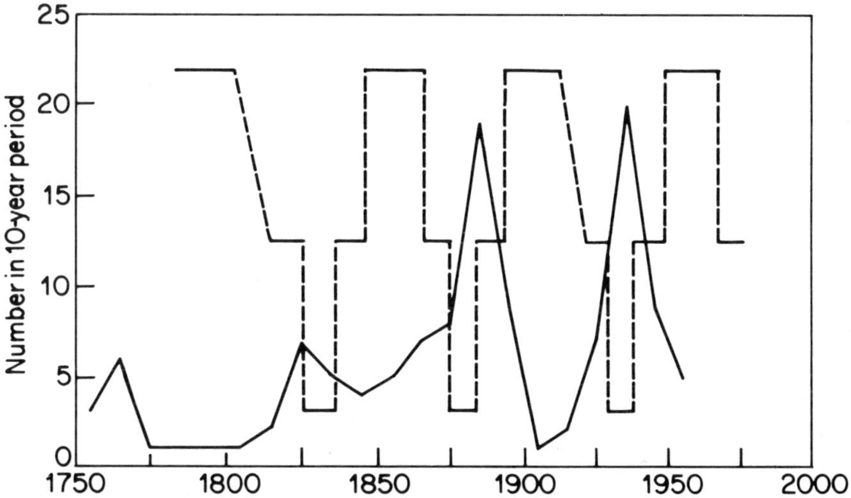

Figure 12.13. Frequency of innovation

Several aspects of this theory are still the subject of debate. There is dispute concerning the definition, selection and dating of the innovations studied and whether clustering in fact occurs. There is disagreement on the possible causes of the clustering: Mensch argues that depression is the trigger; Clark *et al* that it is related to advances in fundamental science and the technological diffusion process; Forrester that recovery, which would occur anyway for other reasons (see following paragraphs), is the trigger.

It has already been noted that it is difficult to observe long waves in industrial activity, at least as measured by aggregate industrial production indices. This does not necessarily invalidate the technological innovation theory. It is a theory that has its foundations in the life cycle of individual industries and if the declining output growth in maturing industries is balanced, or almost balanced, by increasing growth in the new industries there may be little or no evidence of waves in total production.

Energy and mineral production

Data on the growth of world energy production[16] correlate well with the growth of capitalist world industrial production (Figure 12.14). The observed variation provides support for Rostow's theory of long waves. Details of the theory and an analysis of economic history based on it can be found in Rostow and Kennedy.[17] In brief, the theory's proponents argue that alternate periods of over- and undercapacity occur in the sector producing primary goods—food, raw materials and energy. The failure to follow smoothly a steadily rising demand for these products stems from four causes: the long delays that follow the emergence of a profit possibility and the investment decisions designed to exploit it; the long gestation periods associated with the opening up of new capacity (agricultural areas, mines, oil fields, etc); the delays between the completion of an investment and its maximum efficient exploitation; and finally, the lumpiness or non-incremental nature of capacity

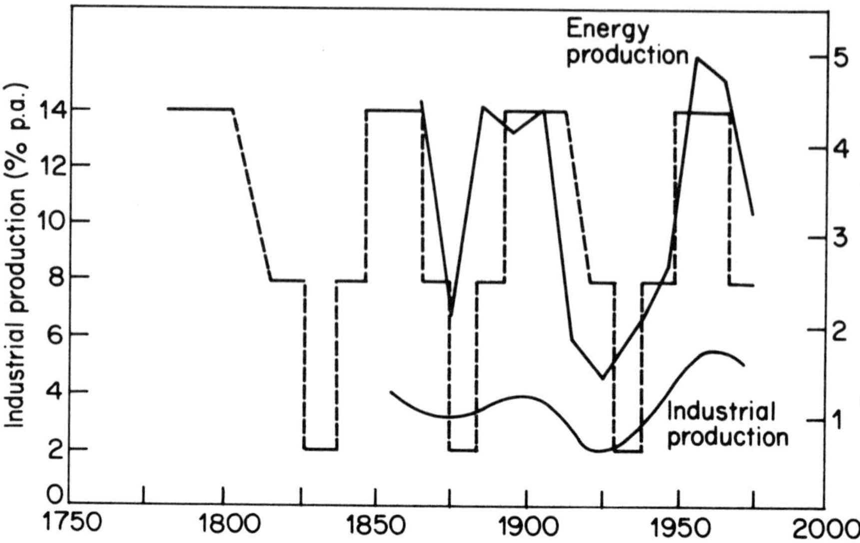

[a]10-year average growth rate in per cent p.a.

Figure 12.14. Correlation between world energy production,[a] capitalist world industrial production and van Duijn's long-wave chronology

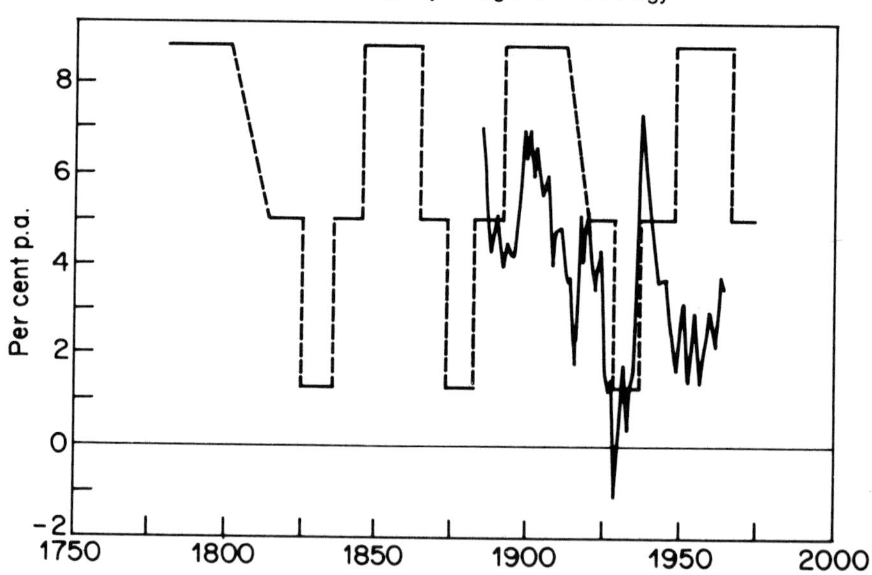

[a]Growth rate in per cent p.a., 11-year moving averages.

Figure 12.15. US mineral production[a]

additions to the sector. The effect of these delays when demand overtakes supply is that supply and demand remain out of balance for several years, the price of primary products rises relative to manufactured goods, and investment is attracted preferentially into the primary sector. Ultimately the new facilities come on stream and overshoot the capacity required to meet the demand. Relative prices now fall and investment swings back to manufacturing until the rising demand once more catches up with the stagnating or declining capacity.

The variations in the index of US mineral production (Figure 12.15) are also broadly consistent with Rostow's theory.

Agricultural production

The smoothed growth of agricultural production in the capitalist world[18] is shown in Figure 12.16. Wars so disrupt this series that little can be deduced from it, one way or the other.

It has been argued that 50-year climatic cycles induce sympathetic cycles in agricultural output and prices which work their way through to the economy as a whole. An early theory postulated sunspot activity as the origin of climatic variations, and evidence is claimed for a 54-year cycle.[19] There is also some direct evidence for regular variations in climate.

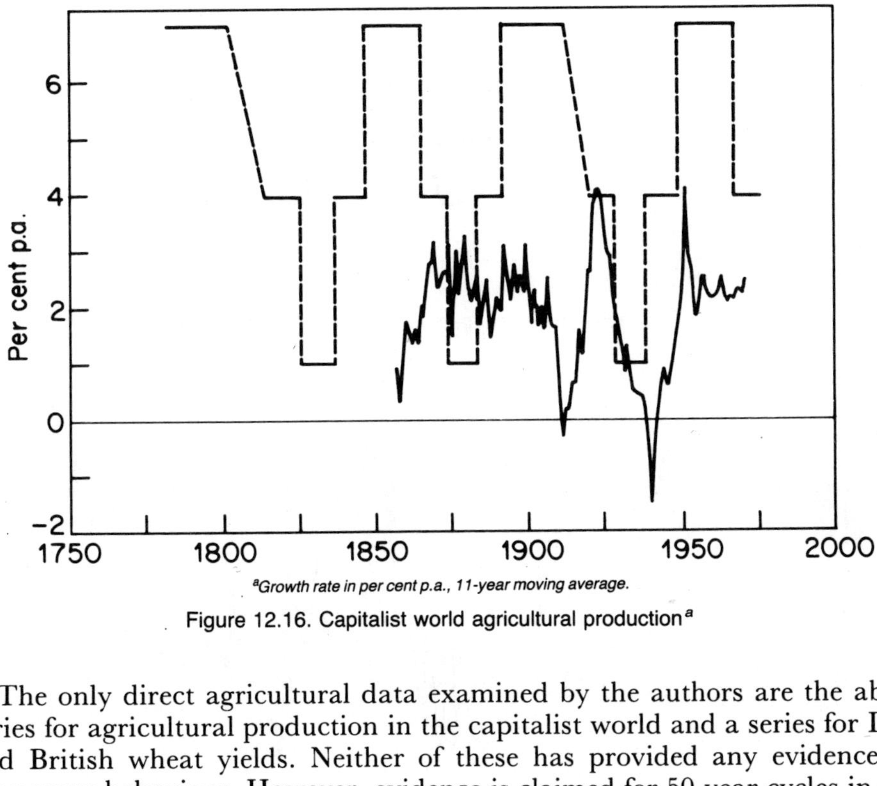

[a]Growth rate in per cent p.a., 11-year moving average.

Figure 12.16. Capitalist world agricultural production[a]

The only direct agricultural data examined by the authors are the above series for agricultural production in the capitalist world and a series for Irish and British wheat yields. Neither of these has provided any evidence for long-wave behaviour. However, evidence is claimed for 50-year cycles in tree rings.[20] Price cycles of course do exist for agricultural products—in common with many other materials and products.

Investment

Several of the theories suggest that investment (rate of capital building) should show cyclical behaviour more strongly than output as a whole. Van Duijn argues that the GB index of production for the producer goods industries does contain a 50-year wave. However, a graph of UK gross domestic fixed capital formation (less dwellings) expressed as a percentage of total national output (Figure 12.17a) shows only the weakest evidence for the peak that would be expected around 1900. The converse is true for the USA. Figure 12.17b demonstrates that the variation in the purchases of capital goods by US manufacturing industries expressed as a percentage of GNP conforms well with the long-wave hypothesis and van Duijn's chronology. Similar results can be obtained for Germany.[21]

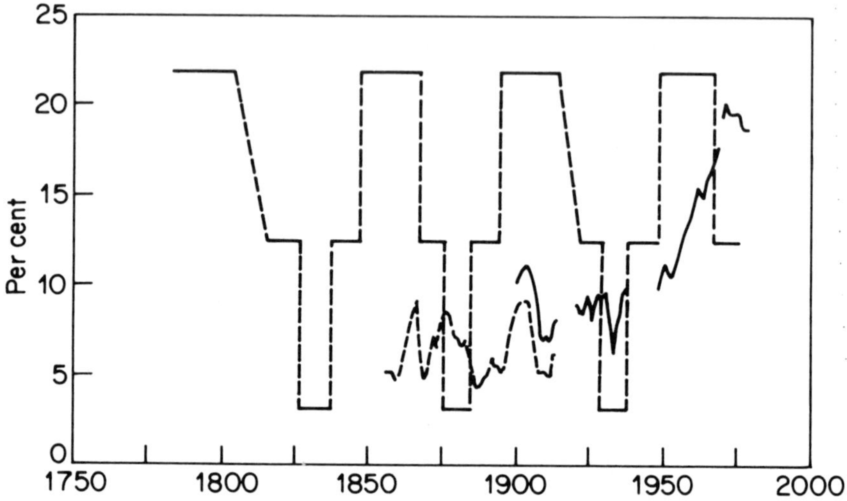

Figure 12.17a. UK gross domestic fixed capital formation (excluding dwellings) as a percentage of total output

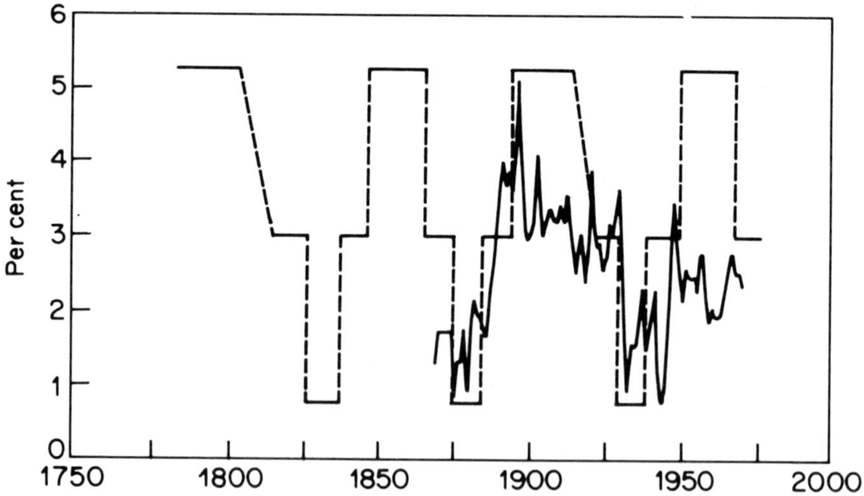

Figure 12.17b. Purchase of capital by US manufacturing industries as a percentage of GNP

One of the theories relying heavily on cycles in fixed capital formation is that of Forrester.[22] Forrester attributes the long wave to alternate periods of over- and underexpansion of the sector producing capital goods—plant, machinery, buildings, etc. The primary cause of the wave is the requirement that the capital goods sector must provide its own capital goods in order to expand. A steadily growing demand for capital goods arising from the consumer sector causes the capital sector to perceive the need for more capacity and it places orders on itself accordingly. This positive feed-back loop, strengthened by several others associated with the managerial response to the consequential delivery delays, escalating prices and profitable investment opportunities, leads to an overexpansion of the capital sector. In due

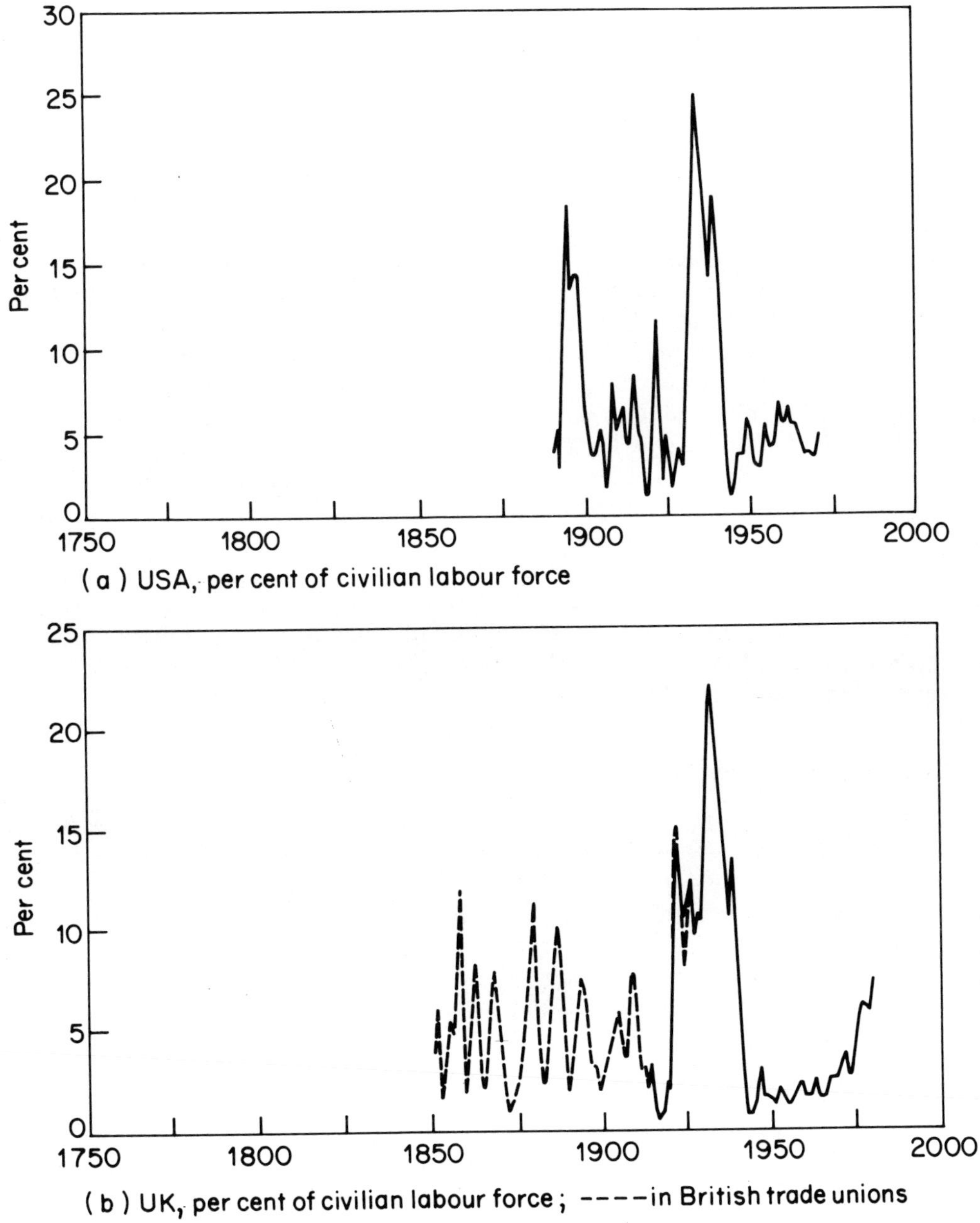

Figure 12.18. Unemployment

course, as orderbooks are reduced, the overcapacity becomes manifest and the feed-back loops operate in reverse. Ultimately, in a growing economy, demand from the consumer sector overtakes the capacity of the capital sector to supply but, because of the long lead times associated with the installation of new capital, a period of undercapacity persists until the new (over-compensating) upswing gets under way.

Unemployment

Unemployment data for the 19th century, where available, are fragmented and not necessarily representative of the total labour force. The data on civilian unemployment for the UK and USA are shown in Figure 12.18. Also included for the UK is a series for unemployment in British trade unions (1851–1926). Both figures clearly illustrate the very high (up to 25%) unemployment rates associated with the Great Depression of the 1930s. While the peak unemployment rate was higher in the US (24.5% in 1933) than in the UK (18.4% in 1932), the duration was considerably shorter—1930–1943 (US) compared to 1921–1942 (UK). Galbraith[23] argues that this was due to the different effects of World War 1 on the two countries, Churchill's insistence on returning to the gold standard, which made British prices uncompetitive in international markets, and the restrictive monetarist policies in the UK in the 1920s, which placed heavy emphasis on productivity. The only other clear peak in unemployment is for the USA in the 1890s. This is coincident with the period of low manufacturing production growth in the USA at that time (Figure 12.11), and may be related to the strong protectionist policies (McKinley tariffs) that were in force, but it is rather late for the van Duijn depression phase of the 1880s. The cyclical peaks in UK unemployment were higher in the 1880s than during the prosperity phase of the early 1900s. However, they were not significantly higher than those in the preceding prosperity phase of the 1850s.

Summary and conclusions

The strongest empirical evidence in favour of the long-wave hypothesis undoubtedly comes from price series. Supporting evidence can also be found in the behaviour of long-term interest rates, world energy production and innovation. Limited and not very convincing evidence can be seen in world industrial production, US mineral production, unemployment and investment. Failure to observe clear cycles in aggregate industrial production may not be particularly significant if the wave is really one of periodic industrial restructuring.

The evidence for the wave can be summarised by reference to the "clock" illustrated in Figure 12.19. The times at which prices and output growth pass their high and low points are taken from the wholesale price and capitalist world industrial production indices. The clock revolves every 50 to 55 years. The periods of rising prices, falling output growth, falling prices and rising output growth come around in the same sequence in each of the three cycles shown. The intervals between the peaks and troughs are irregular. Also

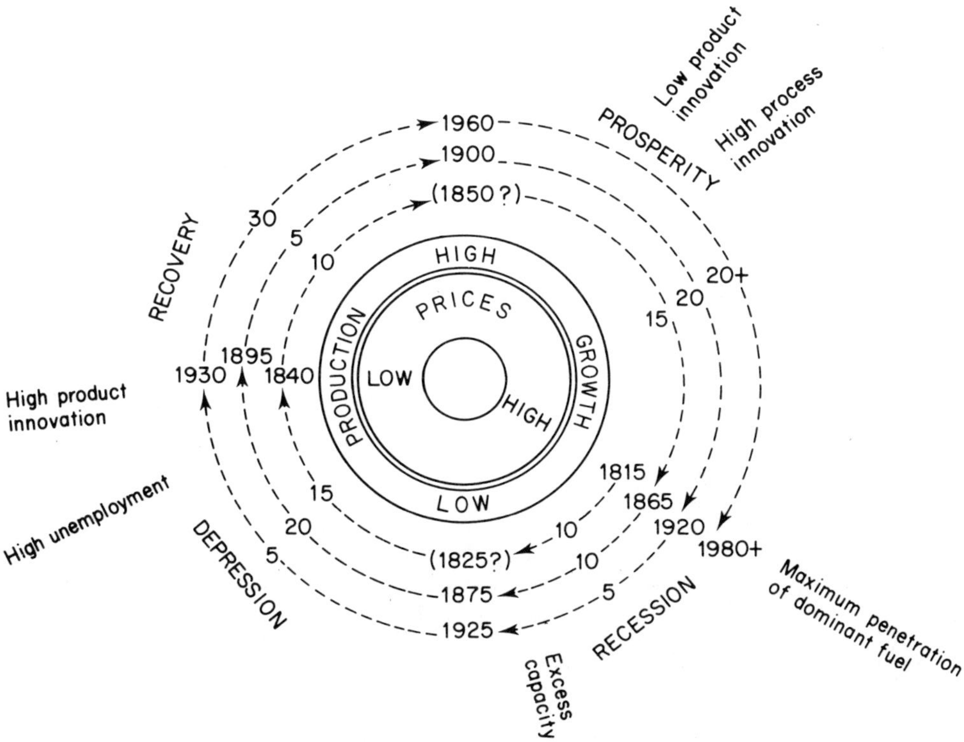

Figure 12.19. A long-wave 'clock'

marked on the dial are van Duijn's estimates for the positions of the prosperity, recession, depression and recovery phases and the appropriate timing of the other economic phenomena referred to in the text.

It appears that the variations in absolute prices may be associated with variations in the money supply, whether brought about by war, gold discoveries or printing. However, it is still possible that these monetary variations themselves are a consequence of what is happening in the real economy. This view is certainly supported by Forrester's model in which the periodic variations in the real economy have their origins in the over- and underexpansion of the capital goods industries. Rostow on the other hand argues that it is the variation in relative prices, between primary and manufactured products, that is important. In this case, waves in the real economy are driven by the over- and underexpansion of the primary sector.

It is not easy to reach firm conclusions on the existence, or otherwise, of the long wave from a direct visual examination of the empirical evidence. The ensemble of time series does suggest that there is some validity in the hypothesis, although no single series on its own carries very much conviction. Statistical analysis, in the authors' experience, strengthens the case, but the shortness of even the longest series reduces the significance of the tests. In a century's time perhaps the evidence will be that much clearer.

Notes and references

1. N.D. Kondratiev, "Die langen Wellen der Konjunktur", *Archiv für Sozialwissenschaft und Sozialpolitik*, 1926, *56*, pages 573–609; N.D. Kondratiev, "The long waves in economic life", *Review of Economic Statistics*, 1935, *17*, pages 105–115.
2. M.G. Kendall, *Time Series* (Griffin, 1973).
3. J.Z. Namenwirth, "The wheels of time and the interdependence of value change", *Journal of Interdisciplinary History*, 1973, *3*, pages 649–683; R. P. Weber, "Society and economy in the western world system", *Social Forces*, June 1981.
4. Many of the time series are taken from: B.R. Mitchell, *Abstract of British Historical Studies* (Cambridge University Press, 1962, 1971); Mitchell, *European Historical Statistics 1750–1970* (Macmillan, 1975); *Historical Statistics of the United States* (US Department of Commerce, 1975); London and Cambridge Economic Services, *The British Economy: Key Statistics, 1900–1970* (Times Newspapers, 1971).
5. Kondratiev, reference 1.
6. W.W. Rostow, *The Process of Economic Growth* (Clarendon Press, 1960); "Kondratiev, Schumpeter and Kuznets: Trend periods revisited", *Journal of Economic History*, 1975, *35*, pages 719–753; *The World Economy: History and Prospect* (Macmillan, 1978).
7. J.K. Galbraith, *Money, Whence it came, where it went* (Andre Deutsch, 1975).
8. J.J. van Duijn, "The long wave in economic life", *De Economist*, 1977, *125*, 4, pages 544–576; "Comment on Van der Zwan's paper" in S.K. Kuiper and G.J. Lanjouw, eds, *Prospects for Economic Growth* (North Holland, 1979), pages 223–233.
9. T. Kuczynski, "Spectral analysis and cluster analysis as mathematical methods for the periodization of historical processes—a comparison of results based on data about the development of production and innovation in the history of capitalism. Kondratiev cycles—appearance or reality?", in *Proceedings of the 7th International Economic History Congress* (Edinburgh, 1978), Vol. 2, pages 79–86; "Have there been differences between the growth rates in different periods of the development of the capitalist world economy since 1850?—an application of cluster analysis in time series analysis", *Historisch-Sozialwissenschaftliche Forschungen* (Stuttgart), 1980, *6*.
10. J.A. Schumpeter, *Business Cycles* (New York, 1939).
11. S. Kuznets, *Economic Change* (London, Heinemann, 1954).
12. G. Mensch, *Das technologische Patt* (Frankfurt, Umschau Verlag, 1975); English edition: *Stalemate in Technology* (Cambridge, Mass., Ballinger, 1979).
13. C. Freeman, "The Kondratiev long waves, technical change and unemployment", in *Structural Determinants of Employment and Unemployment* (Paris, OECD, 1979), pages 181–196.
14. J. Clark, C. Freeman and L. Soete, "Long waves, inventions and innovations", *Futures*, 1981, *13*, 4, pages 308–322.
15. J. Clark, C. Freeman and L. Soete, "Long waves and technological developments in the 20th century", in D. Petzina and G. van Roon, eds, *Konjunktur, Krise, Gesellschaft: Wirtschaftliche Wechsellagen und sozialer Entwicklung im 19 und 20 Jahrhundert* (Stuttgart, Klett-Cotta, 1981), pages 132–169; A.K. Graham and P.M. Senge, "A long-wave hypothesis of innovation", *Technological Forecasting and Social Change*, 1980, *17*, 4, pages 283–311; R. Rothwell, "Technology, structural change and manufacturing employment", *OMEGA*, the Journal of Management Science, 1981, *9*, 3, pages 229–245.
16. C.M. Cipolla, *The Economic History of World Population* (Penguin, 1979).
17. W.W. Rostow and M. Kennedy, "A simple model of the Kondratiev cycle", in P. Uselding, ed, *Research in Economic History* (JAI Press, 1979), Vol. 4.
18. Kuczynski, "Have there been differences . . .", reference 9.
19. E.R. Dewey, *Cycles Journal*, Foundation for the Study of Cycles, June 1970.
20. *Ibid.*
21. H.H. Glismann, H. Rodemer and F. Wolter, "Zur Natur der Wachstumsschwäche in der Bundesrepublik Deutschland", Discussion Paper 55, University of Kiel, 1978.
22. J.W. Forrester, "Business structure, economic cycles and national policy", MIT System Dynamics Group, D-2245-2, 1975; also in *Futures* (London), June 1976, *8*, 3, pages 195–214; "New perspectives for growth over the next thirty years", MIT System Dynamics Group, D-2251-1, 1975; "Economic perspective", MIT System Dynamics Group, D-2667-1, 1977; Graham and Senge, reference 15.
23. Galbraith, reference 7.

CHAPTER 13

INNOVATION AND LONG-TERM GROWTH

George F. Ray

This paper is a revised version of a contribution to an International Institute for Applied Systems Analysis Task Force Meeting on "Innovation and Industrial Strategy". The author gives a historical overview of the impact of innovations on economic growth, pointing out the development arising from breakthroughs in certain areas. The economic mechanism of long waves cannot be described in a one-dimensional way. George Ray also presents a hypothesis on the content of a future upswing of world economy, including in it energy, food, environment and social institutions. In his opinion the microprocessor is only an important tool for technological changes in various fields.

I HAVE recently expressed elsewhere[1] certain views on the part played by innovations in the frequently debated long waves; it is not my intention to repeat here what has been said in considerably more detail there. Let me just mention briefly one or two points that seem to me relevant to the new material that follows.

The long cycles

This is not the place to get involved in the argument about whether long waves in economic activity do indeed exist or not. The controversy started immediately following the publication of Kondratiev's study in the 1920s, has been going on ever since, and will probably continue. More recently, however, as one outcome of the gloomy state of the world economy—including here East and West as well as the Third World, with the sole exception of some oil producers—the study of long waves has again come to the fore and has led to the publication of some valuable views with theoretical and empirical statements by both new converts and old disciples, among them such well known names as Rostow and Forrester.[2]

My interest lies not directly in the theory—or the reality—of the Kondratiev cycle but in its relevance to innovation (or vice versa) and in the way in

George F. Ray is at the National Institute of Economic and Social Research, London.

which innovative activity can influence, indeed give a push to, the whole economic system.

The three long cycles identified by Kondratiev and accepted by Schumpeter in his thesis were allied, in the most simplified form, to the dissemination of steam power, to the railway boom, and to the joint effects of the motor car and electricity. These are obviously crude generalisations, since the "steam power" period cannot be divorced from the development of the coal and cotton industries; the railway boom could not have happened without an iron and steel industry of new dimensions; and the third upswing included—apart from the somewhat loose attribution to the motor car and electricity—the birth of the modern chemical industry, among other things. It is never a simple matter to find a unique cause of historical developments in the economic arena; the single pistol shot at Sarajevo was not the unique reason for the outbreak of the Great War in 1914, nor was Watt's perfection of the primitive steam engine the unique cause of an economic upswing.

Excursion into history

If we start searching for reasons for upswings, and especially if we believe that innovation is, generally speaking, a "good thing" that offers the innovator advantages over laggards, then we can find plenty of evidence pointing to the importance of innovation and to the role played by various innovations in the fortunes of the countries implementing them. The leading role of Britain in the 19th century was in no small measure due to the rapid introduction of steam, to the series of major innovations in the various phases of the cotton industry, as well as to the development of coalmining in that country. It continued with the growth of the coal-based iron and steel industry, which made the railway boom possible. Britain was later overtaken by others, especially Germany—again due to the great German innovations in steel and chemicals. And finally, by taking up both electricity and the motor car before the others, the United States not only arrived on the scene, but overtook everybody else.

The industrial revolution in Britain—and the various innovations leading to it and arising from it—are well known from economic history; to an extent, almost the same is true of the ascendence of the United States. Somewhat less is generally known—especially outside Germany—of the German developments. The German steel industry was indeed thriving on the then new Thomas process: from 1880 to 1900 its output rose tenfold, leaving Britain far behind. The cost of steelmaking in Germany dropped to one-tenth of the level of the 1860s as the combined result of technological progress and scale economies. As early as 1900, Germany supplied 90% of the world production of dyes. The electrical industry owed much to the systematic concentration (especially of Siemens) on invention and to the early large-scale adoption of Edison's patents.

Kondratiev paid great attention to the French economy when he constructed his statistical series supporting his long waves. However, France—certainly the leading country in 18th century Europe, with the largest population—was not among the later industrial pioneers. For a while, during the Revolution and the Napoleonic Wars, it was cut off from the world outside and lost contact with technical progress there; the industrial advance that started in the second half of the 18th century, helped by new techniques imported from England, was halted. For a long time, certainly until the middle of the 19th century, France remained a largely agricultural country. The contribution of a few major industrial innovations was nevertheless important to French (and general) industrial progress—such as Jacquard's loom, Thimonnier's sewing machine, Berthollet's chlorine-bleacher or Girard's flax spinner. The later spread of industrialisation coincided in France with Kondratiev's second cycle characterised by the railway boom.

Among the major European powers of the time, Austria–Hungary lagged far behind the above West European trio. Austrian industry was very slow to introduce steam power and the majority of pig iron production was exclusively charcoal-smelted until about 1870. We cannot even speak of Italy proper before its unification in 1860; by about the middle of the 19th century the leading industry in Piedmont and Lombardy was silk-throwing, and most of these mills worked by water power. Lack of coal characterised the industrial scene: here too ironmaking was based on charcoal. Up to 1860, 1800 km of railways were built. Although railway building was speeded up, industrial progress was slow to 1880 and the real spurt followed in the period 1899–1913, in the third Kondratiev cycle only, initiated by two new products of innovation: automobiles and typewriters.

The history of the economies of smaller industrial countries also provides evidence of the outstanding importance of innovations. The Swiss economy not only survived British competition, which at one time jeopardised the existence of the once flourishing Swiss cotton industry, but by means of innovations—among them, some interesting organisational changes—recovered to become a serious competitor to the British, in both cotton manufactures and textile machinery. From textiles to machinery and then onwards to dyes—this was the self-sustained way of advance of Swiss industry based on industrial and organisational innovations, great flexibility, and that special agility and managerial talent that seems to be a characteristic of the industries of some smaller countries, of which Switzerland is one typical example and Belgium another.

The local availability of coal and iron ore, plus motive power by water, were the bases of early industrialisation of the region that by 1830 had come to be called Belgium. The adoption of steam power was slow, but in 1807 Huart-Chapel invented a reverbatory furnace for melting down scrap iron and introduced several major changes—such as puddling furnaces to convert pig iron into wrought iron (1821). He built the first coke blast furnace in the Charleroi area. New type rolling mills were also added at about the same time. Coming over from England, Cockerill started making spinning machines in Verviers as early as 1799; a number of outstanding innovators helped the Ghent textile industry to prominence, as well as the metal industries in Charleroi and elsewhere. By 1840 the country was highly industrialised, comparable only to Britain, and had relatively the densest rail network in Europe.

Holland, in contrast, did not have any noteworthy industries. The rise of Dutch metal industries paralleled that of the great maritime innovation, the steamship, the first of which appeared in Rotterdam in 1823.

Switzerland and Belgium have been taken as examples of small countries that industrialised early. The situation was very different for Sweden. During the first half of the last century Sweden was one of the poorest countries in Europe. The late start of industrialisation was, however, followed by rapid advance. The first Kondratiev cycle did not touch Sweden at all. Towards the end of the second—the railway boom—Sweden's industry was the fastest growing in the whole of Europe, and in the third cycle Sweden was already in the forefront, especially in the area of electricity, with hydropower and particularly the pioneering work in power transmission. One of the bases of the expansion of Swedish engineering industries was an early invention, the milk separator (1870), which greatly contributed to the creation of an exporting dairy industry in the Nordic countries—apart from its important role in helping engineering to become established. It was followed later by other Swedish innovations in such varied areas as turbines, electrical machinery, gas accumulators and ball bearings.

The country that perhaps benefited most from the Swedish milk separator was Denmark: the new equipment provided a changed technical basis for dairy farming, an important part of the Danish economy, and contributed to a considerable extent to the development of the cooperative movement in Denmark. The latter was instrumental in transforming the export of pigs into the sale of packed pork, another innovation that required technical and organisational change resulting in the significant growth of another part of Danish agriculture.

As in all Nordic countries, industrialisation started late in Norway, with the development of hydroelectricity around the turn of the century. In the five years 1900–5, Norwegian hydroelectric power production rose eightfold. Its beneficial effects can best be illustrated by the growth of the chemical industry: the increasing production of fertilisers, calcium nitrate (based on German and American inventions but further developed by Norwegian innovators, Eyde and Birkeland) as well as carbide (by 1910 Norwegian output amounted to 20% of world production).

Single major innovations can often be seen to have created the basis of whole national industries. For example, the Bessemer steelmaking process offered the opportunity for the large-scale use of Spanish iron ores and for the development of the steel industry in Spain.

Another smaller country whose history is worth studying from this angle is Hungary. Because of historical conditions, industrialisation in Hungary started even later than in Sweden. For a long time the country remained basically agricultural and its industry primarily served agriculture. It was only after World War 2 that industry became the largest sector. This background makes the spectacular development in a few isolated cases even more striking. Among these achievements, probably the most marked were those of two particular electrical engineering firms which acquired international reputations (Ganz and "Tungsram") and developed very rapidly indeed, thanks to the significant innovations of a few highly gifted and successful scientists and engineers in various areas of the then fast-growing electrical industry. This example is of some significance: it underlines the importance of innovations even in a situation where the whole economic climate is not particularly favourable, external economies hardly exist, and the domestic market does not provide any great stimulus to innovatory activity.

Some points arising

Thus, with the benefit of hindsight, it is quite possible to find historical evidence for the importance of major innovations as generators of general or regional economic upswings. It is much more difficult—and perhaps impossible—to answer certain questions that emerge. I will try to sketch them.

- Who was the "basic" inventor? Whoever he was, a long chain of previous inventions helped him. Take Faraday, whose demonstration to the Royal Society in London, anno 1831, is often taken as the birthday of the much later developed electricity industry. He could not have presented his theories without the outstanding achievements of scientists like Benjamin Franklin (1749), Galvani (1791), Volta (1800), Ampère (1822) and others. It was a long way—before Faraday and after him—to the large-scale electricity industry that "created new social benefits, new markets and new jobs" (to use the definition, coined by Mensch,[3] of a major basic innovation).
- Faraday was a scientific genius; so were some other great inventors. But Watt's simple aim was a better engine and Daimler's was a new vehicle he could sell. None of them was thinking, presumably, of his achievement as the start of a new era; nor were their immediate contemporaries thinking it. The great new opening was usually recognised somewhat later by others—such as by Henry Ford in the case of the automobile. With the benefit of hindsight it may be possible to assess the fundamental importance of some major innovation, but it is difficult—or perhaps even impossible—to classify any relatively new development in this super-class without the historical perspective.
- Finally, whilst the significance of a major basic innovation is obvious, from the point of view of its impact on the economy it is not the innovation itself but its *diffusion* across the economy, and the *speed* of this diffusion, that matters most.

These statements may seem to many to contradict the views of Mensch, who argues[4] that one of the main reasons for the decline of the world economy is the dearth of major basic innovations in recent decades. However, I trust that my views do not contradict Mensch's theory at all: they supplement it. There are many innovations "in the pipeline"; some of them may prove to be of major importance but only time will show which one(s)—if any. At present we cannot see the forest for the trees.

The question is nevertheless justified: what is likely to happen? Leaving all other important aspects aside, is there anything on the horizon—in the area of innovations—that may make a marked impact on economic activity or significantly reshape the way of life? Nothing less than the divine inspiration of a prophet is required to answer questions of this kind, and I can offer no more than personal guesses.

Guessing the future

Let us first have the "microprocessor revolution" out of the way. There are many who believe that this great innovation will be the redeemer of the sickly world economy, following in significance the motors of earlier upswings. The importance of micro-electronics can be seen in many areas already. The microprocessor is like a chameleon: it takes on the character of whatever programme has been fed into it. It can direct a guided missile, operate a fuel injector or a coffee dispenser, or control an industrial process. It can be used almost anywhere, in metal machining and in medical diagnosis. It is conceivable that it could be a candidate to lead a technological upheaval, giving the necessary push for a swing out of Mensch's technological stalemate—if we really are in this position.

It is not belittling the significance of the microprocessor when I express the view that I consider it to be the necessary instrument for achieving a really marked change but I expect this change itself to come elsewhere. My candidates for this change are the following areas, in this order: energy; food; environment; social institutions. Some may miss in this short list those natural resources that are neither food nor energy; their omission is intentional. I believe that within the foreseeable future none of these will be in fundamentally scarce supply; but if, contrary to my expectation, shortages do occur, mankind will find a way to live without the scarce material or find out how to replace it. Scientific and technological advance will provide the answer in the case of such an emergency—as it so often has in the past.[5]

The above four areas represent various factors of production (or "resources"), of which three are likely to be in a tight or disturbed position and one in uncomfortable abundance. Energy may become scarce; food supplies inadequate and land for expanding food production difficult to find; and the environment may be disturbed by pollution of various kinds. The fourth area, social institutions, concerns the factor in plentiful oversupply: people.

Energy

With only a very slight stretch of the imagination it can be said that the motors of earlier Kondratiev upswings, as identified by Schumpeterian analysis, were closely connected with energy developments. The first upswing had water power and the beginning of the coal era. The second was clearly based on coal; the third on coal, electricity and the early development of oil. The last upswing (the postwar golden age to 1973) was based on oil and gas, with the birthpangs of nuclear energy.

The energy situation does not require any great analysis here. It will suffice to say that the present key, the oil situation, has already become tight. The price of oil at the producers is now some 15 times its 1972 level in money terms, and increased over this period something like six- or sevenfold in real terms. The world's main source of oil in international trade is the Middle East, whose producers, allied to exporters in other areas, are associated within OPEC; they are—at present—in a position to keep the market tight and set the price by adjusting their supply to demand.

The tight supply position and the high level of the price have, however, a number of implications. On the negative side (which at present—but, hopefully, probably only temporarily—far outweighs any benefits that might come from the positive side as detailed below) are the well-known influences fuelling inflation, reducing domestic output and real income of the oil consumers, and resulting in an imbalance in the international financial position to an extent that from time to time it may seem unmanageable. These are all immediate effects. There is also a positive side to the present constellation, but with effects that can only work in the uncertain future (and some would no doubt add: if at all). Scarcity and the high price are a great stimulus to R and D efforts in three directions: conserving energy, that is raising the efficiency of usage; searching for new sources of conventional fuels; and seeking sources of new types of energy. Mankind will be pressed to progress and it is my belief that solutions will be found; indeed, there are already signs pointing in all three directions. Let me start with the first: conservation.

There are many studies related to the estimation of the price elasticity of energy use. They differ in their estimates but agree, without exception, on one point: whilst the short-term price elasticity may be low, the long-term reaction of demand and usage to price changes is high. The long term may be longer than we would like to see; in the present conditions of very slow growth and reduced investment activity it may indeed be very long.

The three points that have to be borne in mind are these. First, apart from the relatively easy measures that can be taken in the framework of better housekeeping (that is, improved organisation and more careful control of energy use), any more important change that would yield higher energy efficiency requires new investment. That takes time, especially in a period of high inflation and slow growth, since there are many other objectives calling for investment and energy projects compete with others for the limited capital available. Secondly, and this is more important, apart from the once-and-for-all effect of the dramatic rise in oil prices in 1973/74, until recently the financial incentive to invest in energy saving has not been too great. Whilst the macro-economic burden on our economies has increased substantially with the manifold rise in OPEC prices, the cost to the final consumer has risen much less. Up to the closing months of 1979 this rise in real terms was in many cases no more than 30–40%, and in the case of motor spirit even less. We cannot expect too much reaction to such a relatively minor real price change, remembering that energy costs still do not account for a very large share in total production costs (apart from some very heavy users). In this respect I expect a significant change as from about now, when the second price explosion is likely to reach the consumer. Thirdly, it is questionable to what extent elasticity estimates, which are invariably based on historical experience going back to the period of cheap energy supplies, are relevant to the future of a greatly changed energy situation. This is one of the reasons I am avoiding the use of any quantified estimates. Apart from the price, the recognition of possible scarcities may also work in the same direction—that is, to devote much more attention and allocate more resources to energy conservation.

The second effect of the present situation is the search for new sources of conventional fuels. Here we have plenty of evidence practically all over the world, but let me restrict myself to the North Sea. The second wave of oil price rises in 1979 has already induced operators to start building the equipment for the exploitation of smaller oilfields discovered earlier that would not have been touched at a lower price level. Exploration has also received a fillip from the price hike and in May 1980 both Shell and BP announced independently the finding of major new gasfields in the North Sea. Given the fact that so far only a part of that maritime area has been explored, it is not wishful thinking to believe that even more will be forthcoming; and the situation is probably similar in other areas too.

Finally, the search for new sources of energy is also seen in a different light now. In the scientific sense, these sources are not new. It is the engineering that causes problems and the difficulties of producing usable energy from these sources in large, commercial quantities. Whichever of the new sources we take—solar, biomass, sugarcane-based gasohol, oil shales, tar sands, etc—none of them will supply energy cheaply. But whilst so far the difference between the cost of the novel type of energy and that of conventional oil has been very large, with the price-raising race that can be witnessed among OPEC producers the price of oil is coming nearer and nearer to the higher level of the new fuels, helping the latter to reach economic viability within the foreseeable future.

There are, however, some other important points to be remembered. Let me go back in history in order to survey briefly what happened in the past on the energy front, taking just two examples. I will deal with the first only briefly since it has been described in some detail elsewhere:[6] what form did those developments take that helped to overcome earlier, similarly serious energy shortages? The present one is not the first in mankind's history. England was facing serious fuel scarcity in the early 18th century, having burnt up her formerly ample forests; a number of major technological breakthroughs and the following stepwise innovatory developments helped to overcome the problem by converting her economy to the use of the newcomer at that time: coal, of which there was plenty.

My second example requires a little more elaboration. It concerns the United States, the country where (among the more advanced economies) fuelwood remained the energy base for the longest time. As late as 1870, about three-quarters of all the energy used in the US came from fuelwood. By then, the transition to coal was already underway and towards the end of the last century it became the dominant source. The important point in this transition was the change from a severely limited fuel resource to one available in apparently endless quantities. Abundance paved the way for the unimpeded growth of iron and steel production, which, in turn, made it possible to build a railroad network covering the whole enormous continent and to produce the machinery required for the rapidly expanding manufacturing industries. The writings of Rosenberg[7] and Schurr[8] describe how, once the fuel constraint was broken, one development led to another in the various branches of industry and services (and also in agriculture) in a dynamic sequence.

Later on, the US was in the forefront of developing a large-scale electricity generation and supply industry; this removed the limitations imposed on factory processes by the earlier mechanical energy system (which used shafts and belting to transmit power from the in-house prime mover), and led to large productivity increases. The oil period, apart from generally increasing mobility, led to enormous increases in crop yields with the help of tractors and energy-based artificial fertilisers, again removing further constraints—those of the availability of natural fertilisers and animal drawing power. Geographic constraints were also removed: the railways overcame the limits formerly imposed on industrial locations by waterways required for transportation and water wheels for power. In later years, liquid fuels and the ease of transportation by trucks and cars made the limitations imposed by railways and coal disappear. And finally, air conditioning and air transport removed other limitations of climate and distances.

Of course, an endless number of adaptations within the economic system were needed along this long road to changed conditions, but technological advance supplied the answers to the emerging problems. These fundamental changes were all-pervasive; they were caused by the transition from the relative scarcity of one form of energy to the abundance of another, then the emergence of yet newer fuels. There is reason to believe that on the different way that lies ahead similar developments will occur across the board. This way is more difficult because it goes from relative abundance to a tighter situation. The instinct for survival, the "animal spirit", is strong however, and since the energy situation will leave its mark on practically every aspect of present-day life, it is not too naive or too optimistic to expect a similar cascading of technological changes to bring answers to many future problems. We may have some vague ideas about what these changes might be, but we cannot know exactly. Nor did the Americans of 1870.

This is why I believe that major developments will emerge in the area of energy and will radiate and cascade from there. If we think of the very large capital required for a single nuclear power station or one new coalmine, not to mention the billions invested in the North Sea, and if we also remember that new forms and sources of energy will probably require relatively even larger participation from many industries and services, then it must become clear that the impact of these developments will also be fundamental and very great.

I shall dwell only briefly on the other three areas in which—in my personal view—major developments can be expected. All three are in some way—to a greater or lesser degree—connected with energy.

Food

It does not require too much justification to state that food production will have to be increased considerably in order to feed the rising population of the world and to secure its better nutrition. Clearly this is the difficult task of agriculture. The considerable direct and indirect requirements of agriculture (including fertilisers and many other inputs into agricultural production) provide the link to energy supplies. It seems certain that these requirements are going to increase in future, in view of the rising agricultural production.

There are, however, other aspects to be considered as well. Energy materials supply the base to make many of the synthetic, man-made materials the world's industry needs. This base material will be in short supply, hence there will be excess demand for the natural varieties. To limit myself to textile fibres and rubber only: if the present quantities of man-made fibres and synthetic rubber were to be replaced by natural products, the area now serving the production of cotton, wool and rubber ought to be at least doubled and possibly trebled. Because of climatic conditions the possibilities of cultivation are limited to a large extent to those countries and geographic areas where every square metre of good soil is badly needed for the production of basic foods. Thus—apart from some marginal lands—industrial materials may be competing with food for the limited area of land. Given the dual objective, considerable developments will be needed in the technical and the organisational sense to achieve both goals.

Environment

The environment is a vexed subject nowadays, and is further complicated by its association with energy. The production, transformation and consumption of energy are activities which probably cause most of the pollution in the environment. Insofar as one can judge, some new sources and forms of energy may be relatively less polluting (eg solar), but others—such as shale oil—may add markedly to environmental problems in the future. For the present, we have been witnessing the emergence of public opinion against the dangers that nuclear power stations may present to the environment, resulting in demonstration campaigns that have led in some cases to very powerful intervention.

Without going into great detail, I am expecting significant advances to fight the pollution of air and water, since otherwise the emergence of unforeseeable dangers is likely, or at least possible. However, the chances that these developments will become generators of additional economic activity are much less than in the case of the two previous sectors, energy and food.

People

My fourth candidate, changes in the social field, directly concerns people. It is mentioned not so much as a possible source of growth, rather as a necessity. Unemployment is a general problem, indicated both by the high level of registered unemployed, a figure that has refused to return to more normal levels ever since the great recession of 1975, and also by the recent low growth of productivity, which points to overmanning and unemployment within the factory gates. Energy scarcity will hardly help; in all probability it will make the situation more difficult by retarding further economic growth.

Coupled with this is the alienation generally experienced by the worker from his work in an assembly line or other organisational form of modern industry. Technological advance has helped to reduce labour requirements of production and other activities but it has not helped to solve the ensuing problems of the society and those of the individual. With our present institutions, it is likely to be increasingly difficult to keep a growing number of people in employment that is useful and also satisfies them. Social innovators have so far been much less productive and successful than scientists and technologists.

It is a basic truth that man does not live by bread alone. Whether our present social set-up—both the Western and Eastern types—will be able to provide "bread" for the future population now seems doubtful; but it is almost certain that the majority of mankind will not find non-bread, non-material satisfaction, let alone happiness. This is not the right place to go any further in this matter, which is in any case more in the line of disciplines other than economics; whether we need a shorter working week, earlier retirement, some clever reorganisation of industrial and other types of work, the return of the craftsman or something else that cannot even be guessed at—all this would exceed the scope of this paper. But the need for some basic social change will soon be generally felt.

Concluding thoughts

The naturally emerging question is one that asks, if—as it appears—a "next round" is due to begin soon, where is it likely to start? A question obviously impossible to answer; some speculation may, however, not be out of place.

Kindleberger[9] has put forward the elegantly formulated view of the "ageing economy", contrasting the "young countries" with the British climacteric, a condition which may be diagnosed as a kind of general arteriosclerosis attacking the whole system in an advanced period of maturity. This makes the ageing economy lose out "in a field full of innovations". It is an attractive theory, one of its attractions being that (as the author admits with some hesitation), although the past is not easily discarded, "bygones are bygones" and a new start could be made at any time.

Economic history supplies many examples of the rejuvenation of economies that, at various points in time, could legitimately have been considered "ageing". France has emerged dynamically from her long inter-war stagnation; so did Germany from war destruction; Belgium and Holland have successfully recovered from the loss of their colonial empires, as did Austria after her less happy experience in the decades following the exit of the Habsburgs; and the United States gave evidence of their vigour by the lively recovery in 1976–78 from the deepest recession since the war. Industrialisation and developments in many sectors outside industry have reached a stage that is highly developed as compared with the past in the USSR and in the area that has come to be called Eastern Europe. And some of the less developed countries have already progressed rapidly enough to be called "newly industrialised".

One noteworthy experiment in the sociology of industrial work has been undertaken in one of the smaller industrial countries, Sweden (in the Volvo works). A successful new solution for easing energy problems comes from Brazil (in the form of the sugarcane-based "gasohol"). Thus, size or position on the development ladder do not seem to be absolute requirements for discovering and developing new avenues of departure.

The natural and human endowments of various countries are different; it remains true—to some extent—to say that whilst their chances are about equal, they are more equal for some than for others. The differences, however, are relatively much smaller than they used to be in the past. It is not a "handicap" race par excellence any more; it seems to be fairly open.

Notes and references

1. G.F. Ray, "Innovation as the source of long term economic growth", *Long Range Planning*, April 1980, *13*; "Innovation in the long cycle", *Lloyds Bank Review*, January 1980, No. 135.
2. Many references were given in reference 1. Some further and recent works: W.W. Rostow and M. Kennedy, "A simple model of the Kondratiev cycle", *Research in Economic History*, 1979, *4*; J.W. Forrester, G.W. Low and N.J. Mass, *Capital Formation and the Long Wave in Activity* (System Dynamics Group, Sloan School of Management, MIT, Cambridge, Mass., July 1977); J.W. Forrester, "Innovation and the economic long wave", *The McKinsey Quarterly*, Spring 1979; C.J. Stokes, "A long range view based on the Kondratiev cycle", *Business Economics*, January 1980.
3. G. Mensch, *Das technologische Patt* (Frankfurt, Fischer, 1977).
4. *Ibid.*
5. G.F. Ray, "The contribution of science and technology to the supply of industrial materials", *National Institute Economic Review*, May 1980, No. 92.
6. G.F. Ray, "Energy economics—a random walk in history", *Energy Economics*, July 1979, *1*, 3.
7. N. Rosenberg, *Perspectives on Technology* (Cambridge University Press, 1976).
8. S.H. Schurr, "Energy, economic growth and human welfare", *EPRI Journal*, May 1978, *3*, 4. (Also: Resources for the Future Reprint 154.)
9. C.P. Kindleberger, "The ageing economy", *Weltwirtschaftliches Archiv* (Kiel-Tübingen), 1978, *114*, 3.

CHAPTER 14

EXPLAINING LONG WAVES OF CAPITALIST DEVELOPMENT

Ernest Mandel

Although basic innovations are bunched in a 'countercyclical' manner during the depressive phases of long waves, they do not themselves *cause* the transition to an expansionary boom phase. Explanation of the turning points must be sought elsewhere. Long-term changes in the average rate of profit are the main cause of fluctuations in the system, but these are related to other fundamental features of the capitalist mode of production. Whereas the upper turning points from the boom to the depressive phase are determined largely by endogenous factors, especially the rise in the organic composition of capital (growth of capital intensity), this is not true of the lower turning points. Exogenous 'system shocks' of various kinds are needed to propel the system out of the depressive phase. This is the arena of acute social and political struggles, whose outcome is by no means mechanistically predetermined.

It is amusing that the long waves of capitalist development also produce long waves in the credibility of long-wave theories, as well as additional long waves of these theories themselves.

In the mid–1960s, we were somewhat alone in predicting that, towards the end of the decade, the long post-war boom would end and a new long wave with depressive trends would be initiated.[1] The overwhelming majority of economists were of a different opinion: they argued either that no such turning point was to occur, because Keynesian and neo-Keynesian techniques had overcome the system's propensity to produce long depressions; or, in so far as they were addicted to the classical neo-liberal school, especially in its monetarist variant, they saw the long boom ending exclusively as a result of inflation, without trying in the least to tie in this prognosis with an examination of what had happened to the capitalist economy from the wars of the French Revolution until World War 2.

Since the beginning of the 1970s, the situation has changed radically. We have seen the credibility of long-wave theories, which had dropped to near zero

Professor E. Mandel is from the Faculty of Economic, Social and Political Science, Free University of Brussels (Flemish Section), and can be contacted at Jos. Impenstraat 127, 1030 Brussels, Belgium.

during the previous decades, rise dramatically as it had done, for similar reasons, in the inter-war period. When reality stares you in the face instead of you having to dig for it below the surface of appearances, it is hard to stick to the position that its existence remains unproven. So one after another, the pundits of the profession have jumped on the long-wave theories' bandwagon, Samuelson being the latest of a steadily growing list.[2]

In all that undoubtedly creative turmoil, the real problem, however, remains largely unsolved among most academic and professional economists. It is often obscured to the point where it is not even posed. The question is not just to find all kinds of statistical correlations between cyclical and anti-cyclical phenomena operating within time-schedules encompassing several ordinary business cycles. Such discoveries are, of course, useful and, when seriously verified by empirical data, they have to be integrated into a more fully developed theory of the long waves in the history of the capitalist economy. But they cannot, in and by themselves, *explain* why, at certain points in time, a depressive long wave turns into an expansionary one, and vice versa.

Such an explanation can only be structured around *causal* relationships, which have to find their place in the framework of a general theory of how capitalism operates as a system (Marxists would say: of the general laws of motion of the capitalist mode of production). It stands to reason that, in the same way that business cycles, which have repeated themselves more than 20 times on the world market for industrial goods since 1825, have to be related to the system's inner contradictions, so have long waves of capitalist development, which have occurred at least seven times in the same period.

We would wholeheartedly agree with the argument that such a theory, in order to be scientific, must not only be empirically verifiable and falsifiable, but must be constantly checked and rechecked in the light of new long waves, as well as in the light of new data appearing about old ones. But there is something basically different between establishing statistical correlations and establishing causal relationships.

Causes of long waves

Let us take two examples. Gerhard Mensch has argued, convincingly in our opinion, that basic industrial innovations generally occur anticyclically under capitalism. If we take his list of those basic innovations in the first half of the 20th century with a rather long time lag between invention and innovation—never mind whether this list is exhaustive or not—no less than 29 out of 41 such innovations occurred during a depressive long wave (to arrive at that figure, we distinguish here for the 1944–45 period between the USA, where the long boom had already started in 1940, and Europe where it started only in 1948). That degree of correlation is high enough to deduce a general law from it, which, incidentally, fits in well with Marxist business-cycle theory. But it offers no explanation of why and how the turning point between a long depression and a long boom occurs.

Indeed, if it pretended to offer such an explanation, it would contradict itself in a paradoxical way. For if innovations are countercyclical, they certainly do not precipitate in the short term a basic change from a long depression to a long boom. A bunch of innovations occurring in 1930 or in 1935 cannot explain why the long depression ended only in 1948 in Europe and Japan (in 1940 in North America). It is sufficient to note that aggregate investment remained largely stagnant throughout the 1930s in the USA in order to understand that the innovations mentioned by Mensch have not, by themselves, precipitated that basic turn.

The idea that innovation creates more or less automatically an expansionary long wave, if only it is broad and sustained enough, is utterly misleading. Large-scale innovation only creates a long wave of sustained and increased growth if there is simultaneously a significant broadening of the market, and a rate of growth which leads towards a high level of employment. But innovation can lead to increased unemployment and thereby a stagnating market for final consumer goods and stagnant overall investment. According to David Foster, a computer which it takes 350 men/years to produce replaces 4 000 men/years in the service industry.[3] Surely, all things remaining equal, such types of innovation will not lead to sustained high levels of economic growth.

Monetary and psychological factors

The same applies to all those variants of long-wave theory which emphasise essentially monetary and psychological factors ('the general economic climate'), as eg those theories linked with Cassell's gold hypothesis or with Dupriez' generation concept (a generation which is too permissive regarding credit expansion is said to be regularly succeeded by a generation which is too cautious with regard to credit). Here again, empirical evidence seems to confirm an undeniable correlation. Price declines dominated the 1873–93 depression, as they did in the 1913–39 period. Despite appearances, they dominate the present long wave as well (prices expressed in gold, this is, and not in constantly depreciating paper currencies). Conversely, price rises dominated the expansionary long waves of 1848–73, 1893–1913 and 1940 (48)–1970. Abrupt changes in government and central bank attitudes, determined by general changes in bourgeois 'public opinion', have obviously influenced these turning points, as have objective phenomena like sudden ups and downs in the value of gold (the productivity of labour in gold mining).

But it seems clear that these changes are consequences rather than causes of what makes the system as a whole move towards slower or faster long-term growth. It is not because the (bourgeois) public becomes pessimistic that growth slows down. It is because the signs of growth slowing down make capitalists become more pessimistic about their own prospects, and governments suddenly go all out to 'fight inflation'. They did not act in that way for as long as growth was buoyant.

Better still: inflation could stay 'mild' and appear to be 'kept in hand', precisely because growth was smooth and seemed self-reproductive, ie because real surplus-value fuelled capital accumulation. Inflation got 'out of hand', became 'cumulative' and was nourished by 'anticipation' precisely because growth *was* slowing down; larger and larger parts of big firms' investments had to be financed through credit expansion and not through the accumulation of real surplus-value, at the same time as a growing part of consumer expenditure had to be financed through credit, ie inflation.[4]

Profit expectations which govern investment are not purely speculative. They are above all (though not exclusively) a function of *past and present profit realisation*, and of fluctuations in the average rate of profit. So the 'generation cycle', while undoubtedly correlated with the long waves, cannot explain the key turning points in the history of the capitalist economy.[5]

Fluctuations in rate of profit

More generally, as strongly as we reject any 'monocausal' explanation of the business cycle, we reject any 'monocausal' explanation of the succession of long waves in capitalist development. While dissymmetry between the turning point of a long boom into a long depression and the turning point from a long depression into a long boom has to be emphasised,[6] we believe that the expansionary long waves result from an interaction between multiple, partially autonomous variables, leading to a sudden strong upsurge of the average rate of profit and a sudden rapid expansion of the world market.

The advantage of seeing in the fluctuations of the average rate of profit the main cause of the succession of long waves, lies in the fact that profit is, after all, what makes the system tick. Simultaneously, under capitalism it is an excellent synthetic index of the system's overall performance, measured in the light of its own logic. So to see in the long-term fluctuations of the average rate of profit the main cause of the long waves of capitalist development is not to substitute one monocausal explanation for another. Rather, it is to find a cause which is determined by a sufficient number of factors—partially interrelated, but partially autonomous from each other—so as to reflect the system's overall *modus operandi*, while at the same time sufficiently close to the system's heart as to make one understand why changes in *that* factor can precipitate a change in the way in which the system as a whole grows or does not grow.

Turning points in long waves from depression to boom

When we try to demonstrate that the long-term fluctuations in the average rate of profit determine the long waves of capitalist development, we can also solve a parallel problem which remains an enigma to many 'model-builders'—the largely decisive influence of exogenous factors (wars, revolutions, counter-revolutions, the outcome of sharpened class struggles, the sudden discovery of important new gold fields etc) on the turning point from a depressive into an expansionary long wave.

It would indeed be hard to pretend that World War 2 and the ensuing Cold War had nothing whatsoever to do with the 1948–68 boom in Western Europe and Japan, as it would be hard to deny also the obvious relationships which exist between the revolution of 1848 or the full-scale unfolding of the era of imperialism after 1893, on the one hand, and the ensuing long booms following these turning points on the other.

There are essentially two ways in which these exogenous factors succeed, through their influence on the fluctuations of the average rate of profit, in precipitating expansionary long waves of capitalist development.

The bourgeois revolutions of 1848, the discoveries of the large-scale Californian (1848) and Transvaal (1893) gold fields, represent from the point of view of the overall operation of the capitalist system, a sudden broadening of the world market not resulting directly from the previous cycles of expanded reproduction. The system is, so to speak, catapulted into a broader framework, by a 'system-shock' as it were, to use a cultural formula. This leads to a quickening of the pace of capital accumulation.

Likewise, fascism, World War 2 and the Cold War implied a sudden and sharp rise in the rate of surplus-value, only explainable by climactic historical defeats of the working class after decisive episodes of the class struggle. Such a sudden rise in the rate of exploitation of the working class—resulting essentially not from the previous cycles' inner logic of capital accumulation, but from a radical change in the overall sociopolitical environment in which the system operates (destruction of trade unions, elimination of bourgeois democracy, atomisation of the working class, impossibility of collective sale of the commodity labour power etc)—can, if it gets and maintains momentum, reverse radically the tendency of the average rate of profit to decline, thereby initiating a long-term upsurge of that rate, and, likewise, a long-term upsurge of capital accumulation, ie of economic growth.

Certainly, radical changes in the environment in which capitalism operates are insufficient to trigger off a long-term rise in the rate of economic growth (ie a long-term neutralisation of the factors which lead to a decline in the average rate of profit). They must combine with a series of parallel trends endogenous to the system, in order to lead to that result. Elsewhere we have examined how that combination operates in practice,[7] and do not return to that problem here: we only emphasise that all these endogenous factors, however, will not come into their own if they do not receive an initial push from exogenous forces.

The outcome of momentous class struggles—as well as the outcome of big competitive struggles on the world market, inter-imperialist competition etc—is by no means preordained, in the Marxist system of analysis, by the given scope of capital accumulation. Otherwise, there would be little point in involving oneself in working class organisation and working class politics; only completely fatalistic determinism would be justified.

Transition from boom to depression and the falling rate of profit

To take up just one argument: one could defend convincingly the position that, for Marx, the level of wages is *not* determined by any physiological minimum to which wages were supposed to sink, but by the ups and downs of capital accumulation on the one hand (and its reciprocal—the ups and downs of the *conjunctural* reserve army of labour), and by the *structural* reserve army of labour on the other hand (the given historical level of industrialisation). But this is only true if one takes the capitalist mode of production as a totality, and one should add, as a *purely capitalist* totality.

But in real life, capital appears (*has to* appear) as 'many capitals', which in turn are structured into bourgeois (nation) states. Competition between these 'many capitals' and the bourgeois states, as well as the need to produce relatively stable capitalist class relations (maintaining the class struggle within certain bands with which the system can live), determine an international fragmentation of the labour market, ie *relative international immobility of labour*.

This explains why, under conditions of rising rates of capital accumulation, real wages can rise in industrialised countries, when demand for the commodity labour power exceeds supply on the national labour market, and while there are still huge untapped reserves of labour in underdeveloped countries.

Certainly, large-scale international migrations of labour generally accompany expansionary long waves throughout the history of capitalism. They explain why the boom can last long and be self-reproducing for 20–25 years, without wages going through the ceiling, ie without catastrophic declines in the rate of surplus-value. The 'national' reserve army of labour is regularly replenished through such migration. But these cannot go beyond a certain point as long as the labour market is fragmented into national markets and tremendous barriers remain erected against unlimited international mobility of labour. And that is why, towards the end of a long-term boom, there often occurs a stagnation or even a slight decline in the rate of surplus-value, or in any case a slowdown of its rise, which does not permit this rise to neutralise any more the forces imposing a decline in the average rate of profit.

This precise phase in the long waves has often been referred to as the one of the 'profit squeeze', and confused by many authors with the decline in the rate of profit. But for Marx and Marxists, these are two quite different concepts. They tend to be identical only for Ricardians and neo-Ricardians, who eliminate altogether from their calculations the fluctuations of the organic composition of capital, thereby making the rate of profit a straight function of the average level of wages.

For Marx, on the contrary, the decline in the average rate of profit is *not* due to any profit squeeze, ie roughly to any reduction of the share of profits in the national income. It is due to a decline in the ratio of profits to total invested capital, resulting from a disproportionate rise of that part of capital spent on equipment and raw material as against the part spent on wages.

In that sense, Marx's concept of the 'rate of profit' is, together with 'surplus-value', the most specifically 'Marxian' of all concepts Marx uses and, for that very same reason, the most frequently misunderstood one. It is closely tied to the concept of the organic composition of capital and its tendency to rise over time, ie it is based on the assumption of a labour-saving bias in technical progress on a historical scale,[8] an assumption which, in turn, revolves around the function technical progress has to assume under capitalism—not only to cut production costs but also to increase profits by making labour more abundant by reconstituting the industrial reserve army of labour. This assumption should not be seriously challenged in the age of semi-automation, the microprocessor and robotism, which increased permanent unemployment in the West within one decade from less than 10 million to over 20 million.

The asymmetric nature of the two turns—the one from the long boom to the long depression, and the other from the long depression to the long boom—appears clearer in the light of this remark. Given the long-term tendency of the organic composition to rise, and the impossibility for the rate of surplus-value to grow uninterruptedly in the same rhythm as the organic composition of capital, among other factors because of the national fragmentation of labour markets, then what is true for the business cycle remains true for the long waves: the average rate of profit *will* start to decline in the second part of the long boom.

This will eventually lead to the turning point between the long boom and the long depression, for reasons intrinsic to the process of capital accumulation.

On the contrary, a long depression does not in and by itself free forces which precipitate a new long boom. These can only appear as the result of system shocks operating on the process of accumulation of capital, ie of exogenous forces operating on the general environment in which the accumulation of capital unfolds. And this leads to the conclusion that the turn from the long depression into the long boom is essentially *not predetermined*, ie it depends on the outcome of momentous political and social struggles—class struggles, inter-imperialist struggles, wars, revolutions, counter-revolutions—an outcome which only in the last analysis, and very indirectly, can be related to the process of capital accumulation itself, and this in turn only if different time spans are taken into consideration. In a certain sense this outcome depends much more on what occurred during the previous long waves, when these 'subjective forces' bearing on current social struggles were being shaped, than on what is occurring in the current long wave itself.

In that sense, it seems idle to speculate on whether the present long depression of capitalism could lead to a new long boom before the end of this century. In theory, the possibility cannot be denied. *If* the labour movement and the national liberation movement are defeated (and brutally defeated at that) in all the key countries and key struggles of the next decade or decades, *if* the workers let themselves be transformed into drones, *if* huge masses of capital and human beings are destroyed by dictatorships and wars (the last 'adaptation' of capitalism cost mankind 100 million dead—the next might well cost 500 million), then robotism, solar energy, electric motor cars, Huxley's *Brave New World* and a few other 'niceties' would indeed create capitalism's 'fourth age'. But it would look much like barbarism to us, and the cost in the form of regression of human civilisation and bonded labour would be frightening. It would be preferable for the present long wave of capitalism to remain historically as the final one—the one which leads to world socialism.

Notes and references

1. See E. Mandel, "The heyday of neo-capitalism and its aftermath" in *Socialist Register*, 1964.
2. Paul Samuelson, "The world's economy at century's end", *The Japan Economic Journal*, 10 March 1981, page 20.
3. Davis Foster, *Innovation and Employment* (London, Pergamon, 1980), pages 19–20.
4. Technically, one can prove that credit expansion is inflationary if it grows quicker than output.
5. On the asymmetry between the two turning points, see Ernest Mandel, *Long Waves of Capitalist Development* (Cambridge University Press, 1980), pages 51–56.
6. It is interesting to note that during long waves with depressive tendencies, while innovation does occur, funds for *large-scale cumulative innovations* (which make the diffusion of innovations possible) are not available, not because of an 'absolute' scarcity of capital, but because during such depressive long waves capital does not like to take too big risks. In a recent French report (by a banker, a high government official and an MP) the difficulty of financing innovation is attributed to 'comportement des Français préférant les placements immobiliers à leur industrie, manque de volonté d'entreprendre plus que manque d'idée, manque de pragmatisme des innovateurs nationaux" (*Le Monde*, 4 April 1981). But why was such behaviour not predominant during the 1950s and the 1960s? The argument is circular.
7. See Mandel, reference 5.
8. *Economie et Statistique* (of the French National Institute for Statistical and Economic Studies), November 1980, *127*, analysed the French capital/labour balance.

CHAPTER 15

POLITICAL BUSINESS CYCLES AND LONG WAVES IN INDUSTRIAL RELATIONS

Notes on Kalecki and Phelps Brown

Michele Salvati

This paper discusses the concepts of the political business cycle and the electoral business cycle in relation to long and short cycles of capitalist development. In an elaboration and critique of Kalecki's seminal paper of 1943, it is shown that his concept is more relevant to long waves than to the short-term business cycle. Applied in this way it relates closely to the long-term interpretation of British industrial relations suggested by Phelps Brown.

IN THE MID-1970s—and many would choose an earlier date—industrialised capitalist economies entered a phase of slow economic growth, high inflation and acute cyclical instability. This phase has already lasted an unusually long time when compared with previous economic recessions during the long wave of post world war expansion; moreover, no end is in sight and it is predicted that it will continue well into the 1980s. Why this is so is a subject of considerable controversy among economists. But whatever the reason on which individual scholars would put the main explanatory weight—a growing maladjustment in international economic relations, the exhaustion of the expansionary effects of a bundle of basic innovations, etc—the voluntaristic optimism and the short-run focus of the McCracken Report[1] are being increasingly abandoned.

Apart from the economics of the present phase of slow growth and heightened cyclical instability, most social scientists would probably share the view that such a long period of economic difficulties will have serious social and political repercussions. The continuous increase in real disposable income and the absence of significant episodes of economic instability have been important ingredients in the smooth workings of democratic arrangements, whether long prevailing or newly established, in most advanced capitalist countries since World War 2. If this is true, the different behaviour of such variables over a long period of time may make for difficulties and tensions in the politics and societies of many of these countries.

Michele Salvati is Professor of Economics in the University of Turin.

Quite a few social scientists, moreover, would not limit themselves to the recognition of this one-way influence, from the economic to the sociopolitical system. They would probably point to another causal chain, leading from social and political to economic variables, thus adding to the purely economic explanations of the present economic difficulties. To take an example from one end of the ideological spectrum: the influential and vocal proponents of the "overloaded democracy" hypothesis would probably maintain, without denying the relevance of economic factors, that "excessive" demands elicited by a democratic political environment and a growing inability of governments to cope with them have something to do both with the origin of the present economic crisis and with the difficulty of rapidly resolving it.[2]

This double interaction—from economics to politics and from politics to economics—broadly defines the boundaries of the "political business cycle" field. Most contemporary research on the political business cycle, however, is short run in character and electorally oriented in approach. That is, it considers just the short and modest ups and downs in economic activity that have prevailed in most advanced, capitalist economies during the long expansion following World War 2, and examines policy-making through the paradigm of the vote-maximising politician, a paradigm in which only actual and expected outcomes of elections transmit society's impulses to policy-making and economics. Long and deep politico-economic waves are outside the scope of most recent theorising.

In a survey of European industrial relations after World War 2,[3] we have stressed the following points:

- In a rather short time span around 1970, there was a wave of industrial conflict (and of social unrest generally) which was surprising for its intensity, radicalism and simultaneity—surprising, at least, in the light of the theories of industrial conflict prevailing during the 1960s.
- This wave broke at the end of an unusually long period of full employment, although full employment and steady growth had gone hand in hand with a fairly peaceful development of industrial relations.
- The intensity, extent and forms of conflict have been quite different in the various European countries, largely because of different economic situations, politico-institutional arrangements and ideological traditions.
- The same economic, political and ideological factors explain the great variety of responses to the industrial relations and economic crisis during the 1970s.

The ground we surveyed and the time span we covered are those of a long and deep politico-economic wave. In organising the material and in defining the interpretative framework we relied heavily on two brief essays: Kalecki's 1943 essay on the political business cycle[4] and Phelps Brown's 1975 essay on long waves in industrial relations.[5] Returning now to the same ground in a more theoretical vein, we think that an accurate reading and a detailed commentary on these same two essays may be a useful starting point. Kalecki's piece is a deservedly famous one, but now a very old classic. Phelps Brown's is a short piece, written for the general public; most of the research on the influence of political factors on industrial relations has taken place afterwards. If read together, however, and with some patience, they clearly reveal many of the problems that arise when one tries to understand long politico-economic waves and ruptures and rearrangements of long-lasting politico-economic equilibria.

Thus, within the field of social and political waves, this essay is only a preliminary, critical and bush-clearing discussion, pleading at the end for a rather cautious historical approach. It shows that there are mechanisms making for long waves, for ruptures, for recurrent changes in structure. But it does not try either to pin down a periodisation and to test it through various indicators, or to link social and political waves to the rather strong vision of economic waves that characterises some other essays in this book.

Political business cycles (PBC) vs electoral business cycles (EBC)

Kalecki envisages economic policy formation in a capitalist economy *cum* democratic polity as fundamentally shaped by two forces, whose predominance oscillates as the level of activity moves from recession to full employment and then back to recession. Retaining Kalecki's words, we shall call these two forces "the masses" and "the capitalists" (hereafter without inverted commas). We shall start from an economic boom that has lasted long enough to create conditions of full employment. The capitalists, despite the economic gains they make during this period, cannot politically tolerate a full employment economy for long, because of its consequences for work discipline and political stability. They (or their vision) become(s) predominant in government circles as the boom goes on and deflationary policies are enacted, thus leading to a recession. During the recession, a symmetrical and counteracting political development will take place: the masses will not tolerate a less than fully employed economy, and their goals become predominant in government as recession and unemployment deepen. Reflationary policies are enacted and the economy moves back towards a boom.

These are the bare bones of Kalecki's model, the simple mechanics which may lead to a permanent politico-economic business cycle. In the next section we shall delve at length into the two pairs of symmetrical propositions by which Kalecki's PBC is continuously fuelled. Even in this elementary form, however, they are sufficient to understand the radical change in approach that has taken place with the recent EBC theories. Masses and capitalists are not well-defined actors in the political process that leads to economic policy formation, and the process itself—why the masses or the capitalists should become predominant in different phases of the economic cycle—is left institutionally unspecified. EBC theories can be described as an attempt at more precisely defining the actors (political parties in power or in opposition, the incumbent president or his challenger) and at specifying the process (the electoral process) that transmits and mediates feed-back between the economic and political systems.

Though many different theories of the EBC now exist,[6] their common core can be as easily described as that of Kalecki's PBC. Government popularity depends, *inter alia*, on economic variables: in a simple macro model it can be made to depend only on the past values of such variables as real disposable income, employment, inflation and the like (popularity function). Since government parties (or the incumbent president) are seen as seeking re-election (a simple or constrained vote-maximising hypothesis is used here), they will manipulate economic policy instruments in such a way as to maximise their popularity in the run-up to elections (reaction function). Under variously reasonable assumptions about the influence of the selected economic variables on popularity and voting, and about the effects of economic policies, the interactions of these two basic functions may produce jerky, oscillatory behaviour in the economy. The model is simple and points to variables that can be easily defined empirically and to processes that do actually occur and are highly visible in democratic polities. A lot of anecdotal evidence can be brought to bear, and serious efforts have been made to ascertain the presence, relevance and diffusion of electoral cycles in industrially advanced, democratic countries. The issue is still subject to rather radical theoretical and empirical debate, but it is fair to say that the evidence collected up to now is at least congruent with the weak hypothesis that elections lead to economic policy manipulation aimed at increasing the popularity of the party or the president in power, and that such manipulations may account for some of the modest cyclical instability that can be observed in the post-war period.

Whatever the contribution of electorally induced policy manipulations to economic instability, the EBC is such a different animal from the PBC that it is extremely doubtful if the former can be considered a progenitor of the latter. We have just mentioned the rather thorny problems that the PBC left unsolved by simply omitting the link between the social system and the formal political arena: its actors could not be easily detected in the political arena, and no simple, easily ascertainable way of detecting their influence on economic policy-making was given. These problems were solved at the price (or was it considered a further gain?) of completely abandoning the Marxian class analysis of Kalecki's model. In the EBC, the influence of political and social factors on economic policy-making is empirically recorded, but also conceptually sterilised through the parameters and arguments of the popularity function; moreover, with a technically judicious choice, the protagonists of the EBC model are defined as the formal actors in the democratic political system. The protagonists of the PBC, on the other hand, are social classes, though depicted very schematically. This obviously creates all the problems we have just recalled, but adds depth and fascination to the model.

We can illustrate this general point most clearly with reference to the role that elections play in the two models. Elections are the basic cycle-generating device in the EBC model, whereas they barely receive mention in Kalecki's essay. Certainly elections are the single most important mechanism through which the masses, if not the capitalists, bring their goals to bear on economic policy-making, but Kalecki would have found an exclusive concern with pre-electoral manipulations and window-dressing rather misleading. Full employment, on the one hand, and workers' discipline, on the other, are fundamental problems for the entire system and yet are the contradictory aims of a democratic, capitalist society, and their respective prevalence need not be patterned by electoral intervals. First, there are ways other than elections through which the goals of the capitalists—and even those of the masses—may make themselves felt: major shifts in economic policy in both directions, not dictated by immediate electoral aims, are no problem in Kalecki's approach. Second, elections are not taken as the ultimate and unquestioned "revelation" of the people's preferences: they are simply ways in which those goals may express themselves. In Kalecki's model the popularity function may shift according to the phase of the political cycle: though unacceptably schematic, implicit in the model is a deeper, social level of analysis which may explain the vagaries in the mood of the electors—why they vote for a full employment programme in one election and for law, order, discipline and deflation in the next. This can also be formally reconciled with the EBC approach, but, lacking a social level of analysis to explain changes in the preference of the electors, such changes must be introduced through either an exogenous change of the preference function or a constant, but inconsistent (eg "myopic") preference function. Neither solution is very appealing.

The PBC and EBC are theoretical constructs which basically try to answer the same question: why economic policy-making in a democratic, capitalist country tends to generate economic instability. Given the lack of specification of the PBC and the *ad hoc* adjustments always possible with the EBC, the same sequence of events can be rationalised, if not explained, by both. To take the simplest example, think of the two different rationalisations that could be provided of the government's behaviour in expanding the economy before elections and cooling it off afterwards. They descend, however, from different theoretical paradigms—the Marxist and the quasi-neoclassical, post-Downs paradigms of modern political theory—and are profoundly different in structure, despite the superficial resemblance of describing a feed-back mechanism between the political and the economic systems.

In the following sections we shall abandon the comparison between the EBC and the PBC. Starting with a closer reading of the PBC, we shall enquire whether its richer theoretical potential can be developed and its analytical shortcomings eliminated, returning to a final assessment of the two models in the concluding section.

"Political aspects of full employment"

A closer reading

Why should the masses and the capitalists have the goals that Kalecki imputes to them and why should their influence on policy-making oscillate in the way we have just described?

Kalecki was writing towards the end of World War 2 and thought that in democratic, capitalist countries emerging from the war the lesson that unemployment could be avoided by means of appropriate state policies had been firmly learned by the masses: employment had become a political issue, something that can and ought to be influenced by governmental action, and was not a natural event, rising and falling according to its own natural and sacrosanct laws. Memories of the 1930s, and the crucial relevance of employment for the welfare of the masses, would have made long spells of generalised unemployment simply impossible in a democratic polity where the great majority of the voters depended for their livelihood on wage and salary income. As a matter of fact, the necessity that "something must be done in a slump" was agreed to even in capitalist circles

and either under the pressure of the masses, or even without it, public investment financed by borrowing will be undertaken to prevent large scale unemployment. [page 329]

The basic politico-economic goal of the masses and their power to impose it on government, directly or indirectly, through the democratic process are established with this simple argument—a few lines that cut across the haze of social and political intermediations with gordian simplicity.

The fluctuations in the salience of the goal and in the power of the masses to impose it are established with analogous simplicity, though here we find the only hint of the complexity of class structure and the problems of shifting class alliances that can be detected in the strictly "dichotomous" Kaleckian analysis. In a situation in which full employment has been prevalent for some time and inflation begins to develop,

a powerful block is likely to be formed between big business and *rentier* interests, and they would probably find more than one economist to declare that the situation is manifestly unsound. The pressure of all these forces, and in particular of big business—as a rule influential in government departments—would probably induce the government to return to the orthodox policy of cutting down the budget deficit. A slump will follow in which government spending policy would again come into its own. [page 330]

More interesting and speculative are the goals that Kalecki attributes to the capitalists. Full employment and high economic activity are a mixed blessing for industry. From a strictly economic point of view, they can only be advantageous: profits are higher at the peak of the boom than they are in other phases of the cycle, whatever happens to wages and productivity. This statement may be true or false,[7] but is a well-known consequence of Kalecki's theory of the business cycle: prices are determined by a fixed mark-up over prime costs, and such a mark-up is determined by the "degree of monopoly", which is a given for cycle theory. In a somewhat weaker form, Kalecki restates his point in the PBC essay:

profits would be higher under a regime of full employment than they are on average under a regime of laissez-faire; and even the rise in wages resulting from the stronger bargaining power of the workers is less likely to reduce profits than to increase prices, and thus affects adversely only the *rentier* interests. [page 326]

This economic advantage, however, is explicitly contrasted with a socio-political threat:

But 'discipline in the factories' and 'political stability' are more appreciated by business leaders than profits. Their class instinct tells them that lasting full employment is unsound from their point of view and that unemployment is an integral part of the 'normal' capitalist system. [page 326]

Why? Kalecki gives three reasons: the first concerns government interference in the problem of unemployment as such; the second concerns the means that the state is likely to use in order to reach and stabilise a full employment situation; and the third relates to the maintenance of full employment for an extended period of time.

The paragraph that Kalecki devotes to the first—why, if "every widening of state activity is looked upon by business with suspicion . . . the creation of employment by government spending has a special aspect which makes the opposition particularly intense"—is worth quoting in full. To my knowledge, it is one of the first statements, succinct but fully defined, of the theory of the capitalist state that has lately been developed by such authors as Charles Lindblom and Fred Block,[8] the latter in explicit reaction to previous Marxist theories such as those of Milliband and Poulantzas.

Under a *laissez-faire* system the level of employment depends to a great extent on the so-called state of confidence. If this deteriorates, private investment declines, which results in a fall of output and employment . . . This gives to the capitalists a powerful indirect control over government policy: everything which may shake the state of confidence must be carefully avoided because it would cause an economic crisis. But once the government learns the trick of increasing . . . employment by its own purchases, this powerful controlling device loses its effectiveness. Hence budget deficits necessary to carry out government intervention must be regarded as perilous. The social function of the doctrine of 'sound finance' is to make the level of employment dependent on the 'state of confidence'. [page 325]

Just as events in Chile and Portugal had much to do with the recent stress on investment and foreign exchange crisis as a powerful, though indirect, "controlling device" of dominant interests over the admissible span of state intervention, it must certainly have been the Leon Blum experience in France that induced Kalecki to open this path of enquiry.[9] We cannot go into the much-debated question of whether this line of argument can provide a stronger foundation for a theory of the capitalist state than its competitors.[10] Neither can we discuss the preliminary problem of whether this very endeavour—the construction of a theory of the capitalist state—makes sense outside a Marxian approach in which the social level of analysis is considered explanatory and the political level is explained.[11] A simpler *caveat* may now be in order.

Apart from such extreme instances as 1937 France and 1973 Chile, recent history is punctuated by less extreme cases of capital's "strikes" or "flights"; and there is little doubt that one of their main effects has been that of "keeping in line" state intervention with what was thought as admissible or desirable by dominant economic interests. However, rarely, if ever, has the level of state intervention necessary in the recent past in order to obtain full employment in advanced capitalist countries been of such magnitude as to impair the effectiveness of that "powerful controlling device". If we exclude periods of war, private investment elicited by private consumption has always been an ingredient of the full employment activity level at least as important as state expenditure—high enough, anyway, to make the potential blackmail of the state of confidence fully and continuously effective. While the government may have learned "the trick of increasing employment by its own purchases", it has never exploited such knowledge in order to dispense with the vagaries of the state of confidence of the capitalists: in very general terms, the post-World War 2 expansion has been the product of an alliance, much more than of a conflict, between entrepreneurial confidence and state intervention.

The second reason for capitalists' hostility to full employment concerns the means through which the government is likely to pursue full employment, aside from stimulating private investment: ie, state expenditure on investment and "consumption subsidisation", as Kalecki calls welfare expenditure. Having briefly reviewed the main, and more traditional, arguments against public investment, Kalecki states that

> subsidising mass consumption is much more violently opposed by these "experts" [the economists] than public investment. For here a moral principle of the highest importance is at stake. The fundamentals of capitalist ethics require that "you shall earn your bread in sweat" . . . [page 326]

Written in the early 1940s, this statement is strikingly prophetic. As economic conditions worsened in the 1970s, the boundary skirmishes over the proper areas of private and public investment—the boundaries themselves are different in different capitalist countries—were superseded by a massive attack on "consumption subsidisation", and the ideological overtones of the debate fully justify Kalecki's reference to a "moral principle". That people should earn their bread—and provide for old age, sickness and unemployment, etc—through sweat and toil may be justified on the basis of an ethical principle, or defended through efficiency considerations, or simply grounded on the sheer inability of state finances to maintain or further extend the level of today's welfare provisions; without doubt, nonetheless, some "economists" continue to wage their secular war against consumption subsidisation as vehemently as ever.

We have referred to, and commented on, the first two reasons for capitalists' hostility to full employment because of their intrinsic interest and relevance. We should be aware, however, that the case for a political business cycle basically rests on the third.

> . . . even if this opposition . . . were overcome—as it may well be under the pressure of the masses—the *maintenance* of full employment would cause social and political changes which would give a new impetus to the opposition of business leaders. Indeed, under a regime of permanent full employment, the "sack" would cease to play its role as a disciplinary measure. The social position of the boss would be undermined and the self assurance and class consciousness of the working class would grow. Strikes for wage increases and improvements in conditions of work would create political tensions. [page 326] . . . Workers would "get out of hand" and the captains of industry would be anxious to "teach them a lesson". [page 329]

Only in two circumstances, according to Kalecki, could a capitalist economy be run in conditions of permanent full employment. The first is the fascist state.

> The dislike of government spending policy as such is overcome under fascism by the fact that the state machinery is under the direct control of a partnership of big business with fascist upstarts. The necessity of the myth of "sound finance" which

served to prevent the government from offsetting a confidence crisis by spending on public investment or consumption is overcome by concentrating government expenditure on armaments. Finally, "discipline in the factory" and "political stability" under full employment are maintained by the "new order" . . . Political pressure replaces the economic pressure of unemployment. [pages 326–7]

Here capitalists and fascist state managers are presented as united in a common front against the workers, and somehow Kalecki falls short of the more complex, tripartite relationship sketched when dealing with the first reason for capitalists' hostility to full employment. With the experience of German, Italian, Spanish and Austrian fascism in mind, along with that of Deladier and Petain in France, he must have thought that the tensions between authoritarian state managers and the business community were of a second order magnitude compared with the tensions between the latter and the "masses". If successful, a democratic polity is perhaps "the best possible shell"[12] for a capitalist economy; but for many European countries in the inter-war period, such a shell was clearly not successful.

The second case is that of a non-authoritarian, full employment capitalism. But in this case

"Full employment capitalism" will have . . . to develop new social and political institutions which will reflect the increased power of the working class. If capitalism can adapt itself to full employment a fundamental reform will have been incorporated into it. If not, it will show itself an outmoded system which must be scrapped. [page 331]

This second situation was speculative, first because Kalecki could not point to any historical example, second because he doubted whether, had it existed, it would still have been a case of a capitalist economy and society. Republishing his article in 1962, he dropped the last section which dealt both with fascism and with "full employment capitalism". Fascism was not a salient problem by then, and the second case was, in his view, far from materialising.[13]

The major problems

"Political Aspects of Full Employment" is a short essay: our reading has not omitted any relevant part of it, and has actually extended some parts. One further extension would be very much in the spirit of that essay, and should be singled out for its importance: we can safely add a broader set of political, institutional and ideological repercussions, which Kalecki leaves implicit, to the individual and social repercussions of full employment mentioned by Kalecki—those connected to the disappearance of the fear of the "sack". A situation of permanent full employment would mean that interest groups and political parties, *stricto* or *lato sensu* "connected" to the working class, would probably become stronger *vis-à-vis* their counterparts "connected" with business interests. A shift in the political and ideological arena would probably conjoin a relevant shift in the social arena, as Kalecki clearly hints at when talking of "full employment capitalism".

Even an extended and generous reading, however, would confirm the impression derived above from the simple mechanics of the PBC. First of all, the social analysis on which the PBC is based is exceedingly schematic: the brief mention of *rentier* interests is more a sociological translation of an economic category than an analysis of the complex stratification of a modern capitalist society. In particular, the amorphous and shifting "centre" of these societies, so crucial in a political and electoral analysis, is left completely unexplored. The linkage between a social phenomenon, class, and the political phenomena of parties, political organisations and interest groups is by no means so obvious. Second, partly because of the schematisation of the social analysis, but also because the political level deserves an enquiry in itself, the absence of any explicit political science argument deeply weakens the PBC: after all, political cycles depend on economic policies; policy-makers are members of governments; governments are formed by political parties or coalitions thereof; in order to form governments, parties have to win elections. The whole *problematique* of the EBC must be at least a part of the PBC, even if the main stress of the analysis is laid on the class determinants of policy-making.

Beyond a simple restatement of what we had already seen through the "simple mechanics" of the PBC, our closer reading allows us to discover a further difficulty. Though the amplitude and the intensity of the PBC is never precisely specified by Kalecki, many implicit references seem to point towards a rather quick succession of "political" booms and recessions.[14] This is certainly the prevailing interpretation, and it is why it is commonly held that the PBC and the EBC essentially refer to the same set of occurrences: the short, policy-induced swings in economic activity which have characterised the long expansion following World War 2. If such an interpretation is correct, there is a partial incongruence between the nature of these short cycles and the reasons Kalecki gave for their occurrence.

Short interruptions or slow-downs in the growth process and moderate and brief increases in a structurally low unemployment rate simply do not deliver the goods that capitalists expect from them. They deliver different goods. A short, policy-induced recession still succeeds in eliminating or reducing an unacceptable balance of payments deficit, and, in the past, it usually succeeded in dampening wage demands and inflation; even in the 1970s some short-term trade-off between inflation and employment was discernible in many countries. But it does not greatly influence the social and political developments which Kalecki had in mind. As has frequently been observed, people assume that a recession will be short and will not greatly change their long-term job security and income expectations; attitudes and behaviour at the workplace need not be revised. Neither do consumption expectations, relative both to marketed goods and to public goods and transfers, seem to be very sensitive to "short and mild" oscillations in activity levels. If anything, the number of people who do not earn their living through the sweat of their brows increases during a recession. More generally, the growth of "consumption subsidisation" does not seem to have much to do with short cyclical manipulations. Political parties, unions and other organised groups whose constituency is disproportionately composed of manual workers (or salaried employees in general) are often hurt by this kind of recession, but not severely; to change their relative strength in the political arena something more seems to be necessary.

If such a brief characterisation of the effects of a short political cycle is plausible, we are forced to conclude that it is as if most industrialised, capitalist democracies had experienced conditions of permanent full employment, at least from the late 1950s and until the mid-1970s: the short political cycles of the post-war period have not been sufficiently strong seriously to counterbalance the basic reasons for the capitalists' dislike of full employment. There is no necessary contradiction in Kalecki's argument. It could be argued that intentions are one thing, results another: capitalists may have been motivated by the reasons given by Kalecki, and the outcomes have been more modest—mainly economic effects of boom and recession. It could also be added that the conditions of relative strength between the two poles of Kalecki's dichotomous model simply did not allow a full realisation of the capitalists' goals: capitalists have had to be content with the slight dampening of the undesirable economic and political effects of full employment achieved through short and mild recessions.

Nonetheless, a problem remains, which casts doubt on the soundness of Kalecki's approach. Many capitalist democracies have experienced, and within them capitalists have apparently prospered, practically continuous conditions of full employment for a very long spell of time, certainly long enough for the radical tensions Kalecki had in mind to occur. The sociological theories of the "withering away of conflict" and of the "end of ideology", so popular in the early 1960s, did not come out of nothing. It is true that conflict reappeared in the late 1960s/early 1970s, accompanied, as usual, by an outburst of ideology; up to those years, however, we had had a very long period in which the political consequences of full employment had not materialised.[15] Why do full employment and "political stability" seem to have gone hand in hand for so long? There are two main responses to this problem and a series of derivative combinations of the two: (i) such tensions and incompatibilities do not exist, and therefore Kalecki's suggestions lead us astray; (ii) they develop in a different historical time span from the one Kalecki envisaged. Since the first response is not of an "either/or" variety, but depends on the strength of the alleged tensions, there is room for intermediate cases. It could be maintained, for example, that there *are* tensions but they are not as strong as Kalecki may have thought, and they take more time to develop than he suggested.

As we have argued elsewhere,[16] our own view is that this third case fits the post-war experience of industrialised, capitalist democracies more closely than the other two. This means that the time horizon over which the development of the "contradictions" between full employment and political democracy unfolds is much longer than a "short" political cycle; that some of the "undesired" political effects of full employment may not materialise; that they may be weaker than Kalecki thought; and that capitalists may come to terms with them with institutional adaptations that fall short of a "fundamental reform", whatever that might mean.

Short cycles and long waves: the "non-monetarist" view of Phelps Brown

When we argued above that "short and mild" recessions in economic activity do not greatly influence the variables on which capitalists' dislike of full employment depends, we made a fairly uncontroversial statement insofar as wages are concerned. That rank-and-file demands and unions' "pushfulness" are not as sensitive as they once were to market conditions was a common complaint in many capitalist countries during the late 1960s and the 1970s. Quite a few economists would explicitly argue that recessions have not been sufficiently deep or long-lasting to provoke a real reversal in expectations about job security and price behaviour, and would summarise this development by claiming that the "natural" rate of unemployment is now higher than in the past: a much higher level of unemployment would now be necessary in order to lower deeply entrenched expectations and appreciably reduce the rate of wage and price increase.[17]

One does not need to be a monetarist, however, to recognise the element of truth behind such statements, and a "non-monetarist" explanation of the pay explosion of the late 1960s and early 1970s, which fully recognises and develops the expectation argument, has been provided by Henry Phelps Brown.[18] As with Kalecki's, Phelps Brown's essay is short and addressed to the non-specialist; suggestive, rather than analytical and fully documented. It portrays a quasi-cyclical pattern in a bundle of social phenomena—*lato sensu*, industrial relations—which are quite close to those described by Kalecki when he discussed the reasons for capitalists' dislike of full employment. Here, however, the similarities end. Kalecki's fluctuations are short, policy-induced cycles; Phelps Brown's are long waves, of which economic policies are only one in a very complex set of determinants. Kalecki provides a clear, albeit schematic, model, explaining why such fluctuations occur; Phelps Brown merely presents a historical analogy and a few reasons that may account for it. The analogy is striking, however, and the reasons suggestive; both are pertinent to the long-run *vs* short-run problem in which Kalecki runs into difficulties.

The analogy, essentially, is located between two phases of industrial "turbulence" in Britain (with a few references to the continental experience) separated by something like half a century. In Phelps Brown's treatment, the first phase seems to encompass a very long period, stretching between the first and the third decade of the twentieth century: the outbreak of World War 1 temporarily suspended many of the symptoms of unrest, but the ebbing of the long wave of industrial militancy came only after the decisive confrontation of the mid-1920s. The second phase is the wave of industrial militancy and high pay increases that swept Western Europe between the late 1960s and the early 1970s—in fact, it was the "pay explosion" of that period, which is subject to widely different interpretations by economists, that prompted the "non-monetarist"—but, really, *non-economic*—view of Phelps Brown. Before turning to the reasons why these long waves swell and subside, let us show how close Phelps Brown is brought to Kalecki by the width of his perspective.

What economists care about—a disturbing increase in wages—may be only one in a set of connected phenomena; one, moreover, that is highly dependent on the permissiveness of international economic and financial relations and on the attitude of national monetary authorities. It is therefore impossible to understand the nature of a wage push when it goes beyond the pure economic consequences of a sellers' market if attention is not focused on the whole—not just wages, but the change in attitudes and expectations among employees; the extension of trade unionism; the power of the strike; and the clash between industrial power, the rule of the law and the authority of government. [page 11]

The historical experience of advanced capitalist countries allows us, so Phelps Brown maintains, to identify periods in which the whole political framework of industrial relations—not only the rules and the conventions concerning factory work and its reward, but the very roles of workers and their representatives within the industrial and the political systems—is violently called into question. Typical of these periods are wage demands that go well beyond previous standards; demands about changes in work conditions that establish a break in the previous bargaining routine, given the existence of one; a clear shift towards the grass roots of the relevant decision-making levels within the existing trade unions; the coalescence of new collective identities among workers previously excluded from industrial representation and the birth of new organisations; more intensive and more extended conflicts, which assume different forms and expressions; intensive and continuous participation in decision-making, and the prevalence of expressive over instrumental decisions; objectives and goals that tend to go beyond the realm of industrial relations as previously defined and address the political system. In sum: the whole set of explicit laws or implicit rules of the game that for so long defined the role of dependent labour within industry, society and politics is questioned and challenged.

It is difficult to find better examples of what Kalecki's capitalists fear—a loss of discipline in the factories and of political stability—than Phelps Brown's references to the state of industrial relations in Britain, and in Europe, around World War 1 and around 1970. Almost *verbatim* Phelps Brown repeats and extends Kalecki's arguments on the powerful disciplinary role of the "sack", and on the consequences of its disappearance.

Among the factors making for a change in the employees' outlook after the second world war, the most evident is the unprecedented maintenance of a high level of employment. By giving the individual employee less reason to fear that if he withdraws from his job he will not get it back, it will have increased his readiness to strike, and to join and support others who are striking. By giving him more independence of the employer it will have changed the texture of relations, making subordination less acceptable and deference less natural, so removing an invisible but powerful sanction against pressing claims against one's immediate superiors. [page 17]

Kalecki and Phelps Brown clearly have the same social phenomena in mind and even use the same historical examples. Writing in 1973, moreover, Phelps Brown knew very well that the short-run political cycle of the post-World War 2 period did not prevent the realisation of a practically continuous state of full employment. Short cycles may have had some role in retarding the development of the political consequences of full employment, but, in his view, major causes of such delay lie elsewhere.

There is an additional reason why Phelps Brown's analysis is so relevant to our purposes. Kalecki does not empirically examine why workers, once conditions of full employment are established, "get out of hand". The answer, as we shall argue in the conclusion, seems to be directly derived from a paradigm in which workers' strength in the market and political democracy are held to be incompatible with the basic rules of the industrial relations game in a capitalist society; this explains why Kalecki tended to neglect the "timing" problem. For Phelps Brown, why explosions in industrial relations occur, why they take such a long time to mature in the presence of permanent conditions of full employment, why they peter out and a new, partially reformed set of rules of the game is established and then endures, are all empirical questions that he tries to answer through historical analysis.

Obviously, these are not just empirical questions, and theory creeps into the definition of the problems and the selection of the areas of possible answers. For theoretical reasons, the answers Phelps Brown provides are more convincing with reference to the problems of why tensions in the industrial relations system take time to mature and why they subside than to the problem of why they first occur. For the sake of argument, let us accept the general answer that Phelps Brown provides to the question why they occur: in the working of a historically given industrial relations system a set of expectations and behaviour tends to develop, and is, in the end, incompatible with system performance; expectations outstrip potential outputs and behaviour conflicts with the rules.

Why this should take so long to develop out of a previous situation of equilibrium is answered by Phelps Brown with a "generational" argument, which makes crucial reference to the persistence of expectations and preferences and to the customary nature of behaviour. Phelps Brown essentially argues that during the first part of the post-war full employment period, in the 1950s and early 1960s, the majority of the working class was composed of people who had a direct or a very close family experience of the 1920s and 1930s: their basic values, their hopes and fears, were formed in that period and were not greatly susceptible to later change, despite the presence of continuous full employment. A job was always something that could be lost, and attitudes towards work and discipline had become personality traits. Different expectations concerning employment and different attitudes towards work and discipline were introduced only when new generations reached a "critical mass". A great deal of scattered evidence supports this argument, and its plausibility is reinforced when we widen the horizon to institutional actors in the industrial relations or political systems. The institutional counterpart of the persistence of expectations and behaviour is organisational inertia, if one excepts periods of violent social mobilisation: marked changes in power relationships, in the industrial relations or political arenas, may occur long after their economic preconditions are established.

The same argument about the "stickiness" of expectations and habits and about organisational inertia should also be applied in reverse; that is, to the ebbing of a long wave of innovation and tension in the industrial relations system. But here the intensity and breadth of new experiences—a crisis of stabilisation—may well work as a surrogate of time: "the big blows" and the "traumatic experiences" of the 1920s[19] in Britain succeeded in extirpating in a remarkably short span of time the state of unrest in industrial relations that had existed since the first decade of the century. (Both Kalecki and Phelps

Brown recall that much stronger "blows" succeeded in an even shorter time in establishing a "new order" in industrial relations—and in society at large—in countries like Germany and Italy.) Out of the long crisis in the British industrial relations system in the earlier part of this century, a new global arrangement emerged in which some of the demands of workers and the unions were accepted: but the settlement would hardly have endured for nearly half a century had expectations not been drastically lowered by the "big blows" and the "traumatic experiences" of the 1920s.

But why should a historical set-up in industrial relations give rise, over the long run, to a radical challenge from workers and their organisations? This is the crucial first question in Phelps Brown's analysis, since it is from this challenge that a long wave of innovation and stabilisation is set in motion. Phelps Brown's answer is scattered throughout a host of concrete comments and historical references in his essay. In order to simplify the analysis, and to bring subsequent discussion to bear on it, we can re-order his arguments along three analytical dimensions: the first concerns power, the second values, ideology and beliefs, and the third political and social institutions.

Power, ideology, institutions

In his comments on the power situation of unions (or of informal groups of workers) in the context of industrial relations, Phelps Brown refers to both aspects of power that have received extended attention in recent studies of unionism and strikes: power as ability to inflict harm on, and power as ability not to be harmed by, the adversary—offensive and defensive power. He refers to both in his analysis of the situation of the late 1960s and early 1970s. Unions and workers could inflict harm by withholding labour in a boom, when firms are concerned with output and deliveries; in a highly interconnected economy, moreover, small groups of workers can wreak havoc with the activity of very large sectors of the economy. Defensive power is also strong, due to the permanence of quasi-full employment and the underlying pledge of political authorities to maintain it. He mainly refers to offensive power when dealing with the industrial relations crisis around World War 1: a full employment pledge was clearly absent, but there were economic booms, and strikes in crucial sectors, like railways and coal, could stop the working of the whole of industry.

As we have repeatedly underlined, Phelps Brown's analysis is not restricted to industrial relations in the narrow meaning of the term: at its peak, a "long-run" industrial relations crisis is also a political crisis, in which the rule of law is challenged and the whole political context of industrial relations is thrown into discussion. Reference to such a wider context is absent, however, when Phelps Brown deals with power: how market power can enhance the political power of unions and parties "connected" to the working class and how this political power can defend market power—reinforcing the full employment pledge, for example—are never explicitly mentioned; we shall discuss the different arenas in which power can be exercised when dealing with the institutional dimension.

Power gives rise to large-scale conflict and/or change only when it is exercised. In a short-run analysis of pure industrial conflict, we can take the willingness to exploit market power as a historical given, but we cannot do the same in a long-run analysis and in a wider political context; here the motivation and the willingness to engage in conflict and to expand the usual set of issues around which conflict is waged must be taken as variables, demanding explanation. We have just criticised Kalecki on this point, since he seems to take as self-evident the notion that as soon as workers are given a strong defensive power (permanent full employment), a political threat to the basic rules of the game of a capitalist society ensues. For Phelps Brown, who does not rely on a simplified Marxist paradigm, this is a bit of a mystery and, in order to explain it, he is driven into the normative dimension.

What he says, in short, is that, because of the normal workings of a growing capitalist economy or because of exceptional events, a traditional order, which gave everyone his role in the workplace and in society at large, is disrupted. Out of such disruption a situation of incompatible expectations and of normlessness emerges, which may explode in a political crisis. The arguments and the historical examples are familiar: the widening of mass education and increasing information and geographical mobility erode the traditional bonds of loyalty and deference; the usual constellation of reference points through which each worker assesses his status and income position is increasingly disrupted by economic growth and technical progress; the very growth of income and the continuous change in consumption patterns and life styles unleash incompatible expectations. Since the World War 2 period in particular, the very way in which the fruits of technical progress have been distributed—through increases in money wages exceeding the increase of consumer prices—has fuelled a condition of status uncertainty and irritation: keeping ahead of price rises and trying to combat the erosion of real wages has become a national sport. Moreover, above and beyond the effects of the normal workings of a fast-growing economy, wars radically uproot established self-images and codes of behaviour: the work-relations and the social roles to which soldiers returned after the war were hardly ". . . fit for heroes to live in" (page 13).

The arguments we have just summarised largely fit into the process of change in values, beliefs and systems of ideological hegemony at large that John Goldthorpe,[20] examining the non-economic causes of inflation, has defined as "decay of the status order". Against such arguments, every reader of Durkheim would immediately object that the destabilising effects of the decay of the old status order could be counteracted by the formation of new solidarities; more simply, there is no compelling theoretical reason or historical evidence that leads from more education, information and geographical mobility to a radical threat to the established order in industrial relations or in society at large. The boundaries of the normative dimension are wide, and coincide with those defined by the answers to the fundamental question of sociology since the time of Durkheim, if not Toqueville: why a society characterised by a minute division of labour "sticks together"—in particular a capitalist society, where the market works both as a basic organising principle of economic activity and as the fundamental stratification mechanism in society. Not only are the boundaries of the normative dimension extremely wide and include a great variety of specialised fields of knowledge and research, but the terrain is also crossed by profound paradigmatic rifts.

Fundamental legitimation strains, threatening the development of a capitalist society, and the "contradiction" between capitalism and democracy, may be arrived at by different intellectual routes from both conservative and revolutionary thinkers of the past (Burke, Schumpeter, Marx) and may lead to such different ideological and political visions as those of Samuel Huntington and Jürgen Habermas.[21] Almost as ancient, and equally respectable (Weber, Durkheim), is the pedigree of the dominant paradigm, which sees the forces of normative cohesion and integration as prevailing over those of disruption and radical change. In the versions popular in the 1960s,[22] a post-industrial, pluralist society is one in which conflict has withered away, having been successfully routinised and insulated in different institutional contexts, where it tends to become a rather innocuous bargaining game between limited-purpose interest groups. These paradigmatic rifts are obviously mirrored in the approaches to all the problematic fields in which that basic question is relevant: in the theory of social classes, in the sociology of revolutions, in the theory of industrial conflict, etc.

With his few historical observations, and a few plausible analyses of them, Phelps Brown steps into a field that is swarming with never appeased, century-old polemics. His remarks are adequate to his limited purpose: the impression that industrial relations crises of more than normal intensity "have to do" with profound normative maladjustments, with ruptures in the prevailing system of ideological hegemony, is successfully conveyed to the reader. But the sources of such maladjustments, and the varying intensities of industrial relations crises in different periods and different countries, are problems that cannot be answered without explicitly betting on some very general theoretical paradigm and successfully defending it against the criticisms of its challengers.

We have already recalled Phelps Brown's comment on the (limited) role of a new institutional arrangement of industrial relations in explaining the collapse of the big wave in the early part of the century: in his view, the "big blows" and the "traumatic experiences" of the 1920s were the decisive factors. Phelps Brown, however, does not deny the relevance of institutional arrangements in industrial relations and in the political system at large to the nature and the patterns of industrial conflict: institutions may be more or less appropriate to a new set of demands or power conditions. As a matter of fact, he spends a whole page (page 23) of his short essay in a rather appreciative description of the openly corporative arrangement of British industrial relations *and* political institutions proposed by John Mackintosh. He finishes, it is true, with one of his typically wise little remarks ("but it is one thing to devise new institutions, and another to gain acceptance even of the existence of the need they are designed to meet"); his judgement of the needs of the British system comes out clearly, however, and one is left with the impression that, unfortunately but perhaps inevitably, another "big blow" will be necessary to obtain a wider recognition of them.

Starting with the pathbreaking research of Shorter and Tilly,[23] recent years have witnessed an upsurge of studies on the role of political factors, and of political institutions in particular, on the development of industrial relations and especially of industrial conflict. Some have been directly prompted by the big wave of conflict between the end of the 1960s and the early 1970s; others examine different types and intensities of conflict in different national experiences over a longer time span. Some advance highly structured interpretations, others are more in the spirit of ground-clearing, theoretical contributions.[24] There is little doubt that they represent a substantial advance over the narrower traditional studies of industrial relations. Though always relevant for explaining waves of conflict and unionisation, the political element in industrial relations has acquired a new transparency and visibility in many industrialised countries in the post-war period, and a new "political market" has developed in which unions exchange moderation for various kinds of advantages obtained from the state and organised business. It is on this new, wider chessboard that the moves of the different actors of the industrial relations system must be observed and explained. As we have already noticed, for example, power is a dimension that must be taken into account not only in the traditional, collective bargaining market, but also in the political market.

There is little doubt that, together with more traditionally studied aspects of unions and of collective bargaining institutions, the new institutions and practices of the political market are extremely significant in explaining different patterns and profiles of conflict: though not conclusive, the studies of Hibbs, Shalev, Kopit and others go a good way towards establishing this point.[25] At times, however, one has the impression that a stronger claim is being made by some of those who stress the importance of the institutional–political dimension: ie that "power" and "institutions" are *the* hard variables, the ones to which attention may be limited, whereas the factors we have encompassed in the normative dimension are "vague and tautological psychological concepts".[26] As a reaction to the vague and tautological way in which "expectations", "aspiration gaps", "communication contagion" and the like may be used as explanatory variables, this critical point is well taken. If, however, we want to understand the flow and ebb of long waves of conflict in industrial relations and society at large, it is difficult to avoid the normative dimension, however messy and controversial it may be.

In what is still the most subtle analysis of the "new chessboard" of industrial relations, Pizzorno states that "periods of destabilisation and conflict appear when the unions are either unable or unwilling to exchange moderation for power, and either unable or unwilling to prevent new entries and the formation of new collective identities". This is a compact and suggestive description of *how* a crisis appears on the new chessboard: in a world of organisations and representative institutions, destabilisation necessarily implies a crisis of representation. But *why* does a crisis of representation occur? Those "vague and tautological" concepts and factors were marshalled precisely in order to answer a question "why", not a question "how". Nowhere does Pizzorno systematically tackle the former and, insofar as he hints at it, some of the vague and tautological factors belonging to the normative dimension seem to sneak in through the back door.

Conclusions

Having finished our reading of Phelps Brown, and through him of the most recent literature, let us go back to Kalecki and to the problem of political cycles.

In analysing problems of interactions between politics and economics, the time dimension is crucial. We have argued that the engine of Kalecki's political cycle is somehow too powerful and more appropriate to a Phelps Brown wave: if ever full employment unleashes the consequences Kalecki had in mind—a threat to political stability and to discipline in the factories—it will do so over a period much longer than Kalecki assumed. Phelps Brown's arguments about the "stickiness" of expectations and behaviour in the labour market may be extended to the context of organised political actors, where the same viscosity can be seen at work; as they stand, however, they already provide very useful insights into the factors that make for a slow development of the tensions deriving from full employment.

An implication of the above is that short-run electoral cycles, if they exist, normally cannot be interpreted with the big guns of Kalecki's political cycle: they may well be superimposed on a trend or on a long wave that will eventually end up in "political instability" and "lack of discipline in the factories", but they are of a different order, closer to the "normal" workings of the political system. The paradigm from which much of the recent work on electoral cycles stems—that of the vote-maximising politician—may be rather crude: slightly more than common sense cynicism passed off as scientific detachment.[27] It is in this direction, however, that further empirical work must be conducted if one is interested in the short-run interaction between politics and economics. The first priority is to abandon the attempt at formalising the model at any cost; in this field, formalisation often leads to excessive simplification and to rather trivial abstractions. The play may be further enriched with other actors, beyond "government" and "opposition", if they are relevant in the political system under scrutiny: other parties (hence the problem of coalition games), interest groups, etc. The players may be granted slightly more complex utility functions than the common desire to win elections. The country and the time span must be carefully delimited: one must be sure to avoid political developments that cannot be tackled with a "normal" politics model. There is already some work moving in this direction: instead of exploring the properties of a pair of difference or differential equations, the reader will understand something more about the politics and the economics of a given country in a given period.

What about the long period? Will a fully employed society, with a capitalist economy and a democratic polity, inevitably end up in the situation of radical tensions that Kalecki described? As we have already suggested, the answer Kalecki gives to this question was partly a result of the times in which he lived and of the political experiences he had in mind. Basically, however, it was a result of the acceptance of a proposition that can be derived from the Marxist paradigm:[28] the incompatibility of capitalism with full political democracy, ie with universal suffrage. Following this proposition, the basic lack of legitimacy of the capitalist order is taken as given. What are variable, and therefore relevant for explaining conflict and change, are power conditions, both in the political arena (more or less democracy) and in the market arena (more or less full employment). Since the two arenas are strictly connected, power conditions in each are interchangeable and supplementary: they belong to a unique power dimension. Since the normal workings of the system are based on power, so the implicit argument runs, there is a minimum amount of power that the capitalists must have if the system is to function; and since power conditions are interchangeable, it may lie in either arena: a less than fully employed economy, or a less than fully democratic society. The system is necessarily doomed (to a "fundamental reform") if such a minimum amount of power disappears, as results from full employment in a fully democratic society: the (assumed) basic lack of legitimacy of the system is bound to assert itself in these conditions.

We are now better able to account for the difficulties encountered when reading "Political Aspects of Full Employment": the schematism of class analysis and the absence of a properly political level of enquiry. The normative dimension and the politico-institutional dimension—ie how different historical patterns of ideological hegemony, on the one hand, and how different politico-institutional frameworks, on the other, influence the nature, the arenas, the intensity and the outlets of social conflict—are simply absent from Kalecki's model. Their absence is endemic to the Marxist cultural tradition. The great Marxist debates of the first quarter of the twentieth century—the *Bernstein-Debatte* and that on class and class consciousness—tackled the two dimensions of analysis we have just mentioned, but one would look in vain for any admission of the relevance and the (even relative) autonomy of ideological and political factors; one must wait until Gramsci to find more open and creative attention given to both. Even now, and understandably so, it is rather difficult for a Marxist to go far beyond a purely structural approach within the first dimension—whereby "consciousness" is strictly linked to the position in the productive process—and to recognise a rather limited "relative" autonomy of political institutions, within the second.

What is left, then, of Kalecki's *political cycle*? Apparently, *not the cycle*: if an oscillatory pattern can be detected, this is closer to an irregular and rather loose long wave, insufficient for an ambitious, completely endogenous model. Wars, international economic crises and other irregular events have often been crucial causative factors in the few cases in which discipline in the factory and political stability have been seriously threatened, and make the construction of a single-country, self-contained model simply pointless. And *not the politics*, at least in the usual meaning of the word: there is very little politics of this kind within the political business cycle.

This said, and reiterating the criticisms we have made of the schematic Marxist vision underlying Kalecki's model, the impression remains, nevertheless, that Kalecki is saying something very simple, but also quite fundamental, about the political system we live in. The simple thing Kalecki is saying can be better appreciated if we restate his basic assumptions in a weaker and more problematic form:

- The degree of normative adhesion of the subordinate classes to the basic rules of the game of a capitalist society—or, in other words, the degree of ideological hegemony that dominant classes and government elites are able to exert—is variable and precarious.

- Institutions in the industrial relations system and in the political system at large may be more or less adequate to the conditions of power and of ideological hegemony between the great collective actors of society.

Once the first assumption is so restated, the others follow, albeit weakened:

- Therefore power counts, both in the market and in the political arenas.
- Shifts in power conditions may lead to systemic crises and/or profound social change.

These "weakenings" are all research problems. Both the emergence of tensions and conflicts, or, on the contrary, the success of government elites in reaching their objectives of social control even in difficult conditions, are possible outcomes of different historical sets of economic, political and ideological variables. To such historical sets, minutely reconstructed, the greatest attention should be paid: instead of jumping into the forest of some general theory of the crisis (or of long waves), one should be content with the single trees of specific crisis management.

The corollary of such a suggestion—and the response to the obvious objection that one always needs some kind of theory—is to keep a provisionally eclectic and instrumental attitude towards the theoretical instruments that are on sale today: class, pluralist, elitist or others. This suggestion is only superficially anti-theoretical: none of the models just mentioned "does a good job of characterising policy-making in any nation, specific policy sector, or point in time".[29] After an intensive treatment of single trees of crisis management there is some hope of reaching generalisations that could be both wide-ranging and empirically grounded; anyway, this is the only strategy we are able to envisage that may overcome the theoretical impasse in which we are trapped.

Notes and references

1. OECD, *Towards Full Employment and Price Stability* (Paris, 1977).
2. S. Brittan, "The economic contradictions of democracy", *British Journal of Political Science*, April 1975; "Inflation and democracy", in F. Hirsch and J. Goldthorpe, eds, *The Political Economy of Inflation* (London, Martin Robertson, 1978).
3. M. Salvati and G. Brosio, "The rise of market politics: industrial relations in the seventies", *Daedalus*, Spring 1979.
4. M. Kalecki, "Political aspects of full employment", *The Political Quarterly*, October 1943.
5. H. Phelps Brown, "A non-monetarist view of the pay explosion", *Three Banks Review*, 1975, No. 105.
6. The best reviews are those of B.S. Frey, "Politico-economic models and cycles", *Journal of Public Economics*, 1978, *9*, 2; and R.F. Winters *et al*, "Political behavior and American public policy: the case of the political business cycle", in S. Long, ed, *Handbook of Political Behavior* (Plenum Publishing, 1980). Beyond the common core, the differences are remarkable. In the best cases, the works of political scientists are informative and sensitive to the historical and institutional peculiarities of the contexts they analyse: such is, at least, the book of E.R. Tufte, *Political Control of the Economy* (Princeton University Press, 1978); and many of the contributions devoted to the problem by Bruno Frey and his school. The formalised

analyses of the economists are on the brink of irrelevance, if one excepts W. Nordhaus, "The political business cycle", *Review of Economic Studies*, April 1975, *42*, because of his originality. A venomously intelligent critique of the whole literature can be found in B. Barry, "Does democracy cause inflation? A study of the political ideas of some economists", in L.N. Lindberg and C.S. Maier, eds, *The Politics and Sociology of Global Inflation* (Washington DC, Brooking Institution, 1981).

7. *Stricto sensu*, it is false. Recent research on short-run behaviour of wages, productivity and prices has shown that this piece of Kalecki's cycle theory does not strictly correspond to what happens in industrialised, oligopolistic economies. At the peak of the upswing, the growth of labour productivity often tends to slow down, whereas firms cannot, or do not want to, increase prices in proportion with unit costs; the peak of profits and profit margins tends to happen, therefore, before the peak of the cycle. See K. Coutts, *et al*, *Industrial Pricing in the United Kingdom* (Cambridge University Press, 1978); and P. Sylos-Labini, "Industrial pricing in the United Kingdom: a comment", *Cambridge Journal of Economics*, 1979, no. 2. R. Boddy and J. Crotty, "Class conflict and macropolicy: the political business cycle", *Review of Radical Political Economics*, Spring 1975, build their own "Marxist" theory of the political business cycle just on this slow-down of profits at the end of the upswing; in our view, this means to be more "economicist" than the economists and to miss the key contribution of Kalecki.
8. See C. Lindblom, *Politics and Markets* (New York, Basic Books, 1977), especially chapters 13 and 14; and F. Block, "The ruling class does not rule: notes on the Marxist theory of the state", *Socialist Revolution*, May–June 1977, no. 33; and "Beyond relative autonomy: state managers as historical subjects", *Socialist Register*, 1980. A critique and an expansion of this thesis can be read in C. Crouch, ed, *State and Economy in Contemporary Capitalism* (London, Croom Helm, 1979).
9. M. Kalecki, *Selected Essays on the Dynamics of Capitalist Economy, 1933–1970* (Cambridge University Press, 1971).
10. See Block, reference 8.
11. J. Cohen, "Marxism and politics: or trouble in paradise" (mimeo), Department of Political Science, MIT, 1980.
12. B. Jessop, "Capitalism and democracy: the best possible political shell", in G. Littlejohn, ed, *Power and the State* (New York, St Martin's Press, 1978).
13. G. R. Feiwel, "Reflections on Kalecki's theory of political business cycles", *Kyklos*, 1974, *27*, page 39, extensively quotes from the preface to the Polish volume in which the shortened version of the 1943 essay—the same that is included in the English version of the *Selected Essays* (Kalecki, reference 9)—was first printed. Two things are worth recalling here: first, that Kalecki equates his political cycle with the short and mild "stop-and-go" episodes of the 1950s and 1960s; second, that he explains the persistence of exceptionally high average levels of employment by the role of arms expenditure, a kind of public expenditure that capitalists do not dislike. As we shall see in the next section, this is a rather unconvincing way of dealing with the "political consequences" of a long period of full employment.
14. Eg, the reference in Kalecki, reference 4, page 330, to the 1937–38 business cycle in the US economy, and the explicit recognition recalled by Feiwel, reference 13.
15. The best way to capture this extraordinary change in intellectual atmosphere between the 1960s and 1970s is to compare a famous special issue of *Daedalus* devoted to Europe (*A New Europe*, spring 1964; see in particular the essays of S.M. Lipset and R. Dahrendorf) with the two issues (*Looking for Europe* and *The European Predicament*) that the same journal devoted to the same subject in winter and spring 1979. A recent critical assessment of the theories of the end of conflict and ideology—still the mainstream view, if such exists—is contained in W. Korpi and M. Shalev, "Strikes, power and politics in the western nations, 1900–1976", in M. Zeitlin, ed, *Political Power and Social Theory* (Greenwich, JAI Press, 1980).
16. Salvati and Brosio, reference 3.
17. On the natural rate of unemployment and on the trade-off between unemployment and inflation, the most recent review is A. Santomero and J. Seater, "The inflation–unemployment trade-off: a critique of the literature", *Journal of Economic Literature*, 1978, *16*.
18. See Phelps Brown, reference 5.
19. *Ibid*, pages 19–21.

20. J. Goldthorpe, "The current inflation: towards a sociological account", in F. Hirsch and J. Goldthorpe, eds, *The Political Economy of Inflation* (London, Martin Robertson, 1978).
21. S. Huntington *et al, The Crisis of Democracy. Report on the Governability of Democracies to the Trilateral Commission* (New York University Press, 1975); J. Habermas, *Legitimation Crisis* (London, Heinemann, 1976).
22. See reference 15.
23. E. Shorter and C. Tilly, *Strikes in France* (Cambridge University Press, 1974).
24. For two good reviews, see G. Baglioni, "La Pratica dello sciopero nell'azione operaia e sindacale", in G.P. Cella, ed, *Il movimento degli scioperi nel XX secolo* (Bologna, Il Mulino, 1979); and Korpi and Shalev, reference 15.
25. We mainly refer to the moderating influence that the "political market", if it works effectively, may exert on industrial conflict. See, on this topic, D.A. Hibbs, "On the political economy of long-run trends in strike activity", *British Journal of Political Science*, April 1978; M. Shalev, "Strikes and the state: a comment", *British Journal of Political Science*, April 1978; and Korpi and Shalev, reference 15.
26. A. Pizzorno, "Political exchange and collective identity in industrial conflict", in C. Crouch and A. Pizzorno, eds, *The Resurgence of Class Conflict in Western Europe since 1968* (London, Macmillan, 1978), page 289.
27. See Barry, reference 6.
28. Such a proposition always finds supporters when the economic and political development of market economies runs into serious difficulties: to this problem A. Wolfe, *The Limits of Legitimacy. Political Contradictions of Contemporary Capitalism* (New York, The Free Press, 1978), has devoted an enquiry of great interest. Its origin, as is well known, can be traced back to the comments that Marx devoted to the French political life of mid-19th century. See K. Marx, *The Class Struggles in France, 1848 to 1850* (Moscow, Progress Publishers, 1952).
29. See Lindberg and Maier, reference 6.

CHAPTER 16

INNOVATION, INFORMATION AND LONG-TERM GROWTH

André Piatier

Short-term and partial analyses have led to overpessimistic predictions (non-conjunctural unemployment, decline of certain countries, and a slow-down in creativity). A more general approach is based on the idea of industrial revolution: the neolithic—cultivation and domestication; in the 19th century—coal, steel, railways; in the mid–20th century—petrol, cars, etc. A fourth revolution is beginning (electronics, information sciences, telematics, biotics, aquaculture, solar power etc), but this is still in an embryonic phase, whereas earlier revolutions are in their mature stage. An interpretation based on the information sector can lead to a certain pessimism. This can only be avoided by a broader definition of the economy (pattern recognition) in which man, and human investment, have an equal place alongside material investment.

MANY commentaters have drawn rather pessimistic conclusions from contemporary economic difficulties, which in the short term appear unfavourable for innovation.[1] But it is important to assess the evidence of a decline of innovation and of productivity in a balanced and long-term perspective.

In a number of articles, notably in the USA, fears have been voiced that Western countries are innovating less than before. This seems to be confirmed by certain statistics—R and D expenditure, number of scientific and technical personnel engaged in R and D, the demand for patents etc, which show signs of decline.

It would be important to point out this danger even if the possible consequences were not serious. But there is good reason to think that the diagnosis is false.

(1) Observations are taken over a few years only and it is difficult to say if we are dealing with a cyclic or a long-term variation.[2]
(2) The statistics which have been studied are not very definite. They do not cover all areas of innovation. The construction of prototypes, testing and

Professor A. Piatier is the Director of the Centre d'Etudes des Techniques Economiques Modernes at the Ecole des Hautes Etudes en Sciences Sociales, 63 rue Claude Bernard, 75005 Paris, France.

perfecting products are not all covered by R and D; not every discovery is the object of a patent, and so on...

(3) These statistics do not deal with industrial innovation itself, but with operations which affect it. They tell us no more about the production of new products or processes than iron mining statistics tell us about the mechanical engineering industry.

(4) There is no fixed relation between research costs and results. There are long but variable delays between research, invention and innovation.

Furthermore, there is no fixed relation between the number of patents and the number of new products which industry launches: patent statistics deal with very disparate ensembles. So we are only dealing with the symptoms of a possible decline in innovative efficiency in the West. But even such a decline seems less likely, as is shown below.

Indications of a fall in profitability and productivity relate to a different analysis. As far as profitability is concerned, the evidence of cyclic fluctuations is greater than that for Marx's long-term tendency for the rate of profit to fall leading to the disappearance of capitalism. And the internal contradiction in the doctrine between the tendency of the profit rate to fall and monopolistic concentration (whose aim is to raise profits) does not seem to have been studied seriously by Marxists. As for an explanation of the stagnation of US productivity, the arguments deployed are hardly convincing—even if we refer to Campbell R. McConnell's important study.[3] This author invokes the diminution of the capital–labour ratio, the employment of (less skilled) women and youth and the move of the agricultural workforce into industry. He adds in technological progress—although the amelioration of the quality of equipment goods and greater organisational efficiency (a combination of factors) seem to us to be some of the decisive factors in the improvement of productivity! He takes up the arguments of Seymour Melman according to whom the existence of a vast military–industrial sector probably contributes to a fall in the rate of productivity. But European studies tend to show the opposite. The US–French 'model'—with a military sector—brings different advantages, and along a different route, from that of the Germano–Japanese model.

The decline in productivity in industry can, moreover, come from a lack of work enthusiasm (a factor which McConnell wrongly attributes only to radical or Marxist writers). But in the global economy, this decline can derive from the growth of the tertiary sector where increasing returns do not yet exist, and from the growing bureaucratisation and information-related costs of large public and private organisations (see below).

Two long-term schemata for the 1980s

The present evolution of industrialised societies is thrown into an interesting light by two different attempts to insert them into a long-term analysis. The first is based on the idea of an industrial revolution, and the second on the idea of the development of a new sector devoted to information.

Industrial revolutions

When several major innovations appear simultaneously, their combination is an industrial revolution.[4] Until now we have spoken of *the* Industrial Revolution, that of the 19th century based on coal, steel and steam power. This led in turn to the mechanisation of textiles, the expansion of chemicals and other manufacturing industry, and to the widespread appearance of railways.

An industrial revolution is a secular movement which can be schematised by means of a logistic curve. Four "ages" and their corresponding industries can easily be read into it: infancy with weak growth, adolescence with rapid growth, adulthood with slower growth, and maturity with zero growth. Coal, steel, railways, textiles and chemicals grew up together: the crisis of the 1930s, which was neither a conjunctural crisis nor the final crisis of capitalism, marked the end of a big wave of major innovations.

Successive industrial revolutions. It is hard to say if, before the 19th century, there were many other industrial revolutions. Certain historians have had a tendency to see them in the Middle Ages or around the time of the Renaissance. But at these times, there was not so much deep-going technical change as progress by extension and reclamation and then the discovery of new worlds—and by the intensification of urban activity.

Although we shall follow the conventional usage in talking of 'industrial revolution', we tend to agree with the historian Cippola[5] in using the expression 'industrial revolution' for the *neolithic* revolution. The cultivation of vegetables constituted the first major transformation in human life. In effect they replaced the natural converters of solar energy which they found in nature (through hunting and gathering) with a first industry, which consisted in producing these energy converters: animals and vegetables could be produced by man in greater quantities and as a function of his needs. The economic effects of this neolithic revolution have been described elsewhere.[6] The speed at which these long innovatory waves succeed each other is accelerating: it took 2000–3000 years between the agricultural revolution and the first industrial revolution and 100 years between the first and the second industrial revolution, which was based on fossil liquid fuels—the automobile, the aeroplane, the metal sheet, organic chemistry and synthetic materials. All these elements were available and ready between the two world wars, but it was under the hammer blows of the post-war reconstruction period that growth really took off, before slowing down again after 1973–74—when the oil crisis probably served as no more than the catalyst for deceleration of growth which was already destined to be a feature of the long-term evolution of the situation.

The new third industrial revolution (end of the 20th century). This is now visible through the innovations which have already taken place. It can draw on new energy sources, new means of transport and new means of communication (integrated networks of telephones, telegraph, automatic couriers, computers, facsimiles, video conferences etc). Satellites and space rockets will undoubtedly coexist with dirigibles and new sail-driven cargo ships, re-emerging from oblivion on the basis of already visible technical changes. Genetic engineering and automation open immense perspectives for manufacture, in so far as the oceano-economy multiplies the food-producing and industrial potential of humanity: here will be found the elements of prosperity and well-being as well as the source of conflicts.

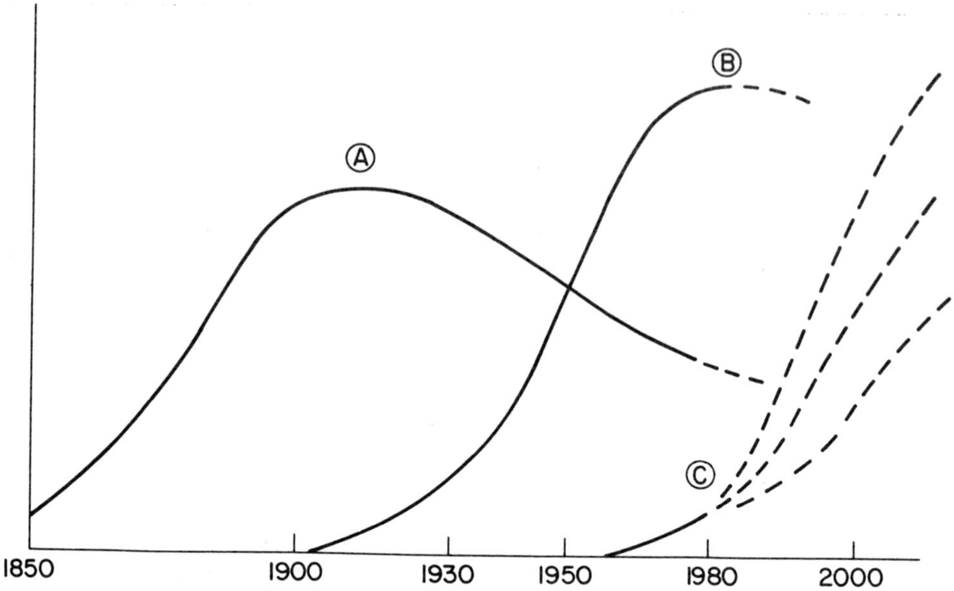

Figure 16.1. The three industrial revolutions.
Key:
Ⓐ Coal, steel, rail, textiles, chemistry.
Ⓑ Oil, cars, steel (sheet), mechanical industry, air transport, organic chemistry, electricity.
Ⓒ New energy sources (including nuclear), oceanic development, biomass, genetic engineering, electronics, telematics, automation, etc.

TABLE 16.1. 'INDUSTRIAL REVOLUTIONS'

	Energy	**Transport**	**Tools**	**Basic production**
Neolithic	Wood Charcoal	The wheel	Hand	Cereals Textiles Stone
19th Century	Coal	Rail	Mechanical industry	Steel ingots Mineral chemicals Manufactures (textiles etc)
20th Century 1st Revolution	Oil Electricity	Car Air transport	Electrical industry	Sheet steel Organic chemicals Plastics
20th Century 2nd Revolution	Nuclear Solar Other new forms	Rocket Telecommunication Dirigibles Sailing cargo ships	Electronics Computing Robotics Office automation Genetic enginering	Synthetic products Aquaculture (neolithic marine) All production

Figure 16.1 (see also Table 16.1) presents secular movements corresponding to the three industrial revolutions. For the third, various predictions are shown as dotted lines, and pose the problem of how rapidly the phases of embryonic and infantile growth can be traversed to lead to the adolescent phase of accelerated growth.

The 1980s : the transition phase. The present period—1975–1985—in this explanation will feature the combination of two movements: the *maturity and decline of* the major industries of the last industrial revolution; the *very weak growth* of new activities of the industrial revolution now beginning and which is only in its embryonic phase. The aggregate effect of these two processes gives rise to very slow overall growth.

To these two long-term movements, which are superimposed on each other, cyclic variations must be added (the two small conjunctural crises of 1974–75 and 1979–81 had deeper effects because of the cessation of growth) and accidental variations (for example the oil crisis).

We have mentioned interpretations of the present crisis which put their sole emphasis on these latter elements of the situation which according to us are secondary. If our diagnosis is accepted, the economic policy to be followed becomes clear:

- accelerate as much as possible the rise of new activities, so that they can soon reach the phase of rapid growth;
- reorient old activities with the aim of making them peak as near as possible to their former maximum level, for example by developing products which are less vulnerable to competition (special steels, high-performance motor cars etc).

If it evolves spontaneously, the onset of the new phase of prolonged growth can be situated around 1990. Only a policy of directed development can reduce the depressed transition period to around 1985.

The minimum content of an industrial revolution. It would appear that for any industrial revolution, there must be a simultaneous appearance of convergent innovations—in energy sources, transport, tools (or the production goods industries) and in 'manufacture'. Hence, an industrial revolution can be defined by this combination of activities and by the prolonged process of expansion which it produces, each new product following a logistic curve. This curve is not so much the 'life cycle' of the new product—a micro-economic notion—as a curve of saturation of the milieu: the new 'population' develops until it has taken up all the terrain open to it.[7] A change of milieu or a change of product can give rise to a new logistic curve. This was the case with steel—in ingots with the first revolution, in sheets with the second. Will it find a new lease of life with the third?

A final indispensable condition for the onset of prolonged global expansion has been, until now, the existence of a reservoir of labour. Marx, who was only able to observe the phenomenon once, spoke of a reserve army of labour. For the first revolution agriculture provided the necessary manpower. With the second revolution the necessary labour force, besides the rural exodus and the surplus labour being expelled from obsolescent industries, was found thanks to immigration. For the third, the principal influx will probably be made up of women.

This schematisation does not take into account the necessary level of skills. Problems of education and formation are dealt with below.

The original character of the third industrial revolution. The long-term expansion which is about to begin has a completely new character: whereas the preceding industrial revolutions augmented man's physical powers, the new one will offer gigantic augmentation of intellectual powers.

Like many other researchers, we have tried to provide some representation of the multiplication of powers which the use of energy and machines brought to humanity. A calculation based on 'mechanical slaves' (total energy production expressed in units of 150 000 calories/year) is quite revealing[8]—the population of Europe and North America in 1650 being scarcely 110 million. In 1950 there were nearly 900 million people. But with their mechanical slaves, they represented the equivalent of 19 billion (10^9) people.

In the rest of the world, where these benefits arrived later and in smaller amounts, there were 1 350 million inhabitants, who together with their mechanical slaves scarcely reached the equivalent of 2 400 million—eight times less than the industrialised world.

The present industrial revolution has totally different dimensions, since the computer revolution offers humanity an augmentation of its *intellectual* forces. We have to measure its impact, for different levels of performance and development of computers and intelligent machines, with reference to an (imaginary) unit of work by the human brain.

The third industrial revolution involves many other original features which were foreshadowed by the green revolution. But during the 1970s these remained undeveloped. Today, with the attention being paid to the biomass (mainly because of the energy crisis), the forthcoming rise of aquaculture, and with genetic engineering now being used, we are seeing a return to humanity's primordial needs, neglected for millennia. Foodstuffs will be one of the first branches of industry to be revitalised, as well as mining activity (that is, the two components of Colin Clark's primary sector).

Until now, innovation has proceeded from primary needs to sophisticated ones. From now on it will cover the complete spectrum of needs and activities—and will also involve the augmentation of both muscle and brain.

In this vision of tomorrow's world we have not referred to the two great precursors, Schumpeter and Kondratiev. It seemed unlikely that writings so far removed from the structures we are now erecting would be important: but our observations of reality would not have been productive without the stimulation provided by their work.[9]

There is an unknown quantity in the scheme which we have put forward—employment. It seems that, until now, most specialists have hesitated to pronounce on this. There is not even a suggested methodology for dealing with the evolution of human labour in tomorrow's world. An attempt must nevertheless be made, but it involves a detour via two contemporary studies—those of Marc Porat and Jean Voge.

Growth of the information sector

Revising Colin Clark's old distinction between primary, secondary and tertiary sectors, Marc Porat[10] has proposed a subdivision of the economy into four sectors: agriculture, industry, material services and information.

Content and evolution of the information sector. This very large sector is subdivided into two subsectors. The primary sector contains organisms which produce or exploit computing resources (measurement and calculation), office equipment, communications, media, including publicity. Porat adds in education, R and D, consultancy services in management organisation, engineering, architecture, accounting expertise and certain financial and commercial services (bank deposits, some legal services), brokerage, exchange agencies, insurance, travel agencies, and agencies dealing with real estate and various other transactions. He is probably wrong to give this group the title of marketing information. The secondary sector comprises activities which provide factory information (planning, study, research, marketing, professional education) and everything else concerning private bureaucracy. To this he adds public bureaucracy (local and national administration and public enterprise).[11]

In the USA (see Figure 16.2), the information sector involves up to 50% of the active population and seems to be reaching a peak. Agriculture reached a ceiling around 1870 before declining. Industry peaked between 1910 and 1940 and declined subsequently. The European countries have gone through a similar evolution, but with a certain delay.

Limits and consequences of development of the information sector. In a series of works[12] Jean Voge draws a number of important conclusions from Porat's study:

- the information sector (in the US case and probably elsewhere) grows about twice as fast as the economy as a whole.
- this growth tends towards an upper limit of 'economic maturity' when the share of the information sector in gross domestic product (GDP) reaches 50%.

These two assertions are backed up by two kinds of 'law':

- N. Parkinson's law, which was enunciated for the phenomenon of bureaucracy but which J. Voge considers applicable to information. According to

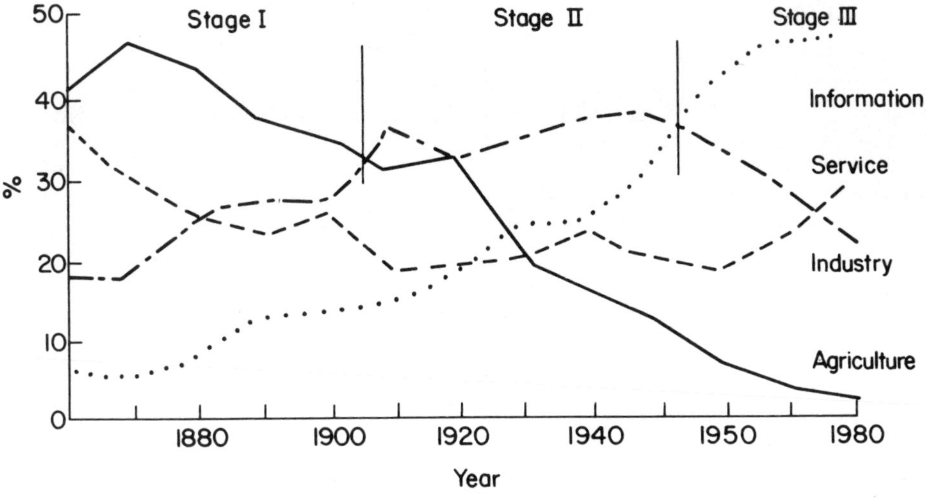

Figure 16.2. Four sector aggregation of the US workforce by percent, 1860–1980, using median estimates of information workers.

this law the volume of information tends to grow as the square of the number of interlocutors. He gives the example of telephone directories, whose mass grows as the square of the number of subscribers, since both the content and the number of copies grow.

- Laws flowing from the principles of thermodynamics, which are valid for numerous physical and biological systems: "frequently, when the power supplied to one of these systems is increased, the power dissipated in the form of heat (thermic and entropic flux) grows as the square of the power supplied. The residual power delivered in the form of useful work or material production thus reaches its maximum when dissipation has risen to half the power consumed".[13]

He concludes that the growth of the information sector leads to a crisis of economic stagnation (and of zero productivity growth) "when one in every two workers is a producer of information and half of all national production consists of information".[13] If the share of the information sector exceeded 50% there would be a regression in material production.

Critique of this theory of stagnation and economic maturity. This pessimistic construction, in which Porat observes that the information sector peaks at 50% of the active population and Voge tries to explain this peak, would be unassailable if (a) the information sector was a homogeneous totality; (b) Parkinson's law applied to all the various activities involved in industrial production; (c) the ceiling of 50% was unavoidable. None of these conditions holds. First, the information sector is a mixture of rather dissimilar things. Some order is necessary. It can be provided by separating out:

(1) activities concerned with the augmentation of intellectual human capital (education and training);
(2) activities dealing with the augmentation of knowledge and exploitable ideas (research, R and D);
(3) activities involving the collection, expression and modification of information in the strict sense of the word;
(4) non-material transport (eg telecommunications);
(5) activities of organisation, arbitrage, decision, division etc;
(6) activities linked to the allocation of rights to goods and services (money, finance, control of property etc).

These six very dissimilar sectors have until now been ignored by economic theory or dealt with marginally by it (for example in the economics of education, of finance, of public service etc...). They must be integrated into a new, general, economic scheme (cf next main section below). For each one, it is necessary to establish a production function, to find out if the returns are rising or falling, to study productivity and confront the results obtained with those from material production, the central (and often exclusive) preoccupation of the economists of today and yesterday.

Second, in economics there are numerous 'laws' which have the same general appearance as the laws of thermodynamics (see Figure 16.3a–d). The oldest is without doubt King's law (above a certain size of harvest, the farmers' income falls).[14] The most recent is that of A.B. Laffer: above a certain level, tax

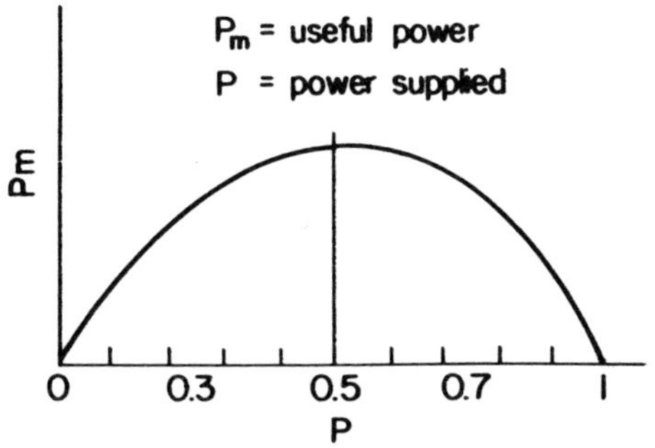

Figure 16.3a. Thermodynamic system with constant potential (J. Voge).

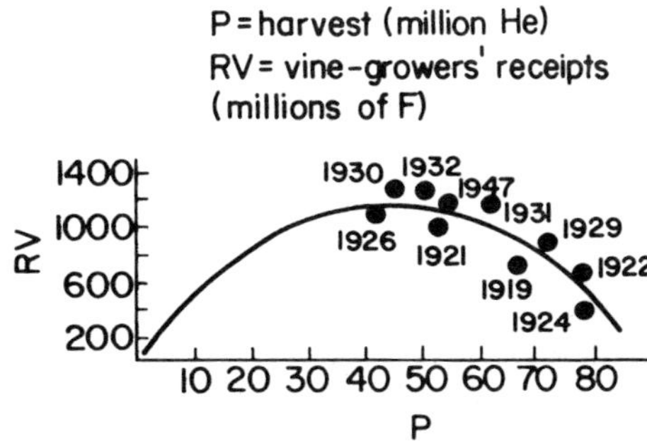

Figure 16.3b. King's law as applied to vine-growing (J. Milhau).

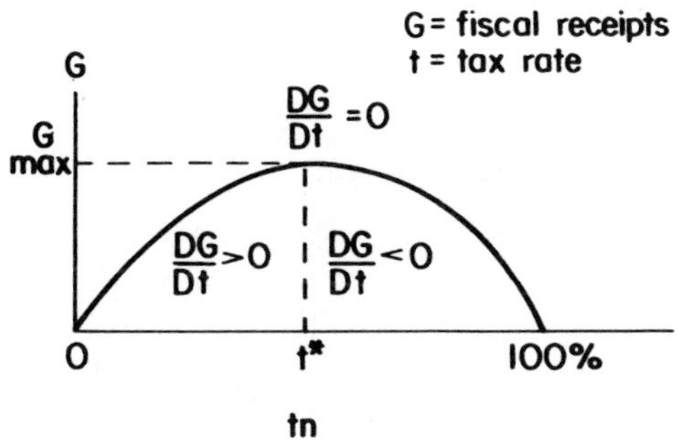

Figure 16.3c. Laffer's curve.

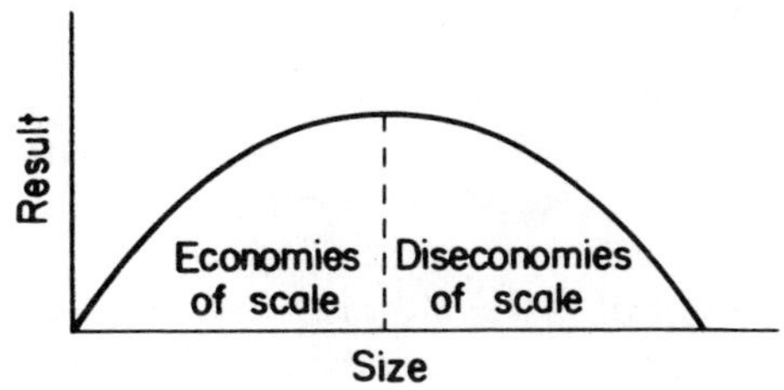

Figure 16.3d. Size of organisations (firms, towns etc).

increases result in a fall in fiscal income.[15] But the highest point of the curve is not fixed once and for all. An austerity plan accepted by the population, or a war effort, displace the curve to the right, whereas a peacetime economy, a recession or other events displace the curve to the left.

The same remark can be made for the last curve in Figure 16.3, relating to economies and diseconomies of scale: improvements in organisation, or system degradation, can displace the optimum size in one direction or the other.

Broader view of the economy

The complexity of modern society cannot be encompassed by the models now available. To take into account the totality of the relations we have been discussing, to introduce innovation into our analysis, and to integrate the use of variables of a less coarse nature than those relating to a global 'information sector', it is essential to construct a broader and better centred economic system. Until now attention has focused on material production, involving marginal humans who play no role other than supplying labour in exchange for revenue, and consuming the results of this activity.

It is not enough to add a non-material production sector, or an information sector, as Porat does. A new, modelable scheme, founded on Fisher's concept of capital, will place man side by side with his milieu. The following enumeration of nine factors of production attempts to take full account of the six sub-sectors of the information sector listed above, which have hitherto been largely ignored by economic theory. Everything taken into consideration in the plans of economic agents, groups, and human societies is capital, either human or material.

Material 'capital' supplies: (1) energy (fossil or renewable); (2) materials (inert or living, that is to say renewable); human 'capital' supplies: (3) manual labour, and (4) intellectual labour, ie ideas and understanding. Ideas and understanding, when they are materialised (patents, reports, publications etc) are added to material capital. Side by side with material capital stands fiduciary capital, consisting of all the titles and rights of possession or disposition, either of capital or of services.

In order to have a genuine production function, and with the reservation that occurs below in relation to investment, the production function will deal first with the outflow of services emanating from capital (manual or intellectual labour, energy) or materials which, being destroyed the first time they are used, possess a value as revenue equal to their capital value: these are what economists normally treat as 'intermediary goods'.

To these four factors of production we must add: (5) the transport factor (displacement and communications)—this is a broader concept than that of transport, including at the very least telecommunications, courier services, and so on; (6) the factor of organisation, arbitrage, decision, coexistence, definition of ground rules and control, etc—this factor deserves a better name than that of public and private bureaucracy which it has been given; (7) the factor of the allocation of the rights of disposition—if there is a state monopoly of money creation, there is a distribution of the responsibility for the determination of remunerations and prices, either through price-fixing mechanisms or through the organisation of 'markets' (the process of transferring rights belongs to this category); (8) the information factor in the strict sense—this corresponds in empirical studies to the famous 'transparency', and in theories to perfection (publicity, media, and information systems should be included here in a limited way); (9) the 'will' factor—nothing will be done with what has gone before unless there is a will to combine these factors in order to produce, and above all to work and think. This willingness has its counterpart in a special process of production concerning culture 'amusement' and everything which, in a given civilisation, is considered useful only because it is pleasant.

Finally, the last term in the production function will deal with the tools which man has created in order to produce more or better. This is the only place where durable fixed capital intervenes. Production, P, is expressed as:

$$P = f(F_1 .... F_n), F(k_1, k_2)$$

But the concept of production has to be made more precise. Here we are only dealing with useful production (cf useful energy in the laws of thermodynamics)—that is to say production which has arrived at the point where it is usable, at the time when it is needed, and in the hands of an agent with the means to use it. Everything else apart from stock must be loss, noise (in the electronic sense) or wastage.

Our societies are riddled with waste—Soviet stations containing mountains of fertiliser at one time, steel production without client industries in the first Indian plans, unsold goods which represent costs which must be counted, along with unemployment, as another consequence of maladjustment.

The means of production are supplied by investment. Investment functions for the various factor categories must be written, distinguishing between replacement and net investment, above all distinguishing between extension of capacity, rationalisation (new processes), and means to make new products.

Investment functions concern: type (1), energy; type (2), agriculture and industry; type (3), health; type (4), education, training and research; type (5), transport and communication; type (6), public and private organisation (bureaucracy); type (7), money, finances, allocation; type (8), information in the strict sense; type (9), culture, leisure, well-being.

Welfare corresponds to functions of types (2), (3) and (9) since, today, economic analysis only takes into account functions (1) and (2), and even here fails to make adequate distinction between renewable and non-renewable forms of energy and materials, the rest being dealt with in a different way. Investment means sacrifice of current production. Thus, at the same time, there is a choice to be made about what must be sacrificed, and towards what end the sacrifice will be made. A decision model will allow comparison of costs (in the broader sense) and the measurement of efficiency in terms of useful production (that is to say in total production less wasted production).

To the extent that the new industrial revolution supplies means of augmenting intellectual labour (computers, robots and so-called intelligent machines) as well as means of augmenting manual labour, some interest attaches to distinguishing between the two categories of investment capital (k_1 and k_2) in the formula above

Conclusions

In this schema, stagnation is not inevitable. Pessimism derives from an over-narrow conception of the social system.

(a) Jean Voge has already suggested a solution through decentralisation. The functions of communication and organisation, for example, would be dealt with more efficiently in smaller units. Progress in computing and telecommunications is moving in this direction for public organisations, enterprises, laboratories, towns, and so on.

(b) The system does not have a fixed potential. The number of men, and human investment, are strategic variables. Many sectors can expect a growth in productivity (teaching machines, the productivity of the bureaucracy, productivity of research). Process innovations will re-initiate a growth in productivity where it tends to fall. The third industrial revolution could be defined as the opportunity to displace the curves of Figure 16.3 to the right.

In the same way that Wagemann[16] has described the economic evolution of societies in terms of successive waves of overpopulation and underpopulation, the reversal of the population trend being the result of productive changes of combinations, we think that economic history is a sequence of expansions–saturations, each new beginning consisting of the appearance of innovations which open new domains of action. But it is essential not to forget that technical progress must precede economic progress which, in turn, must precede social progress.

Notes and references

1. A. Piatier, ed, *Les Obstacles à l'innovation dans les pays de la Communauté Européenne* (Rapports nationaux et rapport général, DG XIII, Brussels, EEC, 1980).
2. The recession phases of the cycle were barely perceptible before 1973, but have become much more marked in a period of relative stagnation and slowdown.
3. C.R. McConnell, "Why is US productivity slowing down?", *Harvard Business Review*, March/April 1979.
4. C. Freeman, "The Kondratiev long waves, technical change and unemployment", in *Structural Determinants of Employment and Unemployment*, Vol 2, reports prepared for Experts' Meeting in Paris, 7–11 March 1977 (Paris, OECD, 1979), pages 181–196; J. Clark, C. Freeman, and L. Soete, "Long waves and technological developments in the 20th century", in D. Petzina and G. van Roon, eds, *Konjunktur, Krise, Gesellschaft: Wirtschaftliche Wechsellagen und sozialer Entwicklung im 19 und 20 Jahrhundert* (Stuttgart, Klett-Cotta, 1981), pages 132–169. I arrived at similar conclusions in the study for the EEC; see reference 1.
5. C.M. Cippola, *Storia Economica dell'Europa* (Bologna, 1974).
6. G. Roupnel, *Histoire de la campagne française*, revised edition (Coll Terre Humaine, Plon, 1978).
   A. Piatier, "Les Réflexions de l'économiste sur la carte des communes de France", *Annales*, July–September 1958.
   A. Piatier, "Les Fonctions de polarisation", *Urbanisme*, 1965, *69*.
7. Colonies of bacteria follow similar logistic curves of growth ending in stagnation, decline, or new development and renewal.
8. A. Piatier, "Les Inégalités de développement démographique et économique", *L'Encyclopédie Française*, Vol. XI.
9. One must add to the names of Schumpeter and Kondratiev that of Izard, whose "transport-building cycle" was an important contribution to the development of the long cycle idea.
10. M. Porat, *The Information Economy* (Stanford, 1976); see also the work of an OECD group of which M. Porat was a member, "Economic analysis of information activities" (Paris, OECD, 1977).
11. However, education and telecommunication are placed in the primary sub-sector.
12. J. Voge, *De l'Education, de la recherche, de l'information et des communications dans la croissance économique* (Paris, Ecole Nationale Supérieure des Télécommunications, 1973–4).
    J. Voge, "Crise économique, production intellectuelle et technologies de la connaissance", *Onde Electrique*, August–September 1975, *55*.
    J. Voge, *Sur un modèle thermodynamique de la croissance et de la maturité economique* (Paris, Ecole Nationale Supérieure des Télécommunications, 1975–6).
    J. Voge, "Société d'information et crise économique", *Télécommunications*, April 1979, *31*.
13. Voge, "Société d'information . . .", reference 12, page 60.
14. An application of King's law was made by J. Milhau to viniculture. Knowing the volume of a harvest, it was possible to foresee the price of wine and the receipts of the peasants.
15. Y.A. Canto, D.M. Hones, A.B. Laffer, *Taxation, GNP and Potential GNP* (San Diego, CA, American Statistical Association, 16 August 1978).
16. E. Wagemann, *Alternationsgesetz der Wachsenden Bevölkerungsdichte-Vierteljahres Heft für Wirtschaftsforschung* (Berlin, 1936).

CHAPTER 17

HISTORIANS AND LONG WAVES

Gerrit van Roon

Some historians have found the long-wave concept extremely fruitful when analysing industrialisation and agricultural development. Particularly important was the work of Braudel and his colleagues associated with the journal *Annales*. Their framework used long-term structural changes (of a century or more) and conjunctural/cyclical changes in the economy, such as Kondratiev cycles. Events of short duration can be far better interpreted and evaluated within this framework. Rosenberg's work provides a good example of the fruitfulness of a method which relates cyclical changes to social and political events.

IN THE PAST, political trends and events have received too much attention in historical research; for instance, the picture of the 20th century has been long dominated by two world wars. The rapid developments since World War 2 demand an analytic historical pattern in which attention is paid to long-term development processes that can be tested for their 'conjunctural sensitivity'. By 'conjuncture', I mean cycles in economic, social, political, and all other sectors of society. Long waves constitute an interesting framework for such an analysis, and this paper makes a plea for a structural-conjunctural approach to history, in which the origins and evolutions of a variety of phenomena can be traced and intimately interrelated.[1]

The drastic extension of the role played by governments, and the spectacular growth in their agencies, is a recent trend in many countries, and is reflected in the absolute and relative rise of government expenditure. Government has become a major producer, is regulating many aspects of social life, and fulfils a redistributive role. Governments influence the industrialisation process and marketing policies by a variety of measures. The objectives of economic policies have widened; governments are mammoth entrepreneurs and mammoth supervisors. The 'nightwatchman state' of the 19th century is now superseded by the 'intervention state' and the 'welfare state'.

At the same time, the increasing scale and concentration of industrial activity has led to positions of huge power, which extend well beyond the economic sector. A far-reaching cooperation has developed between financial and state

Professor G. van Roon is Professor of Modern History and Coordinating Director of the Research Group on Long Term Fluctuations, Free University of Amsterdam, Koningslaan 31–33, Amsterdam, The Netherlands; he is also a Fellow of the Royal Historical Society, London.

institutions. Beyond national borders, the multinationals have become the giants in the world economy. In the planning domain, they have a distinct lead over politicians who, in many respects, have remained "prisoners of geography". Their activities hardly allow any restraint from national laws and regulations.

The problem of concentration and penetration of power, both political and economic, and of internationalisation of industry, is the result of long-term and medium-term processes. The present extent of these problems renders it still more important to distinguish their various components more accurately, and to examine them on the basis of quantitative analyses. By applying a conceptual model[2] of the development of modern industrial society from its origins, we can identify the primary features of this process and put them in a distinct and intelligible context. Similarities and differences in the processes at the international, regional, and national levels can be sought out.

The slow-down in growth, the stagnation and the unemployment of recent years have revived interest in cyclical phenomena. The catchword "growth" replaced the catchword *konjunktur* but, as a result of the longlasting depression, the study of *konjunktur* acquired fresh attention. Thus a working model of the origin and evolution of modern industrial society has to be articulated on the *konjunktur* pattern.

Braudel and the 'Annales' group

In the world of historians, relating the investigation to the *konjunktur* pattern is not new.[3] Historians of agriculture provide a clear example.[4] More generally, the group of French historians of the review *Annales* have been working on these lines since 1929.[5] They pay much attention, usually in regional studies, to subjects formerly pertaining to the 'blind spots' of history: geography, demography, climate, nutrition, diseases, and education. For the group of the *Annales*, now under the leadership of Fernand Braudel, conjunctural evolution is an instrument allowing them to arrive at a better integrated and more concrete historical picture. They concentrate on the analysis of dual, dialectic relationships: those between time and space, and between man and his natural environment. Their first researchers—Marc Bloch and Lucien Febvre—were strongly influenced by geography and were inspired by Durkheim's sociology, in which great attention is paid to the social effects of crisis situations. François Simiand, who had great influence in the *Annales* circle, was a pupil and collaborator of Durkheim.

The *Annalistes* had, and still have, a wide range of interests. They are active at three levels at least: in the historical domain, in interdisciplinary relationships, and in topical problems. In particular, Simiand, Labrousse, and Braudel, have introduced the wave movement into their research. During the recession of the 1930s, the first published his "Les fluctuations économiques à longue période et la crise mondiale" (1932). Labrousse worked along similar lines when tracing the influence of crises in agriculture on the origin of the French revolution, *inter alia* on the basis of price developments.

In 1949 Fernand Braudel, now the senior member of this group, developed for the first time the concept of the segmentation of time. According to Braudel, it is not strictly speaking time itself, but duration, which has to be considered. Braudel sees man and environment as standing in time in a plurality of respects. Historians have to place social evolution in the overall duration of time, that is, to relate it to the movements of structures, conjunctures, cycles and events. Consequently, the segmentation of time is an analytic instrument, more or less a paradigm,[6] with whose aid the overall picture can be reconstructed. In developing this idea, Braudel had the phenomenon of long waves in mind; he sees it as relevant to examinations of both broad and detailed issues.

Braudel understands the structural duration of time to mean evolutions over a period of one century or more, with changes usually being quite gradual. In an article entitled "La longue durée", Braudel fixes the duration of the structural time at two Kondratievs.[7] He likes a broad view of structures; apart from geographical and biological issues, he has in mind technological, economic, social, and cultural evolution. As an example of an economic structure, he refers to commercial capitalism in Western Europe from 1400 to 1800. The fact that he considers the *longue durée* (ie the structural duration of time) as the framework for the course of history is apparent from the following:[8]

C'est par rapport à ces nappes d'histoire lente, que la totalité de l'histoire peut se repenser, comme à partir d'une infrastructure.

There is the conjunctural (cyclical) duration of time, varying in length from a long wave up to cycles of a few years. In Braudel's view, these economic fluctuations can contribute to the explanation of quite different evolutionary patterns and phenomena. The conjunctural duration thus seems to be related to the origin, the course, and the after-effects of great wars. However, this rhythm can also be discerned in cultural, social, and demographic evolutions.[9] While non-economic conjunctures and cycles can be seen as analogous to economic ones, in his view it would be incorrect to reduce them to this. The level of analysis of conjunctural time is important for a model of the development of modern industrial society.

Braudel attaches considerably less weight to the duration of 'events', by which he understands events occurring abruptly, creating a great fuss, but usually having only transient effects. They are surface manifestations of slower, underlying movements, and threaten to mislead the researcher by their short, nervous rhythms. Only when the structures and conjunctures have been plotted out, is it possible to establish the significance of transient events. Thus the aim is to find structures into which such events can be integrated.

The Annalistes do not all agree about the further conceptualisation of segmented time, or on the concepts developed in particular by Bloch, Febvre, and Braudel. But this does not make it meaningless to speak of an *Annales* paradigm. Indeed, interest in the work of the founders is now increasing. The tendency of the *Nouvelle Histoire* in France,[10] initiatives by the Fernand Braudel Centre in New York (under the direction of Immanuel Wallerstein),[11] and activities of the Free University of Amsterdam,[12] are examples of this.

When considering industrial society as a complex integrator of diverging processes, the *Annales* paradigm is of particular importance; it completes and refines the long-wave approach. In particular, the interaction between structures and cycles requires attention. Looking at structures brings to light essential social development processes, while looking at cycles and conjunctures draws attention to changes and shifts brought to the surface by special events. Only in the interaction between the three durations of time can the true intention and significance of the *Annales* paradigm be traced. The influence of the *Annales* school since its establishment in 1929 has been great. Many historians not formally associated with this school have been influenced by it.

Rosenberg's work on waves and social change

The decline we experience daily has roused the historians' interest in declines of the past. As Pierre Chaunu says, "Historiens et économistes cherchent à comprendre ce qui se produit sous leurs yeux".[13] The West German historian Hans Rosenberg may be considered one of the pioneers here. He migrated via England to the United States in 1933, where he lectured in New York and Berkeley and, impressed by the crisis of the 1930s, he concentrated on developing a broad social history, orientated particularly to cycles. This study was affiliated with the work of the *Annales*. Rosenberg pointed out the significance of periods of decline for social processes of development.

A first example is a study undertaken in the early 1930s of the crisis of 1857–1859, which he termed the first world economic crisis.[14] During World War 2, he published an investigation into the social and political consequences for central Europe of the decline in the years 1873–1896.[15] In imitation of certain English authors, Rosenberg classifies these years as the period of "great depression".

A further study was a contribution to the discussion about Bismarck and his time written, however, from the perspective of decline. This book, published in 1967 with the title of *Grosse Depression und Bismarckzeit,*[16] is a remarkable study, going by its classifications and structure alone. As a perspective on a historical period in which economic events gave rise to great social stress, it is still a stimulating book. The slower growth and increasing competition during this period resulted in a drastic fall in prices and a growing uncertainty among producers. New lobby groups were established. Agriculture was outstripped by industry, and the state was required to take protectionist measures. Jews and Slavs were the scapegoats. The middle classes felt hard pressed by the strongly growing labour movement. In such a situation, anti-semitism and pre-fascism could thrive. The changes in the ways of thought are dealt with by Rosenberg in a separate chapter, using the terms of *Klima-Umschlag* and *Gesinnungs-, Glaubens- und Ideenverlagerung*.

In this study, Rosenberg elucidated the importance of this conjuncture period for an explanation of certain phenomena of the 20th century. His work may constitute the starting point for comparative studies in this domain, which include technological development and the role of innovations. On the other hand, it seems to be important not only to investigate periods of decline as isolated items, but also to extend Rosenberg's approach (with the aid of the *Annales* paradigm and of the long-wave approach) to a general diachronical framework, in which the interaction between diverse phenomena and successive periods of high and low conjuncture can be further investigated. Also of particular importance seems to be a systematic investigation of the explanation, the consequences, and the structural significance of so-called "turn-over" phases in terms of qualitative and quantitative changes.

Historians interested in long waves are usually less interested in the origin of these waves. They are fascinated by the structural approach of history, by the question of the patterns of relationship between wave movements in highly different fields, by an ample comparative framework, and by the possibility of thus gaining a more satisfying periodisation of historic processes and periods.

Conclusions

For a more exhaustive study of phenomena in modern history, both synchronically and diachronically, long waves represent an important frame of analysis. Excessive disintegration and segmentation of the past can thus be counteracted as well. In this way it is also possible to examine complex interrelationships more satisfactorily: in a narrower sense the question of relations between waves and economic changes; in a broader sense relations between the economic trend and social, political and cultural developments; and finally relationships between simultaneous wave movements in different sectors. Using such an approach it is possible to study these problem levels in a more systematic way.

With regard to the investigation into aspects of the industrialisation process, a "working model"[17] based on a phase-divided long-wave approach[18] in combination with the stratified, structural–conjunctural approach of the *Annales* has stimulated research in different fields and could lead to much more fruitful research. Thus the social consequences of the accumulation process can be better examined both with respect to concentration, monopolisation and internationalisation in trade and industry[19] and with respect to state intervention.[20] In the conjunctural approach there is a more direct relation between economic development and action by the government. Moreover, shifts in the extent of public expenditure and the expansion of governmental tasks can be better related[21] to social development processes.

Furthermore, this approach may be applied in research into social and cultural movements: in relation to the former, for example, there are studies on social stratification,[22] social mobility,[23] rural changes[24] and shifts in the relation between town and country.[25] The conjunctural perspective could be very important to research into the history of the labour movement, as it allows a deeper insight into the living and labour conditions of the working classes.[26] Many aspects—such as working-class strata,[27] protest and strike movements,[28] prices and wages,[29] interest in revolution,[30] reformism[31] and participation[32]—are highly trend sensitive. Working-class enquiries not only inform us about abuses in boom periods but also contribute concrete data about despair in times of recession.[33]

Although the relationships are very complex, literature and culture, together with ideologies and values, are also affected by the conjuncture. The way in which belief systems and ideas are influenced by crisis has been described in several studies, notably by Hans Rosenberg, who uses the term "Gesinnungs-, Glaubens- und Ideeenverlagerung".[34] Not only was liberalism knocked off its pedestal during the recession of 1873–96,[35] but new aggressive movements with an anti-semitic and pre-fascist ideology also came into being.[36] This "crisis of values" has also been examined by the Dutch historian Jan Romein.[37] It was about the turn of the century, when a period of growth and expansion started, although interrupted several times by short depressions, that cultural pessimism partly gave way to faith in progress.[38, 39] During the 1920s and 1930s people were conscious of the relation between climate of opinion and crisis;[40] feelings of uncertainty as well as of intolerance rapidly increased. Changes in the dominant role played by elements from religious practice or church doctrine have also been studied in a conjunctural perspective.[41]

In studies of countries or areas in the Third World it is also useful to pay more attention to the effects of conjunctural development. Studies dealing with the daily behaviour of the native populations,[42] with the increase of protest and opposition movements[43] and with the politics of expansion of the colonial countries[44] all reveal the influence of the economic conjuncture.

In these different fields of research the approach advocated here should not be focused just on the wave movements as such. It is important to acquire more knowledge about the transition periods between high and low conjuncture, the so-called "turn-over" phases.[45] Interest in demographic phenomena and their effects[46] certainly makes it possible to speak of "highs" and "lows":[47] various studies have dealt with relations between, on the one hand, marriages,[48] population bulges[49] and migration movements[50] and, on the other, the economic conjunctural movement. But too little attention has thus far been paid to these demographic variables as far as an explanation and more exact periodisation of the turn-over phases are concerned. Here it should be noted that this explanation must be multi-causal: not only economic developments should be taken into account but also social, political and cultural developments. Further examination is certainly required to establish the way in which during the transition phases social developments have taken place in a stagnating or innovating society and the variables that played a more initiating or a more secondary part. These turn-over phases are an essential part of the long-wave movement and seem to affect social structures to an important extent,[51] eg slower or faster democratisation or changes in value conceptions. This field of research, which has been almost a tradition in the Netherlands since the articles by J. van Gelderen,[52] is also a fascinating frame of analysis for historians.

Notes and references

1. On this approach, see also G. Barraclough, *An Introduction to Contemporary History* (Penguin, 1969).
2. See G. van Roon, "Arbeitsmodell Industrialisierungsprozess, Wellen wirtschaftlicher Entwicklung und Staatliche Wirtschaftspolitik in den Niederlanden", in D. Petzina and G. van Roon, eds, *Konjunktur, Krise, Gesellschaft: Wirtschaftliche Wechsellagen und sozialer Entwicklung im 19 und 20 Jahrhundert* (Stuttgart, Klett-Cotta, 1981).
3. For example, E. Hobsbawm, "Economic fluctuations and some social movements since 1800", *Economic History Review*, 1952, *5*.
4. See Wilhelm Abel, *Agrarkrisen und Agrarkonjunktur in Mitteleuropa vom 13. bis zum 19. Jahrhundert* (Paul Parey, 1980); B.H. Slicher van Bath, *The Agrarian History of Western Europe, AD 500–1850* (1963); B.H. Slicher van Bath, "Agriculture in the vital revolution", *Cambridge Economic History of Europe*, 1977, *5*, pages 43–132.
5. See T. Stoianovitch, *French Historical Method: The Annales Paradigma* (Ithaca, 1976).
6. See for example, J. Revel, "Histoire et sciences sociales: les paradigmes des Annales", in *Annales—Economies, Sociétés, Civilisations*, 1979, *34*, pages 1360–1376.

7. Fernand Braudel, *Ecrits sur l'histoire* (Paris, Flammarion, 1969), page 48.
8. Braudel, reference 7, page 54.
9. The level of the conjunctures has been especially studied by Ernest Labrousse; see *Conjoncture économique et structures sociales* (Paris, Mouton, 1975).
10. See Jacques Le Goff, *La Nouvelle histoire* (Paris, Retz, 1978).
11. *Review*, the journal of this Center, shows the strong influence of the *Annales*; see "The Impact of the *Annales* School on the Social Sciences" (*Review*, *I*, 3/4, 1978).
12. The projects initiated by the Interdisciplinary Research Group on Long Term Fluctuations of the Free University of Amsterdam.
13. Pierre Chaunu, *De l'Histoire à la prospective* (Paris, Robert Laffont, 1975), page 65.
14. Hans Rosenberg, *Die Weltwirtschaftskrise 1857–1859* (Kleine Vandernhoeckreihe 1396, Göttingen, Vandernhoeck and Ruprecht, 1974).
15. Reprinted in Hans Rosenberg, *Machteliten und Wirtschaftskonjunkturen: Kritische Studien zur Geschichtswissenschaft* (Göttingen, Institut des Sciences Mathématiques et Economique Appliquées, 1978) pages 161–173.
16. Hans Rosenberg, *Grosse Depression und Bismarckzeit* (Ullstein Buch 3239, Frankfurt/M, 1976).
17. Van Roon, reference 2.
18. See A. Spiethoff, in *Handwörterbuch der Staatswissenschaften*, 4th edition (Jena, 1925).
19. See eg the contribution of H. Daems in Petzina and van Roon, reference 2.
20. R. Miliband, *The State in Capitalist Society* (London, 1969); S. de Brunhoff, *Etat et capital: recherches sur la politique économique* (Grenoble–Paris, 1976); F. Marbach, *Zur Frage der wirtschaftlichen Staatsintervention* (Berne, A. Francke, 1950); G. van Benthem van den Bergh, "The interconnection between processes of state and class formation", *Acta Politica*, 1976, pages 289–312.
21. Louis Fontvieille, "Evolution et croissance de l'état français: 1815–1969", *Cahiers de l'I.S.M.E.A.*, Série AF, No. 13, Paris, 1976.
22. See E.P. Thompson, *The Making of the English Working Class* (Pelican, 1965).
23. Georges Dupeux, "L'Étude de la mobilité sociale: quelques problèmes de méthode", in *Conjoncture économique—structures sociales: hommage à Ernest Labrousse* (Paris–La Haye, Mouton, 1974) pages 79–90.
24. Ernest Labrousse, *La Crise de l'économie française à la fin de l'Ancien Régime et au début de la Révolution* (Paris, P.U.F., 1944); Tony Judt, *Socialism in Provence 1871–1914* (Cambridge, Cambridge University Press, 1979); Hille de Vries, *Landbouw e bevolking tijdens de agrarische depressie in Friesland (1878–1895)* (Veenman Wageningen, 1971).
25. Louis Bergeron, "Histoire sociale et histoire urbaine: à la recherche d'une méthode", in *Conjoncture économique—structures sociale*, reference 23, pages 91–96.
26. See Jacques Giele, *Arbeidersleven in Nederland 1850–1914* (Sun Nijmegen, 1979), page 16.
27. Richard F. Hamilton, *Affluence and the French Worker in the Fourth Republic* (Princeton, Princeton University Press, 1967).
28. J. Nadal, J. Vicens Vives and C. Marti, "Les Mouvements ouvriers en Espagne en temps de dépression économique (1929–1939)", in *Mouvements ouvriers et dépression économique de 1929 à 1939* (Assen, Van Gorcum–Prakke, 1966), pages 113ff; Hobsbawm, reference 3; H. Kaelble and H. Volkmann, "Konjunktur und Streik während des Übergangs zum Organisierten Kapitalismus in Deutschland", *Zeitschrift für Wirtschafts- und Sozialwissenschaften*, 1972, *92*, pages 513–544.
29. François Simiand, *Recherches anciennes et nouvelles sur le mouvement général des prix du XVIe à XIXe siècle* (Paris, 1933).
30. Ernst Schraepler, *Handwerkerbünde und Arbeitervereine 1830–1853* (Berlin–New York, Walter de Gruyter, 1972), pages 22ff; Michael Vester, *Die Entstehung des Proletariats als Lernprozess* (Frankfurt/M, Europaïsche Verlagsanstalt, 1975), 3rd edition, pages 20ff., 309ff.
31. Bo Gustafsson, *Marxismus und Revisionismus* (Frankfurt/M., Europaïsche Verlagsanstalt, 1972), Part 1, pages 15ff.
32. "Medezeggenschap en conjunctuur in de jaren na de eerste wereldoorlog", *Tijdschrift voor sociale geschiedenis*, 1975, *3*, pages 287–345.
33. Werner Plum, *Working Men's Reports on the Beginnings of the World of Industrial Labour* (Bonn, Friedrich Ebert Stiftung, 1976); Michelle Perrot, "Comment les ouvriers parisiens voyaient la crise d'après l'enquète parlementaire de 1884", in *Conjoncture économique—structures sociales*, reference 23, pages 187–200.

34. Rosenberg, reference 16, page 66.
35. During the recession period at the end of the 19th century Nietzsche and Spengler wrote their well-known books.
36. Shulamit Volkov, *The Rise of Popular Antimodernism in Germany* (Princeton, Princeton University Press, 1978); Pierre Guiral and Emile Temime, *L'Idée de race dans la pensée politique française contemporaine* (Paris, Editions du CNRS, 1977); M.D. Biddiss, *The Age of the Masses: Ideas and Society in Europe since 1870* (Hassocks, Harvester, 1977).
37. Jan Romein, *Op het breukvlak van twee eeuwen* (Amsterdam, Em. Querido, 1976), 2nd edition, pages 590ff.
38. Maxwell H. Bloomfield, *Alarms and Diversions: The American Mind through American Magazines 1900–1914* (The Hague–Paris, Mouton, 1967).
39. Fritz Stern, *The Politics of Cultural Despair* (1963).
40. Jon Clark, Margot Heinemann, David Margolies and Carole Snee, *Culture and Crisis in Britain in the 30s* (London, Lawrence and Wishart, 1979).
41. A.D. Gilbert, *Religion and Society in Industrial England: Church, Chapel and Social Change 1740–1914* (London, Longman, 1976); Stephen Yeo, *Religion and Voluntary Organisations in Crisis* (London, Croom Helm, 1976).
42. Frédéric Mauro, "Conjoncture économique et structure sociale en Amérique Latine depuis l'époque coloniale", in *Conjoncture économique—structures sociales*, reference 23, pages 237–251.
43. Robert L. Rotberg and Ali A. Mazrui, eds, *Protest and Power in Black Africa* (New York, Oxford University Press, 1970).
44. Jean Suret-Canale, *French Colonialism in Tropical Africa, 1900–1945* (London, C. Hurst, 1971); Immanuel Wallerstein, "The three stages of African involvement in the world economy", in Peter C.W. Gutkind and Immanuel Wallerstein, *The Political Economy of Contemporary Africa* (London, 1976), pages 38, 39, 48–49.
45. I prefer not to speak of turning points; see also Klaus Erich Rohde, *Gleichgewicht und Konjunkturtheorie* (Stuttgart, Gustav Fischer, 1957), pages 170–179.
46. See C. Béhar, "Population et modèles de croissance" (thesis), (Paris, 1972).
47. F. de Roos, *Bevolking en welvaart* (Assen, G.F. Hummelen, 1960), page 3.
48. J. Hajnal, "European marriage patterns in perspective", in D.V. Glass, ed, *Population in History* (London, Edward Arnold, 1965).
49. V. Galbraith and D. Thomas, "Birth rates and the interwar business cycle", *Journal of the American Statistical Association*, 1941; National Bureau of Economic Research, *Democratic and Economic Change in Developed Countries* (Princeton, Princeton University Press, 1960).
50. See B. Thomas, *Migration and Economic Growth* (Cambridge, 1954); M. Walker, *Germany and the Emigration 1816–1885* (Cambridge, Mass., 1964).
51. M. Aymard, "Crise", in *La Nouvelle Histoire* (Paris, CEPL, 1978), page 94.
52. J. van Gelderen, "Springvloed: beschouwingen over industrieele ontwikkeling en prijsbeweging", *De Nieuwe Tijd*, 1913, *18*, pages 253–277, 369–384, 445–464.

GLOSSARY

Conjunctural ("Konjunktur"): Combination of short-term cyclical factors influencing behaviour of the economy.

Echo effect: Some business cycle theories explain economic fluctuations partly or entirely by the replacement of durable capital goods at intervals. This 'echo effect' of the original investment may occur at relatively short intervals for eg machine tools, and at much longer intervals for eg buildings or transport systems, thus creating fluctuations of varying duration. See Tinbergen's paper.

Epidemic models of diffusion: Schumpeter's theory of business cycles stressed the role of innovations and their diffusion through the economy. Following Schumpeter, some sociologists and economists have tried to develop models of this diffusion, drawing on studies of the spread of epidemics. Metcalfe criticises the over-simplification involved.

Innovation: This word is used differently by different authors, but here it generally has the 'Schumpeterian' meaning; he emphasised the distinction between "invention"–the first idea or working model for a new product, process or system–and "innovation"–their translation into actual products and processes used in the economy. Not all inventions become innovations; those which do may have long 'gestation' or development phases. The invention/innovation distinction is important to the theories of Mensch and van Duijn, and to Kleinknecht's paper. Central to Mensch's theory is the idea of shortening this gestation period for some groups of innovations in response to the depression phase of the cycle. This "acceleration" hypothesis is criticised by Clark, Freeman, and Soete.

Juglar cycles: The 'ordinary' medium-term business or trade cycle, which was and is the focus of most 'mainstream' economic theory and research, is central to Keynesian theory and policy prescriptions. Before World War 2, these cycles generally lasted 7 to 11 years, but they have since been shorter. The expression "Juglar" is rare today, used mainly in contrast to longer (Kondratiev) cycles, and shorter (Kitchin) stock cycles. French economist Clement Juglar wrote one of the first major studies of business cycles, in 1889.

Kondratiev cycles (long waves): Economic fluctuations lasting 40–60 years, whose scope and nature (and even existence) are controversial. In the general public debate, the terms "Kondratiev cycle" and "long waves" are often interchangeable, but some authors argue that "wave" has a less mechanistic or deterministic connotation. "Kondratiev cycle" is a misnomer, in that some economists described them before he did; before WW1 van Gelderen developed a long-wave theory, anticipating much of the debate. However, the Russian economist, N.D. Kondratiev, probably did more than anyone to make the idea known in the USSR and the world in general, whilst head of the *Konjunktur* Institute in Moscow in the 1920s.

Schumpeter: Austrian economist, who became Professor of Economics at Harvard, and did more than any other 20th century economist to advance the view that innovation was central to the capitalist growth process, and to the various economic fluctuations associated with it. He popularised the idea of Kondratiev cycles among Western economists, and introduced the notion that they might be largely explained by the irregular appearance and uneven incidence of major innovations and their diffusion. Revival of interest in this key idea underlies this book.

Wolff's Law: A German economist who in 1912 published four "laws of retardation of progress". Essentially he argued that the scope for improvement in any technology is limited, and that the cost of incremental improvement increases as the technology approaches its long-run performance level. Widely referred to in the retardation literature of the 1930s.